Greening the Supply Chain

T0137858

Joseph Sarkis (Ed.)

Greening the Supply Chain

With 45 Figures

 Springer

Joseph Sarkis, PhD
Graduate School of Management
Clark University
950 Main Street
Worcester, MA 01610-1477
USA

British Library Cataloguing in Publication Data
Greening the supply chain
  1.Business logistics - Environmental aspects
  I.Sarkis, Joseph
  658.5

e-ISBN 1-84628-299-3
ISBN-13: 978-1-84996-572-9   e-ISBN-13: 978-1-84628-299-7

Printed in Germany

9 8 7 6 5 4 3 2 1

Springer Science+Business Media
springer.com

Dedicated to my father and mother,

Tanios and Azizi

Preface

Since the Rio Summit in 1992 the paradigm of corporate environmental responsibility has gradually and ever more consistently extended beyond complying with increasingly stringent environmental regulation and also beyond the taking up of proactive initiatives by a few world-class companies. There are multiple reasons for this, not least of which is that current research indicates that the business and financial performance of companies may depend directly on socially and environmentally responsible business practices. Thus, companies are under increasing competitive and other pressures to continuously search for new ideas and methods allowing them to achieve and/or maintain environmental sustainability. Greening the supply chain is one such innovative idea that is fast gaining attention in the industry. This attention may be attributed to many companies realising they have to do something more than applying traditional internal measures of implementing waste-reduction strategies, installing pollution-control technologies, replacing hazardous-material inputs with environmentally friendly alternatives and so on. Many of the positive effects of internal measures have already been achieved by leading corporations. Now the goal has begun to shift to integrating all of the business value-adding operations, including purchasing and in-bound logistics, production and manufacturing, distribution and out-bound logistics, in such a way that activities associated with these functions have the least harmful environmental impact. This is one of the basic tenets of the philosophy of greening the supply chain that is critical for the successful implementation of industrial ecosystems and ecology.

There is also another reason why the greening of the supply chain is gaining popularity with industry. Many world-class companies now realise that customers and other stakeholders do not distinguish between a company and its suppliers. According to Bacallan (2000), 'although they may have nothing to do at all with the problem, companies are often held accountable for the labour practices and the environmental liabilities of their suppliers'. Thus, greening the supply chain helps companies to avoid potential environmental problems that might arise with their suppliers and that in turn could threaten their own environmental performance. Problems arising for organisations can range from actual business continuity and heavy liability issues to perceptual problems of having 'dirty' partners.

As we shall see in this book, there are a number of definitions and philosophies on what a green supply chain may be. These vary from something as small as buying green products from a supplier to the much broader context of an industrial ecosystem. The full material and service value-adding process and its 'closing-the-loop' components will form much of the discussion in this book.

Integrated within the book are ideas and practices of green purchasing, total quality management, life-cycle assessment, environmental marketing, reverse logistics and a long list of other supporting and integral environmental practices. In order to achieve greening at every stage and in every function, we will see that the full involvement and co-operation of suppliers and business partners is essential. This co-operation ranges from the materials, products, components and services they are sharing to the type of energy they are using, and the pollution-control and waste-management technology and services they are providing. Integrating the external relationships of an organisation with internal operations is not only good practice but is a prerequisite for a true green supply chain.

Overview of the Book

The upcoming chapters of this book include work from a number of international authors from theoretical and practical backgrounds, sharing their research and experiences in the field to promote a better understanding of the environmental influence of supply chain management.

To provide a general framework for the topics in this book we have categorised the chapters broadly into methodological, rather than topical, categories for investigating green supply chain management (GSCM). Thus, we first categorise the chapters into those providing more of a conceptual framework evaluation of green supply chains. The concepts primarily have strategic or overview characteristics but also contain topics that have operational aspects related to GSCM. Even though the frameworks are conceptual in focus, they also have strong practical elements, with examples peppered throughout their chapters to help managers better understand, plan, and manage green supply functions. The second part of the book focuses on work that is broad and more empirical. Many chapters here utilise survey instruments as methodological approaches to evaluate green supply chain practice. The third part focuses on contributions that are oriented towards case studies, either to support theoretical frameworks or to explain in more detail how organisations are practically applying green supply chain practices. The final part of the book introduces work in which systemic or technologically based decision support and modeling approaches are discussed. These form the more prescriptive and normative types of techniques for improving GSCM. So, overall, the book's 21 chapters move from broad conceptual and empirical evaluations of GSCM to more specific cases and a tools-based foci.

Part I: Concepts and Frameworks

In Chapter 1 Robert Sroufe discusses strategic environmental sourcing. Sroufe looks at the history of sourcing and relates it to a better understanding of the strategic issues in environmental assessment and selection of suppliers. In this discussion he relates how strategic environmental sourcing provides firms with opportunities to design products that are more environmentally friendly at the same time as reducing waste, risk and costs. In the past, purchasing was considered a clerical function related to manufacturing and materials management. Slowly, it became a tactical function of marketing and manufacturing; now it is considered as part of corporate strategy, helping the firm to be involved in the environmental opportunities found in supply chain management. As part of this strategy, Sroufe discusses supplier quality assurance and assessment programmes, supplier development programmes, supplier measurement programmes and so on.

In Chapter 2, Burton Hamner argues that organisational buyers that use environmental criteria may have an implicit, but mistaken, assumption that imposing these criteria on suppliers will cause the suppliers to move in the direction of environmental sustainability. Hamner presents a typology of strategies used for green supply chain environmental management and outlines various costs to and benefits for buyers and suppliers. Hamner argues that some strategies, in particular the ones most commonly used, are not likely to push suppliers towards more sustainable behaviour. He argues that to achieve significant improvements in environmental performance up the supply chain, buyers will have to make major efforts to work closely with suppliers. This is especially true when suppliers are in countries that have relatively lax enforcement of environmental regulations and less information about the business benefits of environmental management strategies.

In his chapter Hamner mentions some of the work that is taking place under the auspices of the Global Environmental Management Initiative (GEMI). In Chapter 3, one of the central frameworks for evaluating GSCM developed by GEMI is put forward by John A. Harris. The relationship and contribution of GSCM to business value is how the evaluation framework is completed. Within the framework Harris relates five critical topics to questions that businesses need to ask themselves about strategic sourcing and supplier management. These questions include determination of whether there is value in environmental programmes in the supply chain, how to tap this value and what criteria should be used to help add business value. Various issues related to this prescriptive framework are also evaluated in later empirical studies in this book, such as in Chapters 8-10.

Part of the continuous improvement focus of Harris's chapter relates to the next chapter, by Toufic Mezher and Maher Ajam, who, in Chapter 4, begin integrating the concepts of continuous improvement (through total quality management), GSCM and EMSs and how these may all support organisational learning and knowledge management. This very comprehensive framework is valuable to managers to see what aspects of these critical corporate environmental management tools are important in managing an organisation's strategic learning, necessary to remain competitive.

In another framework-based study, Kirsi Hämäläinen, in Chapter 5, considers network dynamics in the supply chain by using the concept of industrial ecology, where the waste of one organisation becomes raw material for another. Managerial implications on the adaptation and other bond-building relationships between partner companies are emphasised in this work. Hämäläinen discusses how partner companies from diverse industries such as energy generation and paper milling can jointly integrate their operations to form a system conforming to industrial ecology where nothing is wasted. She concludes that the development of such a relationship is highly useful for industrial ecosystems and points out how these complementary relationships between mutual parties can play a major role towards attaining environmental sustainability.

A conceptual model that evaluates the distribution or downstream channels of the green supply chain is introduced by Booi Hon Kam, Geoff Christopherson, Kosmas X. Smyrnios and Rhett H. Walker in Chapter 6. They consider green logistics strategies, with a special focus on freight transport and product delivery patterns relating to satisfying customer needs and meeting product deadlines. The conceptual model is evaluated by using three case studies on companies that represent efficient out-bound logistics. In this conceptual model Kam et al. investigate three factors—external influences, internal company characteristics, and strategic goals and the state of available technology and services—that play an important role in strategic product delivery decision-making processes and that have a direct impact on eco-efficiency. This framework provides a useful platform on which to base future research aimed at understanding the effect of changing logistical systems on the environment and the challenges facing logistics managers in the implementation of effective EMSs.

Moving downstream in the supply chain provides discussion on the needs of effective industrial ecosystems that require reverse logistics for used products and materials flowing back into the forward supply chain. An investigation of the concept of reverse logistics within the carpet industry is the goal of Marilyn M. Helms and Aref A. Hervani in Chapter 7. They discuss the complex issues of the supply chain working in reverse to 'reclaim products at the end of their life-cycle and return them through the supply chain for decomposition, disposal or re-use'. Helms and Hervani consider the reverse logistics programme in the recycling of old carpets, discussing how several carpet manufacturing plants are using the take-back programme at different points in the supply chain. An industry analysis for the application of reverse logistics and its implications for businesses is provided in this very comprehensive study.

Part II: Empirical Studies

As the first seven chapters show, there are many concepts related to GSCM in purchasing, production, business operations, delivery and reverse logistics. The authors of the next five chapters use empirical surveys to study how organisations in disparate industries and countries utilise these concepts.

The second part of the book begins with Gregory Theyel's work, in Chapter 8. He discusses the customer and supplier relations for environmental performance, using waste reduction as a measure. Studying firms in the US chemical industry he investigates three types of relationships between customers and their suppliers. These relationships include the stipulation of environmental requirements by the customer to the supplier, the sharing of environmental information and collaboration on improving environmental aspects of products and processes. Theyel observes that closer relations within the supply chain lead to the creation and sharing of knowledge, as also observed by Mezher and Ajam in Chapter 4. Such collaboration can also lead to the industrial ecosystem concept, where conservation of energy and materials is optimised, waste generation is minimised and effluents of one process serve as raw materials to another process (an approach supported by Hämäläinen in Chapter 5).

In Chapter 9, Frances E. Bowen, Paul D. Cousins, Richard C. Lamming and Adam C. Faruk observe that though there is currently a lot of interest in the concept of green supply chains, a gap still exists between the theory and implementation of the concept in industry. Recognising that business would implement green supply chain practices only if it were genuinely to expect to obtain the benefits postulated in theory, Bowen *et al.* conduct a two-phase survey of green supply chain activities in the United Kingdom for a sample of business units. In their research, they identify three types of green supply chain initiatives, involving: businesses that collect environmental information on suppliers and make assessments of the performance of those suppliers; businesses that make changes to the end-product supplied; and businesses that incorporate joint clean technology initiatives with suppliers. They conclude that the most prevailing supplier-related initiative is the product-based strategy, where the company wants to minimise waste, excess cost and resources from the sourcing process at the customer–supplier interface.

The empirical studies in Chapters 8 and 9 are focused on organisations in the USA and United Kingdom. In Chapter 10, Qinghua Zhu and Yong Geng look at environmental issues in supplier selection in the context of large and medium-sized state-owned enterprises in China. As in many other countries across the industrialising world, many enterprises in China are now trying to improve their environmental performance because of pressures from government, communities and non-governmental organisations (NGO) and because of international pressure—particularly from multinational corporations. It seems that along with traditional performance criteria, environmental considerations have also become critical in China because, according to Zhu and Geng, 'it leads to eco-efficiency, cost saving and improved public perception'. In their study, Zhu and Geng conclude that for enterprises producing intermediate products their multinational clients often require ISO 14001 certification. Zhu and Geng also reveal that the other key factors affecting the selection of suppliers by Chinese enterprises are regulatory factors, the costs of purchasing and disposing of materials and the environmental performance of suppliers (similar to the findings of studies focused on the USA. However, many barriers to green purchasing, such as lack of know-how, the necessary tools and management skills, still exist.

Also in Asia, Purba Rao, in Chapter 11, presents an empirical analysis of organisations located throughout South-East Asia, including the Philippines, Indonesia, Malaysia, Thailand and Singapore. Rao finds that customer pressure is the chief factor leading companies in South-East Asia to implement green supply chain measures. Obstacles that hamper the implementation of such initiatives were also evaluated for organisations in these countries. It was found that lack of general interest, manpower and financial resources among suppliers were major obstacles to the adoption of these initiatives.

The final empirical study takes us back to the United Kingdom. In Chapter 12 Lutz Preuss provides results from a study of the practices of procurement managers from 40 manufacturing companies in Scotland. In-depth interviews bring out a number of interesting issues relating to packaging, waste, hazardous materials and so on. With a focus on the way in which customers deal with suppliers, Preuss discusses the passive and active collaborative approaches of customers from the supplier's point of view and concludes that, at the moment, the predominant way of integrating environmental issues in a supply chain is a passive approach, where buyer organisations impose environmental criteria on the supplier, further supporting the findings in previous chapters. Meta-analytically, it seems that globalisation has increased pressures on and responses by organisations, wherever in the world they may be located.

Part III: Case Studies

The third, case-study, part of our book begins with chapters by Jeremy Hall (Chapter 13) and by Frank Ebinger, Maria Goldbach and Uwe Schneidewind (Chapter 14). The common theme between these two chapters is that the authors begin with conceptual frameworks and then use case studies to help support their theoretical concepts.

In Chapter 13, Hall explores why firms invest in supply chain innovations; he hypothesises that they do this to avoid the high levels of environmental risk to which suppliers with poor environmental practices can expose them. But, making an argument similar to one made by Theyel in Chapter 8, Hall states that the effectiveness of environmental innovations depends on an exchange of information through supply chain channels, from the customer to the supplier, and on co-operation and partnership between customer and supplier. Hall's research is focused on a case-study format in which data was collected from open interviews with representatives from a major supermarket chain and from five of its suppliers. The supermarket chain concerned has always taken a very active role in developing its supply chain in order to achieve high-quality product development and to foster a collaborative relationship with its suppliers. Hall introduces a 'sphere of influence' model to aid in the depiction and understanding of the way in which a customer firm influences its suppliers.

In Chapter 14, Ebinger et al. consider the balance between eco-competitiveness (which incorporates economic and environmental objectives) and a competence-based management system. Recognising the growing pressures of globalisation, they emphasise that economic optimisation alone cannot satisfy market and community pressure—there must be a parallel optimisation of economic and environmental objectives in the supply chain. Applying a competence-based approach to the German mail order business OTTO, they arrive at a very interesting conclusion that in addition to resources such as technology, knowledge and management capabilities intangible resources such as co-operation, communication skills, good supplier relationships and supply chain control tools make the ecological success of the supply chain a reality.

In Chapter 15, James P. Warren and Ed Rhodes review the environmental supply chain context of the smart™ automobile. The smart™ vehicle is an efficient car having a lower impact on the environment regarding emissions and a reduced use of non-renewable resources. Warren *et al.* describe how this environmental design has been made possible by the co-operation of suppliers. They highlight: the use of modularity in product design and production facility layout; an emphasis on partner participation from product creation to after-sales service; and the use of a highly customised build-to-order product system to green the entire supply chain. In their case study, Warren and Rhodes compare the process characteristics employed at the smart™ car factory, called 'smartville', to the more 'traditional' approaches of vehicle manufacture, in an attempt to establish the actual or potential reduction of environmental impact in the three stages of vehicle life—manufacture, use and disposal—including the role of main suppliers in this process.

Still within the automotive manufacturing industry, in Chapter 16 Annette von Ahsen provides a case study supported by an empirical analysis of the environmental supply chain and practices at BMW. Ahsen notes that over the past few years, car manufacturers have stepped up their efforts to green their operations and their supply chains, partly in reponse to increasingly stringent environmental legislation and partly in response to increasing customer expectations concerning the minimisation of fuel consumption, the reduction of harmful emissions and the increase of recycling. In this case study, Ahsen shows how BMW influenced the environmental management of suppliers by pushing forward the ecodesign of its products. She points out that it is essential to find suppliers who proactively assist in designing sustainable products. For example, if fuel consumption has to be reduced, the vehicle design must be more lightweight, which requires those who supply the components to source and deveelop appropriate materials. In many cases, automotive manufacturers such as BMW now include an environmental dimension in their vendor rating systems, systems that previously would have focused solely on performance measures such as quality and cost. In this case study one is shown how BMW works jointly with suppliers to optimise not only quality and cost but also environmental performance. The chapter also includes an overview of a survey of automotive suppliers to assess the state of environmental management in the supplier operations.

Philip Trowbridge, in Chapter 17, provides some insights into green supply chain practices in the electronic industry by detailing efforts at Advanced Micro Devices (AMD). He discusses how this effort grew from the strategic integration of supplier relationships with environmental programmes. Trowbridge considers the case of AMD's environmental strategies and organisational linkages. Some detailed operational and managerial characteristics spanning green supply chain practices are presented in this very practical industrial case study.

Part IV: Tools and Technology

The final part of our book is composed of four chapters that provide prescriptive models, tools and technological solutions to help manage and make decisions within the green supply chain.

Meindert H. Nagel, in Chapter 18, discusses the supply chain from the perspective of an original equipment manufacturer (OEM), introducing the concept of environmental quality in the supply chain. He utilises this concept to develop a quantitative model to assess the environmental performance of suppliers. In the case of an OEM the management of supply chain is a complex activity because of the vast number of different components. The usual decisions required to be made involve the performance of suppliers on price, delivery aspects, service, quality and technology. In common with the arguments in previous chapters in this book, Nagel's argument is that, in addition to the above criteria, suppliers should also be ranked and classified in relation to their environmental aspects. By using environmental load elements to form the basis for a supply chain management model and by introducing an environmental performance tool, Nagel arrives at a tangible measure for the environmental performance of suppliers that aids in ranking and classifying suppliers. He provocatively argues that suppliers be penalised through price reductions for poor environmental performance.

In Chapter 19, Hsien H, Khoo, Trevor A. Spedding, Ian Bainbridge and David M.R. Taplin, utilising another decision-making tool and considering an integrated view of supply chain management, with upstream internal and downstream aspects, present a simulation model to examine a case study for a supply chain dealing with the distribution of aluminium metal from raw material processing to the end-user market. In their simulation exercise Khoo et al. consider optimal location, conservation of energy between different plants, promotion of scrap metal recycling activities and costing of pollution as major factors in evaluating simulation runs. An illustrative example of the simulation examines how a supply chain with four plants can reduce transport pollution and energy issues while still meeting customer demand.

Next, in Chapter 20, Severin Beucker and Claus Lang introduce a technologically driven concept of computer-aided resource efficiency. They discuss their work with CARE (Computer Aided Resource Efficiency Accounting for Medium-Sized Enterprises). The objective of the CARE project is to develop a financial and environmental information system for managing and decision-making in companies. The approach uses the concept of resource efficiency accounting (REA), which combines cost accounting with environmental impact data on production processes and the product life-cycle. They explain the importance of such systems within the general context of e-commerce relationships and technology.

The final chapter of the book, Chapter 21, brings together a number of the concepts discussed earlier with a look toward the future and builds on the e-commerce context discussed in Chapter 20. Steven V. Walton and Chris Galea examine the purchasing aspect of an electronic supply chain for service-oriented businesses. By using various case studies they also identify issues relating to broader supply chain management topics, including distribution and logistics. They argue that although e-commerce is potentially a powerful way to improve the performance of environmental purchasing there are some pitfalls with the practice and management of this technology and some directions that it could take in the future to be avoided.

Summary

Overall, the book illustrates that there are a wide variety of practical and theoretical problems and solutions that arise in the study of GSCM. Many organisations in many different sectors will find that some of their problems may already have been solved by other projects included here. Of course, the field of GSCM continues to evolve—and evolve quickly. We hope the interdisciplinary work included here serves as an important piece of this evolution as well as inspiring the serious pursuit of additional research and practice in the future.

This book project would not have come to fruition without the help of numerous people. I would first like to thank all the contributors for their patience and effort. I would like to thank the editorial staff, especially Kate Brown and Anthony Doyle at Springer for their guidance in putting together a high quality document. I would also like to thank the Greenleaf Publishing editorial staff, and John Stuart, for helping me to initiate this project.

Joseph Sarkis
Worcester, Massachusetts, USA

Reference

Bacallan, J.J. (2000) 'Greening the Supply Chain', Business and Environment 6.5: 11-12.

Contents

Part II Empirical Studies

Part III Case Studies

Part IV Tools and Technology

List of Contributors

Maher Ajam
American University of Beirut
Faculty of Engineering and
Architecture Engineering Management
Program
PO Box 11-0236
Riad El Solh
Beirut 1107 2020, Lebanon
mrajam@dgjones.com

Ian Bainbridge
Cooperative Centre for Cast Metals
Manufacturing
University of Queensland
Queensland 4072, Australia

Severin Beucker
Fraunhofer Institut fuer
Arbeitswirtschaft
und Organisation Competence Center
Innovationsmanagement
Nobelstr. 12,
70569 Stuttgart
Germany
severin.beucker@iao.fraunhofer.de

Frances Bowen
Haskayne School of Business
University of Calgary
2500 University Drive NW
Calgary, AB, T2N 1N4
frances.bowen@haskayne.ucalgary.ca

Geoff Christopherson
School of Management
Business Portfolio
RMIT University
Melbourne,
Australia
geoff.christopherson@rmit.edu.au

Paul Cousins
School of Management and Economics
Queen's University Belfast
25 University Square
Belfast
BT9 1NN
UK

Frank Ebinger
Albert-Ludwigs-Universität Freiburg
Institut für Forstökonomie
Tennenbacher Str. 4
79106 Freiburg
Germany
f.ebinger@ife.uni-freiburg.de

Adam Faruk
Ashridge
Berkhamsted
Hertfordshire,
HP4 1NS
United Kingdom
adam.faruk@ashridge.org.uk

Chris Galea
Department of Business
Administration
St. Francis Xavier University
P.O. Box 5000
Antigonish, NS B2G 2W5
Canada
cgalea@stfx.ca

Yong Geng
School of Management
Dalian University of Technology
Dalian
Liaoning Province (116024) P.R.C.

Maria Goldbach
Institute of Business Administration
Chair for Production and the
Environment
Carl von Ossietzky University
Oldenburg
PO Box 2503, 26111 Oldenburg
Germany
maria.goldbach@uni-oldenburg.de

Jeremy Hall
Haskayne School of Business
University of Calgary
2500 University Drive N.W.
Calgary
Alberta, Canada T2N 1N4
hallj@.ucalgary.ca

Kirsi Hämäläinen
University of Jyväskylä
School of Business and Economics
PO Box 35
FIN-40014 University of Jyväskylä
Finland
kirsi.hamalainen@tampere.fi

Burton Hamner
Director
Cleaner Production International
5534 30th Avenue NE, Seattle,
WA 91805
wbhamner@cleanerproduction.com

John Harris
Eli Lilly and Company
Health, Environmental, and Safety
Department
Lilly Corporate Center
Indianapolis, IN 46285
j.harris@lilly.com

Marilyn M. Helms
Sesquicentennial Endowed Chair and
Professor of Management
Division of Business Administration
Dalton State College
213 N. College Drive
Dalton, GA 30720
mhelms@daltonstate.edu

Aref A. Hervani
Chicago State University
Dept. of Geography, Anthropology,
Sociology, & Economics
SCI-321, 9501 S. King Dr.
Chicago,
IL 60626-2186
ahervani@csu.edu

Khoo Hsien Hui
Department of Industrial & Systems
Engineering
National University of Singapore
1 Engineering Drive 2
Singapore 117576
g0203686@nus.edu.sg

Booi Hon Kam
School of Management
Business Portfolio
RMIT University
Melbourne, Australia
booi.kam@rmit.edu.au

Richard Lamming
School of Management
University of Southampton
Southampton
SO17 1BJ, UK
R.C.Lamming@soton.ac.uk

Claus Lang-Koetz
Institute for Human Factors and
Technology Management (IAT)
University of Stuttgart
Nobelstr. 12
70569, Stuttgart
Germany
claus.lang-koetz@iao.fraunhofer.de

Toufic Mezher
American University of Beirut
Faculty of Engineering and
Architecture Engineering Management
Program
PO Box 11-0236, Riad El Solh
Beirut 1107 2020, Lebanon
mezher@aub.edu.lb

Menno Nagel
Delft University of Technology
Faculty of Mechanical, Maritime and
Materials Engineering
Mekelweg 2
2628 CD DELFT
The Netherlands
m.h.nagel@3me.tudelft.nl

Lutz Preuss
School of Management
Royal Holloway College
University of London
Egham Hill, Egham
Surrey TW20 0EX
United Kingdom
Lutz.Preuss@rhul.ac.uk

Purba Rao
Asian Institute of Management
123 Paseo de Roxas
Makati City 1260
Philippines

Ed Rhodes
The Faculty of Technology
Centre for Technology Strategy
The Open University
Walton Hall

Milton Keynes, MK7 6AA
United Kingdom
e.a.rhodes@open.ac.uk

Joseph Sarkis
Graduate School of Management
Clark University
950 Main Street
Worcester,
MA 01610-1477
USA
jsarkis@clarku.edu

Uwe Schneidewind
Institute of Business Administration
Chair for Production and the
Environment
Carl von Ossietzky University
Oldenburg
PO Box 2503
26111 Oldenburg, Germany
uwe.schneidewind@uni-oldenburg.de

Kosmas X. Smyrnios
School of Management
Business Portfolio
RMIT University
Melbourne,
Australia
kosmas.smyrnios@rmit.edu.au

Trevor A. Spedding
School of Management and Marketing
Faculty of Commerce
University of Wollongong
Australia
spedding@uow.edu.au

Robert Sroufe
Carroll School of Management
Boston College
Operations and Strategic Management
Department
140 Commonwealth Avenue
Chestnut Hill,
MA 02467
Sroufe@bc.edu

David M.R. Taplin
Visiting Professor of Systems
Engineering
School of Engineering
University of Greenwich
Chatnam Maritime, ME 4 4TB
United Kingdom

Gregory Theyel
College of Business and Economics
California State University
25800 Carlos Bee Boulevard
Hayward, CA 94542
USA
gregory.theyel@csueastbay.edu

Philip Trowbridge
Advanced Micro Devices, Inc.
5204 E. Ben White
M/S 529
Austin, Texas 78741
USA
Philip.trowbridge@amd.com

Anette von Ahsen
School of Business Administration
Chair of Environmental Management
and Controlling
University of Essen, P.O. Box
45117 Essen, Germany
anette.von-ahsen@uni-essen.de

Rhett H. Walker
School of Business and Technology
La Trobe University - Bendigo
PO Box 199
Bendigo, Victoria 3552
Australia
rhett.walker@rmit.edu.au

Steve V. Walton
Goizueta Business School
Emory University
1300 Clifton Road
Atlanta, GA 30322-2710
USA
Steve_Walton@bus.emory.edu

James P. Warren
The Faculty of Technology
Centre for Technology Strategy
The Open University
Walton Hall
Milton Keynes, MK7 6AA
United Kingdom
j.p.warren@open.ac.uk

Qinghua Zhu
School of Management
Dalian University of Technology
Dalian
Liaoning Province (116024) P.R.C.
erinzhu@hotmail.com

Part I

Concepts and Frameworks

1

A Framework for Strategic Environmental Sourcing

Robert Sroufe

Carroll School of Management, Boston College, Operations and Strategic Management, Department, 140 Commonwealth Avenue, Chestnut Hill, MA 02467, Sroufe@bc.edu

As competition among businesses intensifies on a global scale, companies will continuously be looking for ways to reduce waste and its associated costs, maintain a flexible corporate strategy and improve their position in the marketplace. Managers are finding environmental practices being employed more widely as a result of changing business conditions that emphasise environmental and financial performance (Cordeiro and Sarkis 1997; Hart and Ahuja 1996). Consequently, meaningful and effective frameworks for change are increasingly important because of the cost of environmental options and the need to comply with regulatory pressures and address the concerns of consumer groups. Additionally, the adoption of voluntary environmental initiatives—such as the Business Principles for Sustainable Development of the International Chamber of Commerce, and international standards such as ISO 14001 from the International Organisation for Standardisation (ISO)—have impacted firms in recent years (Corbett and Kirsch 2001; GEMI 1997). These voluntary initiatives are causing many firms to emphasise environmental programmes that are both internal and external in scope. Basically, the stage has been set for the extension of environmental management to the supply chain and to associated environmental sourcing strategies. For the purpose of this chapter, sourcing as a function includes supply-base management, the controlling of total costs, the creation and exchange of long-term value and the creation of value partnerships with suppliers (Handfield and Nichols 1999).

Given the increased amount of attention recently to greening the supply chain (Beckman et al. 2001; Bowen et al. 2001; Carter and Ellram. 1998; Green et al. 1998; Min and Galle 1997; Narasimhan and Carter 1998; Seuring 2001; Walton et al. 1998), there still remains a lack of comprehensive frameworks and a dearth of information as to how the purchasing function can simultaneously integrate environmental initiatives into functional and strategic level decision-making processes.

Strategically, buyers should work with suppliers to explore the mutually beneficial results of solving environmental issues. Environmental issues represent an opportunity for purchasing to further influence supply chain management. For the purposes of this chapter, supply chain management involves the systematic, strategic co-ordination of the traditional business functions and the tactics used across these business functions within a particular company and across businesses within the supply chain, for the purpose of improving long-term performance of the individual companies and the supply chain (Mentzer *et al.* 2001). The greening of business processes can be a catalyst for finding ways to advance other areas of purchasing and supply management strategy. Environmental issues and imperatives can spur new ways of thinking and acting on total quality principles and the concept of continuous improvement (Baker 1996).

A different and new image of environmental sourcing is emerging—one that is cost-driven (Seuring 2001) and strategy-driven, economically justified and integrated with the corporate and product and/or process decisions. If done correctly, purchasing is part of an environmentally conscious management philosophy, which can be defined as a system that integrates product and process design issues with the issues of manufacturing production planning and supply chain management (Handfield *et al.* 1997). Purchasing can help to identify, quantify, assess and manage the flow of environmental waste through the system with the goal of reducing waste and maximising resource efficiency. The role of purchasing is to assess the available sources of supply and provide supply strategy. One criterion of this strategy is to evaluate the suppliers' capabilities regarding cost, quality, lead times, flexibility and the environment. Additional supplier evaluation criteria include distribution, safety, incidents, health records and adherence to environmental regulations and/or involvement in voluntary environmental programmes.

Supply base sourcing strategies and practices have evolved from a typically non-competitive, overlooked element of strategy before the 1970s, to a synergistic and integral part of corporate competitive advantage today. Being part of a firm's competitive advantage means keeping one step ahead of the competition. For firms who are considered to be innovators and early adopters (Moore 1991) there are many challenges and hidden opportunities to recognising and integrating the critical function of purchasing and green supply chain management. For some time now, research has shown that suppliers are critical to the competitive success of firms (Monczka *et al.* 1993). The fact that future supplier performance is expected to improve continuously adds to the complexity of the environment and the importance of purchasing.

Borrowing from Dobler *et al.* (1990), I define strategic environmental sourcing as being concerned with the development of a firm's environmental plans for its long-term material, component or system requirements. This is in contrast to a firm's plans for foreseeable, near-term requirements. Strategic environmental sourcing helps management to focus attention on long-term competitiveness and profitability rather than on short-term, bottom-line considerations. Accepting this definition means understanding the importance of the trade-off of short-term environmental cost to obtain long-term performance goals and benefits. Strategic sourcing therefore involves an action plan designed to achieve and enhance the purchasing manager's specific long-term goals and objectives for his or her function. Strategic procurement has had, and is having, a growing impact on firms' competitive stances in the marketplace.

Given the growing importance of supply chain management and the additional environmental concerns that purchasing mangers are facing, the purpose of this chapter is to briefly review the history of sourcing strategy and to highlight the schools of thought that have helped shape the strategy and practices of the sourcing function. Additionally, the aim of this chapter is to posit frameworks to aid practitioners and academic researchers to gain a better understanding of strategic environmental sourcing and supplier selection, by addressing the following questions:

- How has purchasing evolved to meet the needs of the environment?
- What is strategic environmental sourcing?
- How should suppliers be evaluated and selected?
- What metrics are available to assess suppliers?
- Where are the opportunities for future research?

The chapter is arranged into the following sections:

- Section 1.1: background
- Section 1.2: strategic planning
- Section 1.3: strategic environmental sourcing and framework
- Section 1.4: supplier development programs
- Section 1.5: a model of supplier selection for environmental sourcing
- Section 1.6: supplier assessment metrics
- Section 1.7: managerial implications, and directions for future research

The background section provides insight into the body of knowledge surrounding sourcing strategy, and is followed by subsequent sections that help to identify and highlight a framework for strategic sourcing. The following sections present an incremental model integrated into a new conceptual framework. This is followed by a discussion of the processes and steps involved in a strategic sourcing and supplier assessment model. Finally, there is a discussion of performance metrics and potential areas for new research in this growing field.

1.1 Background

Purchasing strategy first achieved a general level of recognition and interest in the mid-1970s (Farmer 1978; Rajagopal and Bernard 1993; Spekman 1981). Until that time, conventional corporate planning for a firm's long-range planning cycle began with the analysis of its products, markets and its competition—and then worked back through various operating and staff departments. Typically, purchasing or materials management departments were seldom included in this process until the long-term plans were agreed by other departments and translated into annual operating plans. This process worked satisfactorily as long as the materials were readily available at competitive prices. The importance of the purchasing function was taken more seriously when arguments were made that a function, which spends 50%–70% of a firm's revenue, should have more input concerning corporate strategy. Owing to the recessions of the 1970s, the scarcity of some resources, intensive worldwide competition and increasing international and external variables, a marked change in evolution of the purchasing function's involvement in strategy formation was necessary. Not only was the involvement of the purchasing function in strategy formation necessary, it was also critical to the growth and development of the sourcing function and to supply chain management.

If recessions of the past were a catalyst in the recognition of the importance of the purchasing function and sourcing strategy, then perhaps the timing is now right to take sourcing strategy and supply chain management to the next level. The current recession and the exponential growth since the 1970s of US regulatory requirements—such as the Occupational Safety and Health Act, the Resource Conservation and Recovery Act, the Environmental Protection and Community Right to Know Act and the Clean Water Act—and other international legal requirements have put manufacturing firms in a precarious position. Add to this the release of voluntary environmental standards—such as ISO 14001, Green Lights, the 33/50 programme of the US Environmental Protection Agency (EPA), Green Seal and many companies are realising the importance of the supply chain and the need to manage hundreds or thousands of external processes in addition to internal processes (Montabon et al. 2000). Given the increased external pressures on firms, how do firms go about integrating environmental practices into supply chain management and strategic sourcing?

Over twenty years ago, Spekman and Hill (1980) found purchasing personnel, especially at higher levels, do not spend a sufficient amount of time and energy on such important strategic activities as external monitoring. Unfortunately, this is still true in some firms today. Unless high-level purchasing personnel concentrate to a greater degree on these external relationships they will not be able to have a positive impact on a firm's strategic planning process. The point is that being an effective purchasing manager of day-to-day activities is not enough. Strategic purchasing demands that managers become adept at: (1) monitoring the external environment, (2) forecasting and anticipating changes in relevant purchasing related factors and (3) communicating and sharing purchasing-related information with suppliers as well as with internal stakeholders.

Many firms find themselves facing a significant list of critical materials that may not be readily available at competitive prices. Consequently, it is necessary to identify those potentially critical materials early in the design and planning process—and to analyse each issue thoroughly enough to determine whether a serious problem exists. In many firms this situation has led to modification of the corporate long-range planning process. The process now brings together product-demand and material-supply considerations early in the design process.

If integrated successfully, strategic environmental sourcing provides firms with opportunities to design products that are more environmentally sound, to reduce waste, to lower costs and to reduce risk. The unique characteristics of strategic environmental sourcing are the focus on the impact that changes in external environmental issues have on a firm's future material needs and supplier policies. Spekman's generalised model of the strategic procurement planning process highlights the significance and interactive nature of this characteristic. The scope of the purchasing function can be characterised in three ways: by a clerical function, by commercial activity or by a strategic business function (van Weele 1984). In the past, purchasing was looked at as either a clerical arm of manufacturing and materials management or as a tactical function of marketing and manufacturing. Currently, purchasing has become recognised as a cross-functional player in the process of strategy formation; top management recognises the importance of this function and the need for purchasing strategy to be linked to corporate strategy (Monczka *et al.* 1993). Additionally, environmental sourcing strategies are emerging; investing in the supply base through supplier development programmes is seen as an investment in the future, and purchasing managers need to think strategically and be aware of the environmental long-term opportunities found in supply chain management.

When making the transition from a clerical function to that of a strategic business function, purchasing managers will be given greater autonomy and responsibility for identifying external environmental issues and for aligning purchasing goals and firm-level goals. Sourcing strategies require the buying organisation to determine which suppliers are best positioned to provide long-term competitive advantage, what number of suppliers is most appropriate, when orders should be established with suppliers and for how long (Bowersox *et al.* 1985). The purchasing manager should understand and interpret the firm's strategic posture before deciding on the specific techniques to use in acquiring intermediate products from suppliers.

To date, the purchasing function has evolved to meet the strategic challenges of sourcing. Given this evolution, frameworks are still needed to take strategic planning to the next level and integrate strategic environmental sourcing processes.

1.2 Strategic Planning

Although strategic planning is entrenched deeply in the minds of corporate managers and market planners, strategic environmental sourcing concepts need to be diffused throughout the organisation and especially throughout the purchasing function. Before strategic environmental sourcing can affect long-range decisions at the corporate level, purchasing managers must first understand, develop and implement strategic planning more effectively at the department level. The corporate planning process must incorporate more effective integrative and co-ordinating mechanisms among the various components of the strategic planning process. The end result must ultimately enhance corporate profitability.

In order to realise the opportunities of strategic environmental planning, communication barriers must be eliminated. In some organisations corporate planners tend not to communicate well with purchasing personnel. In other companies, the interaction between the two departments is missing because purchasing managers typically do not think strategically. Alternatively, purchasing managers may not be *allowed* to act strategically. Bales and Fearon (1993) found that chief executive officer (CEOs) and presidents do not see the purchasing function as adding much value to anything but the firm's bottom-line profits and its production and operations. Additionally, purchasing is typically not expected to take risks, innovate or assure external customer satisfaction. In developing a strategic approach to environmental sourcing, purchasing managers must themselves be strategically oriented. Communication with all levels of management, and communication with the key environmental personnel within the firm, is important throughout the planning process.

It is important for the purchasing manager to recognise that the planning process is repetitious. This repetitive nature can be seen in the process of strategy development. This process is a hierarchy of strategies and plans, beginning at the corporate level, filtering down to the division level and finally to the department level. At the corporate level, the primary strategic concern is, 'What business, or businesses, should we be involved in?' At the division level, the strategy question is redefined as, 'How should we compete in a given business?' Finally, at the departmental or functional level, the strategic focus converges on integrating the various activities into the total corporate scheme and on designing strategic programmes that are aligned closely with current and anticipated environmental changes.

Environmental strategy formulation and implementation results from the synthesis of corporate and purchasing goals with the various external constraints on the entire system. The process of strategy formulation forces the purchasing manager to select from among a number of options and to focus on a more manageable array of alternative courses of action. Operationally, strategic development will be a progressive, stepped process. Successful attainment of lower-level strategies will be necessary before higher-level strategies can be enacted.

There are three levels to environmental strategy hierarchy and implementation. The lowest levels are performance-related strategies. These strategies focus primarily on managing purchasing resources, on reducing costs and on providing internal services. Mid-level hierarchies contain 'systems'-related strategies. These strategies involve issues relevant to vendor analysis, and other strategies that co-ordinate organisational sub-functions. At the highest level are the competition-related strategies. These strategies typically focus on the buyer's bargaining power, which generates purchasing leverage and, as a result, facilitates the improvement of the firm's competitive market position.

Strategy evaluation brings the planning process full circle and forces managers to confront the appropriateness of potential environmental alternatives. The objective is to understand both the process and the results of the strategic planning process. The following list of criteria is recommended for the procurement planning process:

- Internal consistency:
  Are the procurement strategies mutually achievable?
  Do they address corporate and/or divisional objectives?
  Do they reinforce each other? Is there synergy
  Do the strategies focus on crucial environmental procurement issues?
- External consistency:
  Do the purchasing strategies exploit external opportunities?
  Do they deal with external threats?
- Resource fit:
  Can the strategies be carried out in light of resource constraints?
  Is the timing consistent with the ability of the department and/or business to adapt to the change?
- Communication and implementation:

Are the strategies understood by key implementers?

Is there organisational commitment?

The dominant concern throughout the process is the focus on the effective allocation of corporate and purchasing resources to meet external constraints and opportunities.

As noted by Farmer (1978: p. 12), 'supply strategies do not stand on their own'. The purpose of developing the strategy is to help give the buying firm a competitive advantage. Companies who ignore the opportunities of sourcing strategies forego this potential advantage. The task is to develop a purchasing strategy in conjunction with other corporate strategies and the company supply base.

1.3 Strategic Environmental Sourcing: A Model

Although Spekman's model of strategic procurement planning helps to develop a foundation for the planning process, a better understanding of strategic 'environmental' planning is needed. Working with the definition that strategic environmental sourcing is concerned with the development of a firm's plans for its long-term material, component or system requirements, Rajagopal and Bernard (1993) provide a conceptual framework for the development of a competitive environmental sourcing strategy. The incremental development of an environmental sourcing strategy model has several important implications (see Fig. 1.1). First, the increasing amount of complexity of the purchasing function has contributed to the need for a systematic way to design a strategy. Second, the framework can serve as a guideline to meet the unique requirements of organisations. Third, the framework provides guidelines to help identify environmental needs and thus enable a shift in strategy. The development of this procurement strategy is divided into three phases, involving information, identification, integration, and intent:

- Phase 1: input analysis, including information-gathering and appraisal
- Phase 2: initiation of activities, including identification and integration of processes
- Phase 3: sustaining a competitive edge, including communication and implementation of strategic intent by all parties involved

Phase 1 focuses on the need to be continuously aware of the internal and external environments affecting the firm. Part of procurement's environmental operations should include the collection of information on suppliers, assessment, operational reviews, prioritisation and listing of opportunities and the dissemination of information (Peterson 1996). Information-gathering and processing is crucial to risk identification and is reflected in the ability of the decision-making unit to be proactive or reactive. Mintzberg (1994) suggests that firms with a planning office, or designated strategic planners, should involve such planners in the dissemination of data and in helping to motivate managers to think strategically. Purchasing is ideally positioned to help in this process by monitoring and gathering the required information about the supply base. In this first step of

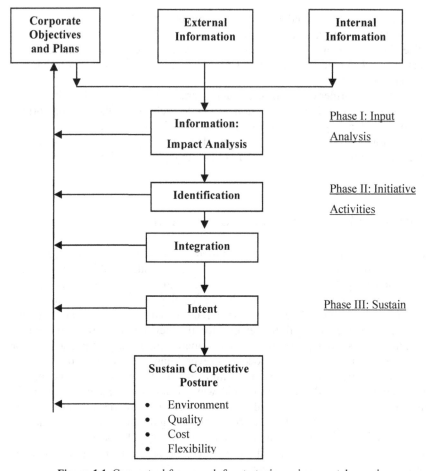

Figure 1.1. Conceptual framework for strategic environmental sourcing

the model, purchasing personnel should concentrate efforts on: assembling a team and gaining organisational commitment and support for that team; identifying metrics; setting environmental goals; and establishing systems and processes (Peterson 1996). As with so many processes today, the need for an information system, or environmental management system (EMS), to accommodate data and performance metrics will be a critical vehicle to disseminate information accurately and in real time. In developing a purchasing information system, the emphasis should be on supplying information that assists the purchasing function in making decisions that are in line with the firm's competitive strategy.

Phase 2 of the process involves the initiation of activities. Here, one finds feasibility analysis, consisting of identifying relevant information and then performing technical and economic evaluations of the data. The feasibility analysis will lead to the understanding and selection of environmental opportunities (Peterson 1996). During this phase it is important to identify relevant issues and then to integrate those issues into the purchasing plan. This includes the identification of:

- Objectives
- Materials, and the classification of those materials
- The supplier base, and the management of that supplier base
- The company's bargaining strength and the strength of critical suppliers
- A strategic decision-making hierarchy
- An organisational structure to facilitate strategy development
- The process elements in strategy development
- Behavioural elements in strategy development
- Available strategic tools

Purchasing strategies are the basis for supplier selection and development. Some of the most important factors required to formulate a company strategy involve the integration of data and information. Integration of suppliers and all functional areas has been a prominent theme in literature for some time now. The establishment of cross-functional relations with top management and environmental experts is essential. With this integration typically comes the need to formalise lines of communication and the protocols for interaction among the various functions. It is here that Mintzberg (1994) warns that caution should be used so as not to go over the formalisation edge. Rajagopal and Bernard (1993) suggest the need for some formalisation. Six modified steps have been identified as essential to the implementation and integration of a sourcing strategy within a firm. These are the integration of:

- Metrics, information and identified goals
- Objectives and policies in order to develop an environmental sourcing strategy for each supply market
- Internal functions in order to develop and implement environmental sourcing strategy
- Selected suppliers with the company in order to develop long-term strategic relationships and mutually beneficial advantages
- Logistics operations in order to execute environmental sourcing operations effectively
- Long-term strategic environmental sourcing plans with corporate planning activities and outputs

In phase 3, sustaining a competitive posture helps to ensure that firms will not become laggards. The assertion is that firms cannot become complacent, or satisfied with only meeting minimum environmental requirements. The goal is to develop a long-term plan to achieve a sustainable, competitive advantage. To obtain this goal, purchasing managers will have to justify projects and obtain funding, install equipment and implementation procedures, and evaluate performance (Peterson 1996). Basically, there must be a strong strategic environmental sourcing *intent* involved in the allocation of resources, specification of individual objectives, and in the motivation and training of those involved. Garrambone (1995) noted that, 'One of the biggest challenges for purchasing will be how to turn an individual accustomed to transactional purchasing into a skilled strategic sourcing professional'. This transition of purchasers to strategic sourcing professionals will not be done overnight, or without continually reviewing and upgrading of the firm's procurement processes and management systems (Burt 1989).

With a model of strategic environmental sourcing developed, attention now turns to supply chain issues that impact strategic sourcing. Additional information is given to help operationalise the strategic environmental sourcing model through supplier integration, including co-operative relationships, and supplier quality assurance and assessment programmes.

1.3.1 Supplier Integration

Long-term strategic advantages can be developed by working with suppliers (Rajagopal and Bernard 1993). Supplier integration is the evolutionary process used to form long-term co-operative relationships with suppliers. Ideal suppliers assist their customers with co-operation to promote product development, life-cycle analysis, performance metric development, risk assessment and timely delivery.

A co-operative buyer–seller relationship uses a supply base that consists of one or a few preferred suppliers. This will maximise bargaining power and achieve economies of scale. This is the opposite of the open-market bargaining model and clerical perspective of purchasing that attempted to sustain a competitive environment by maintaining many suppliers. By managing the relationship with suppliers, a purchasing manager will be better able to contribute to a firm's strategic success (Landeros and Monczka 1989). The characteristics of a co-operative relationship include:

- A supply pool of preferred environmental customers
- An alliance incorporating a credible commitment between buyers and sellers
- Joint activities aimed at environmental problem-solving
- An exchange of environmental information between firms
- Joint adjustments to marketplace conditions

A credible commitment to a long-term relationship is maintained because there is a concentrated joint effort to improve quality and productivity and to reduce waste and overall costs. Disputes are resolved in such relationships by working jointly on the problem instead of taking hard positions in which the outcome may depend only on power. Also, in order to develop a mutual response to changes in the marketplace, the buying and selling firms use joint problem-solving efforts. Last, there is a greater amount of data-sharing in a co-operative relationship. The goal of a successful supply chain is to trade off information for inventory whenever possible, holding inventory in the locations, quantity and form that is optimal for the entire supply chain.

1.3.2 Supplier Quality Assurance and Assessment Programmes

The importance of quality cannot be overlooked. Although it has taken the quality movement decades to become installed in US business, management has awakened to the knowledge that quality has replaced price as the key to increased market share and higher profit margins (Garvin 1983). Instead of trying to inspect quality into a product, managers have learned to design and purchase quality into the product (Burt 1989). Quality and total quality environmental management (TQEM) begin with the accurate description of the item being obtained. This requires co-ordination with the marketing and engineering functions in order to establish the item specification, environmental attributes, quality, quantity and timing.

The selection of a source for an item is the next step. The supplier must have the capability to provide the item and services the buyer requires. This often requires a co-operative relationship and a close look at the supplier's operations to determine whether it can ensure the necessary level of quality. Supplier development is a process that encompasses these activities. Monitoring of the supplier's process control data may be necessary to maintain ongoing adherence to environmental quality standards. Performance summary data will help in this effort and should be aligned with the objectives of the assessment. It is through the assessment that a better understanding of the supplier's processes comes about. It is also here that supplier development programmes can target potential problem areas in the supplier and work with that supplier to move toward a mutually beneficial long-term relationship.

1.4 Supplier Development Programmes

The development of suppliers is important to maintaining a purchasing strategy. In order to compete effectively in global markets, a company must have a competent supply base. Suppliers must be able to produce high-quality parts and materials at an acceptable cost and deliver these on a timely basis. A supplier development programme can link purchasing strategy with a firm's corporate competitive strategy. With the increasing introduction of new technology, just-in-time (JIT) practices and the global scale of the marketplace, a supplier development programme can turn into a competitive advantage. Without a competent supplier network, a firm's ability to compete effectively can be hampered significantly.

The basic objective of the purchasing function is to secure competent supplier sources that will provide an uninterrupted flow of required materials at a reasonable cost. A supplier development programme can be defined as any systematic organisational effort to create and maintain a network of competent suppliers. At a micro level this involves seeking out new suppliers for new 'greener' products and materials. At a macro level, a supplier development programme includes activities that help suppliers continually improve quality, reduce waste and bring about a better understanding of the long-term mutual benefits to both parties.

Today, supplier development programmes are viewed as a complex organisational activity requiring formal and active involvement from a number of functional areas. Although each organisation tends to approach supplier development decisions differently, Hahn *et al.* (1990) have developed a generalised conceptual model depicting the organisational decision process. Key steps on the process involve programme initiation, programme organisation, supplier evaluation, consensus development plans and, finally, implementation and evaluation of the development programme.

1.5 A Model of Supplier Selection for Strategic Environmental Sourcing

There is a need for procedures to help understand the concept of a firm's environmental fitness. Figure 1.2 shows a new framework for purchasers to use when assessing suppliers and making decisions involving supply chain management. The basis for this model is a collection of information from Hahn *et al.* (1990), Rajagopal and Bernard (1993) and Peterson (1996). Whereas Hahn *et al.* and Rajagopal and Bernard supply a good framework for strategy and supplier development decisions, Peterson outlines the building of an environmental programme for purchasing, but with no real framework or detailed supplier assessment. Figure 1.2 is an attempt to pull together critical elements from each of the above models. This new model highlights the importance of having strategic environmental and corporate goals to focus the purchasing organisation, as well as the long-term benefits obtained from the selection of suppliers.

By breaking this model down into its components purchasers are able to understand the strategic interrelationship between competition and suppliers. To begin, support from top management is key to any type of program. The need for an environmental programme is recognised through management's desire to improve the firm's competitive position or in response to specific threats to the firm. In some cases, top management will initiate the programme directly. In other cases, ad hoc groups already working on supplier issues initiate the need for a programme. The recognition of the need for a programme is then transformed into a set of corporate and environmental objectives. These objectives should be broad-based and flexible.

By using the framework of corporate and environmental goals, performance measures for issues such as quality improvement, cost reduction and waste reduction will be created during an information-gathering phase (Peterson 1996). The process of collecting information will also include the evaluation and selection of relevant data. Environmental decision-support systems (Frysinger 2001) or environmental management systems (EMSs) become critical elements in the process of evaluating and assessing supply-base data.

Input sources for this data and information will come not only from the external marketplace and competition but also from existing suppliers and any new suppliers being considered, resulting in a supplier assessment programme. This evaluation typically will involve the measurement of efficiency, quality, cost reduction and on-time delivery and will include specific measurements regarding the environmental practices of the supplier, such as ISO certification (e.g. within the ISO 14000 series), involvement in pollution-prevention and waste-reduction programmes, hazardous waste management, and the meeting of environmental performance measures. Thus, more information will be available to make the supplier-selection decision. If a supplier does not perform well on the assessment but is still included in the supply base, the option to implement a supplier development programme can be used.

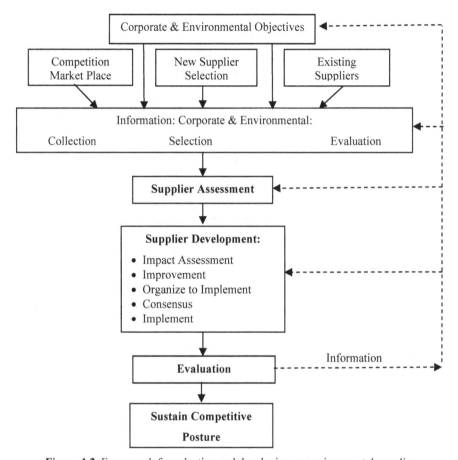

Figure 1.2. Framework for selecting and developing an environmental supplier

When the buying firm is not satisfied with the performance of a supplier, it may initiate selected supplier development activities with that supplier. To get to this step, typically, the supplier assessment has triggered the need for improvements to some aspect of the supplier's performance. The supplier evaluation is an integral part of the supplier development programme. The results serve as a guide to those areas of the supplier's performance that need the most improvement, or the least amount of attention.

The formulation of a supplier development team will need to be developed to analyse the impact assessment and set forth objectives for the programme. The organisation of this type of programme falls on cross-functional teams among U.S.

automotive manufacturers such as Ford and General Motors, on permanent supplier development departments for Japanese automotive manufacturers and on a permanent department utilising several ad hoc teams among some Korean firms (Hahn *et al.* 1990). The development team can next be organised by the material to be purchased or by the supplier to be developed.

The next step is to identify areas for improvement in corporate and environmental performance. The purpose of this phase is to find the specific causes of the problem. Supplier performance problems can be classified in terms of required supplier capabilities—technical, manufacturing, quality, delivery, financial or managerial. This classification helps to narrow the area(s) to be investigated. At this point in the process, the supplier's managers should be invited to participate in the analysis; the objective is to obtain a consensus diagnosis involving both the supplier and the buyer. It is very important that early involvement of the supplier in the analysis is critical for successful programme implementation (Hahn *et al.* 1990).

Once the causes of the problem have been identified, an implementation team with the appropriate expertise must be organised. This group then designs the plan and time-schedule for the programme. During this phase the team must determine the degree of emphasis on each developmental area. There should be a consensus between the buyer and the supplier at this point. Through working with the supplier the development plan is implemented.

The final stage of the suppler development programme is the evaluation of the results. The consensus development plans, developed by both the buyer and the seller, have been implemented. When the implementation is complete, the results should be evaluated for developmental objectives as well as for specific environmental, technical, quality, cost and delivery capability objectives. It is recommended that if the programme works properly over the short term, participating suppliers should qualify for long-term 'certified', or 'preferred' status. Suppliers who do not achieve this status should be eliminated from the supplier base (Hahn *et al.* 1990).

Supplier development programmes are only as good as the metrics by which the suppliers are assessed. But what metrics can firms use to assess their supply base? Capturing attributes of supplier performance is necessary to measure, monitor and manage sourcing and supply chain performance. In the following section, environmental metrics are suggested for use in the strategic environmental sourcing model, supplier assessment and supplier development.

1.6 Supplier Assessment Metrics

The growing importance of both internal and external environmental reporting places even more emphasis on understanding and using new performance metrics. The experiences of the author and previous research projects are now used to highlight performance metrics and look at how managers perceive some environmental metrics. In Handfield, et al. (2002) a review of the literature identified a large number of potential environmental performance indicators (EPIs). For the sake of this chapter, a sample of EPIs from the same study is presented in Table 1.1, the method of data collection briefly reviewed, and then EPIs are discussed in relation to the proposed supplier selection model in Figure 1.2. An assessment of EPIs in the previous study included identifying a number of Fortune 500 companies considering integrating environmental decisions into the supplier selection and supply chain management processes, including Daimler-Chrysler, Baxter, AT&T, General Motors and others. A number of managers at these companies were interviewed by telephone and, in some cases, actual supplier evaluation manuals were obtained from the companies. These companies maintained active databases on a number of performance indicators for their major suppliers, yet were struggling with the development of a systematic method of integrating EPIs into the supplier evaluation and selection decision. One of the biggest problems encountered by purchasers at these companies involved how many metrics to maintain, and which metrics are the most important. For the purposes of this chapter, the metrics used by a firm should help to measure the extent to which a supplier has integrated environmental management systems that positively impact environmental and operational performance. When done effectively, these management systems include management support, communication, measurement, monitoring, and reporting for internal and external stakeholders. When working with new environmental performance measurement, many purchasing managers can be unsure as to what performance metrics reflect actual supplier performance. To help overcome this potential problem a Delphi group (expert panel) was used to assess important environmental information for supplier assessment.

1.6.1 The Expert Panel

The objective of most Delphi applications is the reliable and creative exploration of ideas or the production of suitable information for decision making. In order to develop a framework of the many different types of environmental criteria shown in Table 1.1, a group of supply chain managers was assembled to conduct a Delphi group study. Managers were chosen because they were known as experts in the field of environmental management and had primary responsibility for waste reduction and materials management within their organisations. A primary task of the group was to assess the overall rankings of the criteria individually, then integrate the perspectives to create a single collective framework in which to make sense of the results (for a detailed discussion of applying the Delphi method to operations issues, see Malhotra et al. 1994).

For many firms getting started in the assessment of a supplier's environmental performance, new metrics will be necessary. For managers, the list of EPIs in Table 1.1 can be a good starting point. For performance indices to be useful, they should be easy to obtain, and assess. Difficulties of performance measurement and assessment include how to introduce and integrate new performance metrics, and how to collect data regarding the metric. Purchasing managers, for some time now, have been predicting an increased use of supplier's external reports. To this end, corporate environmental reports are becoming more useful to supplier assessments. The Delphi group expressed concern as to how environmentally conscious a supplier should be. The more relevant aspect of the supplier's environmental performance is that the organisation is improving and working toward waste minimization within its own plant and also working with its suppliers to do the same. The "greenness" of a supplier may only become clear when data is available to compare one supplier to the rest of the supply base.

Table 1.1. Examples of environmental performance indicators

Biodegradable / compostable (%)	On Environmental Protection Agency (EPA) 17 hazardous chemicals list
Commitment to periodical environmental auditing	Ozone depleting Chemicals
Contains no ozone depleting substances	Participation in voluntary EPA programs
Emissions and Waste (per unit of product)	Pre/post consumer recyclable content (%)
	Public disclosure of environmental record
Energy efficiency label	Received any EPA/(RCRA non-compliance fines
Environmentally-responsible packaging	
Global application of environmental standards	Resources and Energy (per unit of product)
Hazardous air emissions	Second tier supplier environmental evaluation
Hazardous waste	
Involvement in Superfund site	Secondary market for waste generated
International Organization of Standards (ISO) 14000 certification	Solid waste
	Take-back or reverse logistics program
Landfill – tons of waste per year	Third party certification (eco labeling)
Longer shelf life than industry standard	Total energy used
Number of hours of training on environment per employee	Toxic pollution
	Volatile Organic Compound (VOC) content (%)
Use of less hazardous alternative (% of weight/volume)	

One task for the Delphi group included going over the list of EPIs to indicate if a metric was relevant, easily assessed and important to their corporate environmental strategy. All indicators were deemed relevant, but that data for some indicators would be more difficult to obtain. For the purpose of this chapter, the most important indicators are discussed in relation to the proposed supplier assessment framework. The 10 most important indicators are shown in Table 1.2.

The performance indicators chosen as the most important by the expert panel help to demonstrate the need for purchasing personnel to understand long term environmental impacts and to not focus only on short term costs. Managers want to emphasize corporate environmental reports, second tier supplier assessment, and documented processes for managing hazardous waste, pollution, environmental management systems, and reverse logistics. As would be expected, there is a relatively large amount of focus on hazardous materials and risk management related to the important EPIs. The difficulty comes in collecting this new environmental information when firms may have little understanding of the internal processes of their suppliers. This is where purchasing personnel need to work with existing suppliers and new suppliers to identify and collect new data. If suppliers produce external reports, especially corporate environmental reports, information in these reports can more easily be used in supplier assessment processes. The evaluation, selection, and development of suppliers will come only after a systematic process is in place to evaluate all relevant performance metrics. Here again there is a need for organizations to pay attention to their own internal communication, monitoring, and reporting to make sure an environmental management system is in place to help facilitate the selection and development of suppliers. It is also important for this same environmental management system to have visibility outside of the purchasing function to help with corporate alignment of strategic and environmental objectives. Those firms considered leaders in environmental management have systems that demonstrate the internal benefits of waste reduction, and important environmental metrics. This same information is then used to help assess, and improve suppliers. The information collected for these systems is continuously updated, accurate, easy to access, and reviewed at multiple levels of management so that decisions regarding suppliers are made with the best available data.

Table 1.2. Top indicators of supplier environmental performance

Top Ten Most Important Indicators	
1. Public disclosure of environmental records: such as corporate environmental reports, and Toxic Release Inventory data	2. ISO 14000 certified: have third party certification of Environmental Management Systems
3. Second tier supplier environmental evaluation: evaluation of suppliers beyond the first tier (direct contact) firms	4. Reverse logistics program: a system is in place for the recovery of products or packaging from the consumer, or supply chain members
5. Hazardous waste management: process in place to document and manage hazardous materials and waste	6. Environmentally friendly product packaging: use of recycled materials for packaging, or waste minimization of packaging materials
7. Toxic pollution management: process in place to document and manage toxic pollution	8. Ozone Depleting Substances: substances that when used contribute to the degradation of ozone
9. On EPA 17 Hazardous Material List: does the firm use any of the chemicals on the Environmental Protection Agencies Hazardous Materials list?	10. Hazardous air emissions management: process in place to document and manage air emissions

For practitioners and researchers, the information and models put forth in this chapter give only a limited focus on purchasing regarding the complexity of environmental metrics, and the issues involving environmental assessment of suppliers. The information in Table 1.1 helps to set the groundwork for what metrics practitioners can include in supplier assessment and in the internal auditing of their own facilities and processes. It is hoped that the information from the Delphi study discussed here will help academic researchers to operationalise the concept of strategic environmental sourcing and to highlight the need to obtain this type of environmental performance information. In the next section I review the questions motivating this chapter, further managerial implications and opportunities for future research.

1.7 Managerial Implications and Directions for Future Research

With increased attention given to managing the supply chain in environmentally conscious ways there is a need for systematic development and validation of strategic models and frameworks for operationalising strategic environmental sourcing and environmental supply chain management. The development of frameworks or models is only the first step in theory development. Two models are posited in this chapter to provide guidelines and potential environmental metrics to practitioners struggling with environmental sourcing issues.

Motivations for this chapter can be found in the search for answers to several questions concerning sourcing, strategy and the integration of environment management into sourcing processes. To start, has purchasing evolved to meet the needs of the environment? It appears the stage has been set for the strategic involvement of the purchasing function in supply chain management. Purchasing is no longer a clerical function. Instead, purchasing managers and personnel have the opportunity to leverage their positions within firms to impact material acquisition and supply chain management processes through strategic long-term planning.

An additional motivation for the information presented in this chapter was a call for a better understanding of strategic environmental sourcing. Strategic environmental sourcing is concerned with the long-term material, component or systems requirements of a firm. This definition has been placed within the context of strategic procurement planning and supported with a framework for selecting and developing environmental suppliers. While the strategic environmental sourcing framework calls for collecting and using information to make long-term decisions regarding supply strategy, having a strategy is not enough. Key to any successful sourcing strategy is the need to integrate suppliers and important metrics into processes management.

Information and frameworks posited in this chapter deal with how firms should evaluate and select suppliers. In order to integrate strategic environmental sourcing into the long-term planning processes of a firm, supply base considerations need to be addressed. Supplier assessment, development and integration are important mechanisms for developing and adjusting the strategic environmental sourcing plan. Guidelines and environmental management systems that aid in the assessment and selection of suppliers who understand environmental goals are critical to the success of any strategy. Working closely with suppliers to meet environmental goals and sharing information and will help ensure a firm's supply base is a strong resource contributing to a sustainable competitive position in the marketplace.

In reviewing the literature and existing schools of thought, opportunities for future research abound. Opportunities not only take the form of frameworks for supplier decision-making when dealing with strategic environmental issues but also are evident in the measurement and assessment of suppliers.

Suggested future research involves the following questions:

- At what level of strategy formulation is it best to have environmental information?
- What are the cross-functional impacts of environmental metrics?
- What 'industry-specific' metrics are involved in the role of purchasing in supplier assessment of environmental corporate practices and products?
- How can we identification and collection of the major variables and measurement standards to test supplier development programme effectiveness.
- What are the major performance metrics that result in a better understanding of the effective role of procurement in the design process?
- What are the major factors prohibiting the integration of environmental issues into strategic planning decisions?

There are tremendous opportunities for research in strategic environmental sourcing and supply chain management. Future research should address how firms go about strategy formation, determine the short-term and long-term impacts on the firm's environmental and operational performance, look at scale and construct development and investigate barriers to environmental practices.

References

Baker, R.J. (1996) 'Let's Think Green', *Purchasing Today* (July 1996): 48.

Bales, W., and H. Fearon (1993) CEOs'/Presidents' Perceptions and Expectations of the Purchasing Function (Tempe, AZ: Center for Advanced Purchasing Studies).

Beckman, S., J. Bercovitz and C. Rosen (2001) 'Environmentally Sound Supply Chain Management', in C.N. Madu (ed.), *The Handbook of Environmentally Conscious Manufacturing*, Kluwer Academic Publishers: 317-39.

Bowen, F.E, P.D. Cousins, R.C. Lamming and A.C. Faruk (2001) 'The Role of Supply Chain Management Capabilities in Green Supply', *Production and Operations Management* 10.2: 174-90.

Bowersox, J.D., P.L. Carter, and M.R. Monczka (1985) 'Materials and Logistics Management', *International Journal of Physical Distribution and Logistics Management* 15.5: 27-35.

Burt, N.D. (1989) 'Managing Product Quality through Strategic Purchasing', *Sloan Management Review* (Spring 1989): 39-47.

Carter, C., L. Ellram (1998) 'Environmental Purchasing', *International Journal of Purchasing and Materials Management* 34.4: 28-39.

Clean Water Act (1977), Washington, DC: US Government Printing Office.

Corbett, C., and D. Kirsch (2001). 'International Diffusion of ISO 14000 Certification', *Production and Operations Management* 1.3: 327-42.

Cordeiro, J.J., and J. Sarkis (1997) 'Environmental Proactivism and Firm Performance: Evidence from Security Analyst Earnings Forecasts', *Business Strategy and the Environment* 6: 104-14.

Dobler, D.W., D.N. Burt and L. Lee Jr (1990) *Purchasing and Materials Management* (New York: McGraw-Hill).

Emergency Planning and Community Right-to-Know Act (1986), Washington, DC: US Government Printing Office.

Farmer, D.H. (1978) 'Developing Purchasing Strategies', *Journal of Purchasing and Materials Management* (Autumn 1978): 6-11.

Frysinger, S.P. (2001) 'Environmental Decision Support Systems: A Tool for Environmentally Conscious Management'in C.N. Madu (ed.), *The Handbook of Environmentally Conscious Manufacturing*, , Kluwer Academic Publishers: .317-39.

Garrambone, L.J. (1995) 'Sourcing Strategy', *American Management Association* (Member Spotlight May 1995): 62.

Garvin, D. (1983) 'Quality on the Line', *Harvard Business Review* (September–October 1983): 64-75.

GEMI (Global Environmental Management Initiative) (1997) *ISO 14000:Measuring Environmental Performance,* (March 1997: Washington, DC: GEMI).

Green, K., B. Morton and S. New (1998) 'Green Purchasing and Supply Policies: Do they Improve Companies' Environmental Performance?', *Supply Chain Management* 3.2: 89.

Hahn, C.K., C.D. Watts and K.Y. Kim (1990) 'The Supplier Development Programme: A Conceptual Model', *Journal of Purchasing Management*, Spring: 2-7.

Handfield, R., and E. Nichols (1999) *Introduction to Supply Chain Management* (Upper Saddle River, NJ: Prentice-Hall).

Handfield, R., S. Walton, L. Seegers and S. Menlyk (1997) 'Green Value Chain Practices in the Furniture Industry', *Journal of Operations Management* 15.4: 293-315.

Handfield, R., S. Walton, R. Sroufe and S. Menlyk (2002) 'Applying Environmental Criteria to Supplier Assessment: A Study of the Application of the Analytical Hierarchy Process', *European Journal of Operations Management* 141: 70-87

Hart, S.L., and G. Ahuja (1996) 'Does it Pay to be Green? An Empirical Examination of the Relationship Between Emission Reduction and Firm Performance', *Business Strategy and the Environment* 5: 30-37.

Landeros, R., and R.M. Monczka (1989) 'Co-operative Buyer/Seller Relationships and a Firm's Competitive Posture', *Journal of Purchasing and Materials Management* (Autumn 1989): 9-17.

Malhotra, M., D. Steele and V. Grover (1994) 'Important Strategic and Tactical Manufacturing Issues in the 1990s', *Decision Sciences* 25.2: 189-214.

Mentzer, J., W. DeWitt, J. Keebler, S. Min, N. Nix, C. Smith and Z. Zacharia (2001) 'Defining Supply Chain Management', *Journal of Business Logistics* 22.2: 1-26.

Min, H., and W.P. Galle (1997) 'Green Purchasing Strategies: Trends and Implications', *International Journal of Purchasing and Materials Management* 33.3: 10-17.

Mintzberg, H. (1994) 'The Rise and Fall of Strategic Planning', *Harvard Business Review* (January–February 1994): 107-14.

Monczka, M.R., J.R. Trent and J.T. Callahan (1993) 'Supply Base Strategies to Maximise Supplier Performance', *International Journal of Physical Distribution and Logistics Management* 23.4: 42-54.

Montabon, F., S. Melnyk, R. Sroufe and R. Calantone (2000) 'ISO 14000: Assessing its Perceived Impact on Corporate Performance', *Journal of Supply Chain Management* 6.2: 4-16.

Moore, A.G. (1991) *Crossing the Chasm: Marketing and Selling Technology Products to Mainstream Customers* (New York: Harper Business).

Narasimhan, R., and J. Carter (1998) *Environmental Supply Chain Management* (Tempe, AZ: Centre for Advanced Purchasing Studies).

Occupational Safety and Health Act (1970), (Washington, DC: US Government Printing Office.

Peterson, J. (1996) 'Stop It and Shrink It', *Purchasing Today* (July 1996): 28-31.

Rajagopal, S., and K.N. Bernard (1993) 'Strategic Procurement and Competitive Advantage', *International Journal of Purchasing and Materials Management* (Autumn 1993): 13-20.

Resource Conservation and Recovery Act (1976), Washington, DC: US Government Printing Office.

Seuring, S. (2001) 'A Framework for Green Supply Chain Costing', in J. Sarkis (ed.), *Greener Manufacturing and Operations: From Design to Delivery and Back* (Sheffield, UK: Greenleaf Publishing): 150-60.

Spekman, R.E. (1981) 'A Strategic Approach to Procurement Planning', *Journal of Purchasing and Materials Management* (Winter 1981): 2-8.

Spekman, R.E. and R.P. Hill (1980) 'Strategy for Effective Procurement in the 1980s', *Journal of Purchasing and Materials Management* (Winter 1980): 2-7.

van Weele, A.J. (1984) 'Purchasing Performance Measurements and Evaluation', *Journal of Purchasing and Materials Management*, Autumn: 20, 3, 16-23.

Walton, S., R. Handfield, and S. Melnyk (1998) 'The Green Supply Chain: Integrating Suppliers into Environmental Processes', *International Journal of Purchasing and Materials Management* (Spring 1998): 2-10.

2

Effects of Green Purchasing Strategies on Supplier Behaviour

Burton Hamner

Director, Cleaner Production International, Producer, http://www.CleanerProduction.Com, 5534 30th Avenue NE, Seattle, WA 91805, wbhamner@cleanerproduction.com

Environmental management of purchasing and the supply chain (green purchasing) is now relatively common among larger companies and appears to be increasingly used as a corporate practice. For example, a 1995 survey of 1000 buyers of office equipment and supplies (Avery 1995) showed that 80% of respondents were taking part in environmental initiatives within their organisations. In 1993, just 40% of respondents responded this way (Avery 1995). Most readers will themselves know of organisations that are using environmental criteria of some sort in purchasing.

The practice is becoming common enough that academic efforts are being made to develop typologies of motivations and strategies. Drumwright (1994) proposes a framework explaining *why* organizations engage in green purchasing. She differentiates organizations into two general categories. The first category holds organizations for which green purchasing is a deliberate outcome of articulated strategies of socially responsible behavior. In Type I organizations, green purchasing is an extension of the *founder's ideals*. In Type II organizations, green purchasing is *symbolic* of the corporate mission. The second category holds organizations in which green purchasing is motivated by basic business reasons. Type III organizations see green purchasing as *opportune*, while Type IV organizations engage in it because of external *restraints*. Drumwright also proposes strategies for vendors who seek business from the four types of organizations.

Other investigators have studied groups of companies to identify *how* they engage in green purchasing. Lamming and Hanson's (1996) literature review and investigation of five major UK companies led them to propose five basic types of strategy used by companies for green purchasing:

- Vendor questionnaires
- Use of environmental management systems
- Life-cycle assessment
- Product stewardship
- Collaboration and relationships

Bowen *et al.* (Chapter 9 in this book) developed a more detailed categorisation of green purchasing strategies based on a survey of 24 business units in public UK companies:

- Product-based green supply: Participation in recycling initiatives that require cooperation with a supplier; Collaboration with a supplier to eliminate packaging; Efforts with suppliers to reduce waste.
- Greening the supply process: Building environmental criteria into the vendor assessment process; Use of a scoring system to rank suppliers on their environmental performance; Use of a supplier environmental questionnaire; Use of environmental criteria in the selection of strategic suppliers; Presentation of supplier environmental awards; Requiring suppliers to have an environmental management system;
- Advanced green supply: Use of environmental criteria in evaluation of buyer performance; Use of environmental criteria in risk-sharing and reward-sharing agreements; Participation in a joint clean technology programme with a supplier.

Lloyd (1994) proposed an even more general typology of purchasing strategies, with only two categories:

- External certification of suppliers
- Questionnaire and audit approach

The Global Environmental Management Initiative (GEMI) (see Chapter 3 of this book for details about this program) has published an Environmental Self-assessment Programme which is based on principles established by the International Chamber of Commerce in its Business Charter for Sustainable Development (ICC, 1991). Principle 11 of the Charter asks signatories:

> To promote the adoptions of these principles by contractors acting on behalf of the enterprise, encouraging and, where appropriate, requiring improvements in their practices to make them consistent with those of the enterprise; and to encourage wider adoption of these principles by suppliers.

GEMI has developed a four-level typology of strategies or performance in working with suppliers that proceeds from compliance, to systems development and implementation, to integration into general business functions, to a total quality approach:

- Performance Level 1: Compliance - Company reviews and gives preference to suppliers that comply with environmental, health and safety laws and gives preference to suppliers that match the company's environmental policies and standards.
- Performance Level 2: Systems Development and Implementation - System exists to evaluate potential suppliers' environmental policies. Suppliers who do not comply with environmental policies are dropped.
- Performance Level 3: Integration into General Business Functions - Supplier selection models are integrated with environmental priorities

evaluation system. A coordinated approach for evaluating suppliers is followed by all business units.

- Performance Level 4: Total Quality Approach - Corporation gives preference to suppliers who accept and implement ICC principles. Supplier evaluation system considers their environmental management quality improvement systems and suppliers are continuously being evaluated for consistency with the corporation's environmental policies. Company collaborates with suppliers to identify and implement appropriate improvements in the corporation's and suppliers' Environmental Management Systems.

The categorisations described above, however, are incomplete, perhaps because the literature has tended to focus on companies that are already known for their leading-edge practices. Smaller companies and government agencies usually have much simpler green purchasing strategies. These usually focus on product content, such as use of recycled paper or avoidance of products with toxic chemicals.

For organisations that seek to promote sustainability beyond their own operations (type-1 or type-2 organisations) the key question include: What impact does the chosen supplier management or green purchasing strategy have on the behaviour of suppliers? Does the supplier simply provide technical solutions, or does it change its own behaviour towards more sustainable practices? Although there appears to be a common belief among all green purchasing advocates that the practice will encourage broad sustainability in the long run, that is not necessarily the case. Some strategies are more likely than others to promote sustainable behaviour among suppliers. The strategies also vary greatly in the cost and effort needed by the buyers.

2.1 From Supplied Products to Supplier Behaviour

The concern about impact of purchasing strategies on suppliers stems from the recognised need for environmental management throughout the full supply chain of a product. Suppliers can produce 'greener' products' without necessarily becoming green themselves. For example, companies worldwide have stopped using chlorofluorocarbons (CFCs) in production as a result of bans on those materials imposed by buyers and regulations, but there is no evidence of a corresponding worldwide wave of companies becoming green as a result. Suppliers can make technical changes to products or production practices without changing their management behaviour significantly. This is especially true in developing countries where environmental regulations are not strictly enforced and where the competitive advantages of environmental management are not recognised. The German ban on textiles dyed with azotropic dyes has caused thousands of textile producers to change their dyestuffs to more 'friendly' types but has generally not caused them to reduce pollution or improve their environmental management practices, yet this is what is necessary for a sustainable supply chain.

Ideally, green purchasing strategies should cause suppliers to develop good environmental management practices and pass on similar requirements for improved environmental performance to their own suppliers, but, because this takes time and money, suppliers will not do it unless they also adopt environmental management as a behaviour paradigm. To use Drumwich's typology, the suppliers must move from type 4 (green as a result of restraints imposed by buyers) to type 1 or 2, green because of intrinsic motivations of corporate leaders (idealistic or symbolic of corporate commitments).

To investigate this further, a more comprehensive typology of green purchasing strategies is proposed. Each strategy can be considered for its effect on changing supplier's behaviour in the direction of reduced environmental impact and sustainable development. Each strategy can also be considered for the cost and effort it requires. Buyers will need to consider the trade-offs between impact on supplier behaviour and the cost and effort for the buyer within the framework of its own motivations and goals for green purchasing.

2.2 Strategies Used in Green Purchasing

The strategies listed below are ordered by the relative level of effort required by the buyer to implement them. Also discussed is the relative impact of the strategy on the supplier's environmentally sustainable behaviour. The ranking of strategies and impacts is, of course, subject to debate, and there are many variations possible within each strategy.

2.2.1 Product Content Requirements

Here, buyers specify that products must have desirable green attributes. This is perhaps the most common type of green purchasing and is exemplified by the many thousands of organisations that make it a policy to purchase paper with recycled content. The cost to buyers is usually not much higher than that of 'normal' purchasing practices.

The impact on supplier behaviour tends to be low and predominantly technical. The suppliers look for recycled stock to include in their products and may invest in special facilities for producing recycled stock, but there is no obvious incentive for the supplier to adopt sustainability strategies beyond those required to maintain market share.

2.2.2 Product Content Restrictions

In this case, buyers specify that products must not contain environmentally undesirable attributes. This is also a very common strategy. Bans on CFCs or other chemical content, on plastic foam in packaging and on solvent-based coatings are among the most common examples. Buyers may have higher costs because the elimination of product ingredients may require them to adjust their own production or product design. However, costs often are lower because the buyer avoids problems associated with using toxic chemicals or with disposing of excessive solid waste.

The impact on suppliers is again technical but is more likely to have positive environmental effects than are product content requirements. This is because the elimination of the use of toxic chemicals often reduces the supplier's own environmental impacts, after initial capital investments for production changes are made. The supplier may need to pass on the need for alternative chemicals to its own supplier, which increases the likelihood of a change in supplier behaviour.

2.2.3 Product Content Labeling or Disclosure

In this strategy, buyers require disclosure of the environmental or safety attributes of product contents. In the USA this is in fact common in the provision of material safety data sheets (MSDSs) with commercial products. However, MSDSs address only safety, not environmental effects. Other kinds of labels in use include environmental 'seals of approval', such as Green Seal, and indicators of relative environmental impact, such as scientific certification systems, offered by various commercial organisations.

The cost to buyers of requiring this information is very low because the buyer does not commit to actually buying a different kind of product and thus may not have to make any production or design adjustments as a consequence. However, the cost to suppliers is higher because they have to develop or obtain the label, which requires either internal research or fees to outside organisations. The impact on the suppliers may also be higher, because the exploration of environmental impact will at least be educational for management and perhaps more far-reaching than simply developing a technical solution such as eliminating a chemical ingredient or including recycled stock in the product.

Some studies support the observation that technical product standards are much more common than any other strategy. A survey of UK companies by *Supply Management* magazine showed that less than half used environmental performance to assess suppliers, but a far greater proportion claimed to use various sources to assess the environmental credentials of the raw materials themselves (Tyler 1997).

2.2.4 Supplier Questionnaires

Here, buyers ask suppliers to provide information about their environmental aspects, activities and/or management systems. The effort by buyers is higher than in dealing with products because the questions being asked must be related to the management goals and policies of the buyers, which requires internal management decision-making for the buyer. It is assumed that the buyer is ready to make the technical adjustments to production that are indicated by the match between supplier response and management goals; thus the cost of surveying suppliers is in addition to the technical costs discussed for product-based purchasing strategies.

A survey of 300 small to medium-sized businesses in the United Kingdom showed that over 40% had been asked about their environmental performance by their customers (Barry 1996). However, the survey also showed that most of the businesses were unconvinced of the need to improve their environmental performance, and 84% were unaware of the duty-of-care regulations about disposal of waste.

The cost for suppliers is in the development and provision of information. Suppliers who already have good environmental management information will find responding to questionnaires relatively easy. Others without the information to hand will have to develop it. The impact of this strategy on suppliers is questionable. For those suppliers who must develop new information, the discovery process may have an impact on management. However, without a clear indication from buyers that certain kinds of answers will result in negative action (reduced purchasing preference) there is no reason to assume that suppliers will take action beyond providing the information.

2.2.5 Supplier Environmental Management Systems

2.2.5.1 Uncertified Suppliers
In this case, buyers require suppliers to develop and maintain an environmental management system (EMS) that generally conforms to one of the recognised international standards, such as the British Standard 7750 (BS 7750), ISO 14001 from the International Organisation for Standardisation (ISO), the European Union Eco-Management and Audit Scheme (EMAS) and the Responsible Care initiative of the US Chemical Manufacturers' Association (CMA). However the buyer does not require the supplier to have the system certified as fully compliant with the appropriate standard, either through self-certification (allowed for BS 7750, ISO 14001 and Responsible Care) or third-party certification (required for EMAS).

The US automobile manufacturing industry, among others, is adopting this approach. Ford Motor Company is considering the use of ISO 14001 as a benchmark for EMSs to be used by its suppliers, although the company has not decided to require certification of its suppliers to the standard (Bergstrom 1996).

The cost to buyers of imposing this requirement on suppliers is quite low; buyers can simply demand that suppliers have an EMS. The cost to suppliers, of course, is much higher if they have to develop an EMS if they do not have one already, or if they have to modify their EMS to meet whatever specification the buyer requires. The impact on supplier behaviour is also higher than the impact of technical requirements as the supplier will have an organised approach to environmental management.

2.2.5.2 Certified Suppliers
Here, buyers require suppliers to have an EMS that is certified. This also is of low cost to buyers and of even higher cost to suppliers to ensure compliance with EMS specifications, especially since the cost of certification by third parties may be expensive. The impact on supplier behaviour would be greater than for uncertified EMS requirements because additional management resources would need to be allocated by the supplier.

All three of the major international EMS standards—ISO 14001, BS 7750 and EMAS—require that the EMS address suppliers' environmental aspects or their compliance with the buyer's environmental policies. However, the standards do not specify what the suppliers should do or whether the suppliers themselves should have EMSs or even be in compliance with environmental requirements.

It is critical to recognise that an EMS alone does not guarantee a significant improvement in supplier environmental behaviour (Hamner 1996). An EMS may be wrapped around a basic compliance assurance system that does not address any issues of sustainability other than minimal compliance with regulations. Thus the impact on supplier behaviour of an EMS, without additional requirements for specific moves towards sustainability, must be considered in general to be relatively low.

2.2.6 Supplier Compliance Auditing

In this strategy, buyers audit suppliers to determine their level of compliance with environmental requirements. This requires a significant effort by buyers and appears to be feasible only for larger organisations that already make a practice of close inspection of supplier operations. Buyers may use professional environmental compliance consultants to supplement their own capabilities for determining compliance status. Under this strategy, suppliers would be strongly motivated to achieve compliance with environmental requirements and would also engage in high-level dialogue with buyers about environmental issues; thus the impact would in general be higher than with the strategies outlined above.

2.2.7 Supplier Environmental Management System Auditing

In this case, buyers audit not only the compliance status of the suppliers but also their EMSs. This increases the buyer's efforts and may also entail the use of additional experts in EMS. Since the scope of supplier investigation is higher, the impact on suppliers would also be higher. Again, there remains the question about whether the supplier EMS is addressing sustainability, or only compliance assurance.

2.2.8 Buyers Set Their Own Compliance Standards

Here, buyers develop their own standards for environmental compliance and require buyers to meet these standards. Buyers conduct their own inspections to determine the level of compliance with the standards.

Members of the US apparel industry are notably using this strategy. The non-profit group Business for Social Responsibility is managing a project called 'Greening the Supply Chain in the Apparel Industry'. Members of the project include leading US apparel retailers such as Levi-Strauss, Nike, Gap, Eddie Bauer and others. The group of companies developed its own standards for waste-water discharge and is requiring their suppliers worldwide to comply with these standards. The motivation for this effort appears to be a recognition that environmental standards vary widely, especially in developing countries, and that simply requiring compliance with local standards may not result in effective environmental protection. Suppliers who do not meet the standards within a time-frame set by the buyers are dropped.

The effect on suppliers has been obvious. According to comments made by representatives of the apparel companies, suppliers are moving quickly to install pollution-control systems that will meet the group standards for waste-water discharge. Some of the buyers are also educating suppliers about waste minimisation and cleaner production approaches as ways to meet the standards, but it appears that, in order to meet the deadlines set by buyers, the suppliers are moving quickly to the proven technologies of pollution control. Thus, this approach is changing supplier behaviour away from deliberate non-compliance but does not itself lead to supplier programmes for long-term environmental sustainability.

2.2.9 Product Stewardship

In product stewardship, buyers take responsibility for managing the environmental effects of products throughout the product life-cycle. Companies such as IBM are notable for their efforts to engage all the producers in their supply chain in discussions regarding environmental sustainability (Lamming and Hampson 1996). The Asset Recycle Management programme at Xerox focuses on managing all Xerox product materials from 'cradle to grave', with the objective of recovering asset values and reducing costs (Bhushan and Mackenzie1994). This involves working with suppliers throughout the supply chain to manage recovery of materials as well as extensive application of design for environment (DfE) tools. The Responsible Care initiative of the CMA includes the Product Stewardship Code, which specifies that members must 'evaluate HSE [health, safety and environment] programmes of suppliers and require them to provide HSE information'.

Product stewardship requires a very high level of effort from buyers, but the effect on suppliers' sustainable behaviour is not obviously direct. Suppliers may make numerous technical changes to accommodate product stewardship demands, but may not themselves develop the management commitments to sustainability that are needed to ensure their own continuing efforts towards total quality environmental management. This obviously depends on the level and type of dialogue held between buyers and suppliers.

2.2.10 Education and Collaboration

In this approach, buyers educate suppliers about environmental issues and environmental management strategies and work closely with suppliers to solve environmental problems. A major focus of the education of suppliers is on the economic benefits of improved environmental performance. In 1991, the S.C. Johnson Corporation held an International Suppliers' Day Environmental Symposium at its headquarters during which the company's objectives and concerns were presented and breakout sessions were held to discuss solutions to technical problems (Makower 1994).

The Nike Corporation's Environmental Action Team (NEAT) is another example of education and collaboration with suppliers. In 1996 Nike held an environmental summit at its headquarters, attended by representatives from many of its contract manufacturers around the world as well as Nike country representatives. A significant part of the summit was educating the participants about the benefits and strategies of pollution prevention. Nike's chemists are working closely with suppliers to implement the use of water-based adhesives in place of solvent-based ones and the company has hired a full-time pollution prevention expert to visit and work with the suppliers.

This approach seems to require the same high level of effort from buyers as product stewardship, but the impact on supplier behaviour is much more specific and targeted. A major emphasis is placed on educating the supplier's top management about the economic benefits of cleaner production and pollution prevention, within an overall mission of contributing to sustainable development. Even if the suppliers do not embrace sustainability as their own mission, under this strategy they are more likely to move their behaviour towards sustainability than they are under any other.

2.2.11 Industrial Ecology

In this strategy, buyers work with suppliers and with customers to develop a fully integrated system for recycling and re-use of materials within an industrial ecology framework. The only notable example of industrial ecology in practice is the famed Kalundborg industrial area, which evolved over time rather than through deliberate planning (KCIS 1996). In this system, organisations are both buyers and suppliers to each other.

Companies such as AT&T have embraced the concept of industrial ecology (see the words of Robert E. Allen, chairman of AT&T; Allen 1995), but putting it into practice remains a formidable obstacle that may never be fully realised on a large-scale basis. The impact on suppliers is equal to the effort to that on buyers, because effectively they become interchangeable. With many buyers and suppliers, any imbalance of effort can significantly impair the metabolism of an industrial ecosystem.

2.2.12 Overview

Figure 2.1 shows the relative positions of the green purchasing strategies outlined above against axes of impact on supplier behaviour, and buyer's level of effort. The positions of the strategies can vary as the strategies are modified or combined, but Figure 2.1 does indicate that there is a clear trade-off between the buyer's desire to promote supply chain sustainability through green purchasing and the level of effort required.

Recent research supports this generalised set of relationships. In a survey of US chemical firms, Theyel (see Chapter 8) found that a reciprocal learning process between customers and suppliers occurs as firms exchange information to set and meet environmental requirements. Firms that collaborate with customers tend to collaborate with their suppliers similarly, with the greatest successes in waste reduction occurring in firms that meet their customers' environmental standards and in turn set standards for their suppliers.

In the United Kingdom, Charter et al. (2001) of the Centre for Sustainable Design conducted a survey of major corporations to evaluate the implementation of sustainable supply chain management (SSCM). They also found that the firms with the most impact on their suppliers' environmental behaviour were collaborating closely with those suppliers and making significant efforts over time. Charter *et al.* reported that key factors that have influenced successful SSCM strategies have been

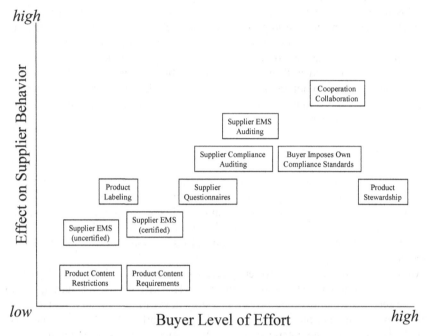

Figure 2.1. Green purchasing strategies: level of buyer effort in relation to impact on supplier behaviour

the power of companies over the supply chain and the role of business risk drivers in forcing companies to manage risk more effectively in their supply chains. The key measure of the success of SSCM tools appears to be the amount of buy-in from senior management (Charter *et al.* 2001).

2.3 International Purchasing Concerns

The assumption that green purchasing promotes better environmental performance from suppliers is particularly questionable when suppliers are located in countries or locations where environmental regulations are lax or unenforced. In such places there are few or no other drivers for improved performance and usually very little information available to suppliers about the competitive benefits of environmental management. My experience visiting manufacturing companies in South-East Asia has revealed numerous examples of companies that have changed their product contents to meet Western buyer requirements, but the companies have not adopted environmental management practices or in some cases have not even installed pollution control systems necessary to meet local environmental protection standards.

The few instances we have seen of top management commitment to environmental performance have been in companies where the Western buyers have engaged the management in serious dialogue about the need for and benefits of environmental management. Even with the buyer intentions and goals clear, it has been necessary for the buyers to provide numerous examples of how environmental management has improved the profits of companies *in developing countries*. Examples of Western companies saving money through environmental management do not impress managers in developing countries.

Fortunately, there are now many published examples of environmental management success stories from developing countries, and these can be used to enhance discussions with suppliers. But personal, repeated contact appears to be critical in getting supplier interest, and the buyer may be the only source of information available about environmental management strategies that can improve the supplier's business position.

Even so, the impact of green purchasing strategies on suppliers in developing countries is generally going to be very limited, for a simple reason: buyers can have an effect on their immediate suppliers, but it is very unlikely that the suppliers will in turn try to have an effect on their own suppliers. It usually takes a great deal of effort simply to respond to the buyer's immediate concerns about compliance and product contents. Buyers will probably have to work closely with suppliers if there is to be any transfer of environmental management practice or concern further down the supply chain.

2.4 Recommendations for Managers

Before becoming focused on green purchasing, managers need to decide what outcomes they hope to achieve. If the objective is to reduce the environmental concerns created by excess packaging, toxic ingredients and so on then firms can include environmental specifications in the product requirements, but if the objective is to actually improve the suppliers' own environmental performance it will be necessary to have a more extensive interaction with them directly.

Managers should begin with an evaluation of their own firm's capacities. Changing supplier behaviour to improve performance takes a lot of work. Does the firm have significant influence with the supplier? Is it prepared to engage the supplier at top-management levels in discussions about the environment? Is the firm's own top management committed to long-term environmental collaboration with suppliers? Is strategic supply chain management a focus of the firm and is it building capacity in this area in general?

Next, it is important to determine specific goals for the supplier's environmental performance. These are often focused on product designs, in which case the collaboration is to be a technical one. But if the goal is for the supplier to have improved management of environmental issues the collaboration will need to be at a managerial level, and possibly include capacity-building or mentoring for the supplier. The goals should also be in line with the firm's own strategic directions. For example, a firm seeking a low-cost position should focus on improving the supplier's efficiency of resource use and on waste minimisation, which can reduce the supplier's costs and thus the sales costs. If a firm is seeking a differentiation strategy or a high-value strategy the goals should be to improve the environmental quality of the firm's products, in collaboration with the supplier.

The firm should then begin building the capacity of its own purchasing managers. They are the contact points and interfaces with the suppliers and they need to be well-trained in concepts of environmental management, strategic supply chain management and collaboration. This is no small undertaking in many cases. Many purchasing managers have relatively adversarial relations with suppliers. They are always trying to get lower prices and faster delivery and better quality, but not in a collaborative way. To promote environmental improvement, purchasing managers need to act as mentors and advisors. Environmental staff should train purchasing staff, and vice versa.

Several excellent new tools have been developed to help train purchasing managers. The Global Environment Management Initiative, a consortium of major US corporations, has produced the publication, *Strategic Sourcing: Environment, Health and Safety: New Paths to Business Value*. This source is designed to train purchasing managers about the business benefits of promoting environment, health and safety to suppliers and contractors and provides specific methods (Harris covers this project in detail in this book; see Chapter 3). The US EPA (2000) has produced a manual, the *Lean and Green Supply Chain*, which also provides training. Many other organisations have also produced training materials. There is no shortage of resources, many of which are available for free.

Finally, all should participate in an exploration of the potential benefits to both buyer and supplier from environmental performance improvement. Green purchasing in a collaborative framework can have significant long-term benefits, but these need to be clearly identified and understood by all parties. It may be possible to estimate cost savings, but it is more important for all parties to understand that long-term improvement in environmental performance is a result of better management, which has benefits across the board that are both tangible and intangible.

2.5 Conclusions

Green supply management is becoming a major component of corporate environmental management strategies. When the motivation for green supply management is for business opportunity or to respond to external restraints, then buyers are not likely to be concerned about the impact of the strategy on the suppliers' environmental behaviour. But if the motivation for green supply management is based on leadership commitment to sustainable development or the desire to promote sustainable development generally, then the question of impact on supplier behaviour becomes very important. The more directly the buyer is involved with the supplier, and especially with the top management of suppliers, the more likely it is that buyer commitment to sustainability will have an effect on the supplier's behaviour. In many cases it will be necessary for the buyer to make the business case for environmental management and to educate the supplier in methods for reducing environmental impacts.

Organisations that seriously want to promote environmental sustainability will need to recognise that green purchasing is only a limited tool. For it to have a significant multiplier effect the organisation will need to commit the resources necessary to engage suppliers in sustained dialogue and education. The purchasing department will need to become a centre of excellence in 'train the trainer' for environmental management.

References

Allen, R.E. (1995) 'Foreword', in T. Graedel and B. Allenby (eds.), *Industrial Ecology* (Englewood Cliffs, NJ: Prentice Hall).

Avery, S. (1995) 'Buyers Go Green: Slowly', *Purchasing* 119.4: 43-45.

Barry, A. (1996) 'Buyers Start to Spread the Green Message', *Purchasing and Supply Management* (February 1996): 21-23.

Bergstrom, R. (1996) 'The Next "Quality" Job at Ford: Getting Green', *Automobile Production* 108.11: 54.

Bhushan, A., and J. Mackenzie (1994) 'Environmental Leadership Plus Total Quality Equals Continuous Improvement', in J. Willig (ed.), *Environmental Total Quality Management,* (New York: McGraw-Hill): 71-93.

Charter, M., A. Kielkiewicz-Young, A.Young and A. Hughes (2001) *Supply Chain Strategy and Evaluation* (London: Centre for Sustainable Design, University College).

Drumwright, M. (1994) 'Socially Responsible Organisational Buying: Environmental Concern as a Noneconomic Buying Criterion', *Journal of Marketing* 58.8: 1-19.

Hamner, B. (1996) 'A Strategic Approach to ISO 14001', *Corporate Environmental Strategy* 4.3: 46-52.

ICC (International Chamber of Commerce) (1991) *Business Charter for Sustainable Development* (Paris, France: ICC).

KCIS (Kalundborg Centre for Industrial Symbiosis) (1996) *Industrial Symbiosis: Exchange of Resources* (Kalundborg, Denmark: KCIS).

Lamming, R., and J. Hampson (1996) 'The Environment as a Supply Chain Management Issue', *British Journal of Management* 7 (Special Issue, March 1996): 45-62.

Lloyd, M. (1994) 'How Green Are My Suppliers? Buying Environmental Risk', *Purchasing and Supply Management* (October 1994): 36-39.

Makower, J. (1994) *The E Factor* (Oakland, California: Tilden Press).

Tyler, G. (1997) 'Blueprint for Green Supplies', *Supply Management* 2.7: 36-38.

US EPA (US Environmental Protection Agency) (2000) *Lean and Green Supply Chain* (Washington,   DC: US   EPA).

3

New Paths to Business Value: Linking Environment, Health and Safety Performance to Strategic Sourcing

John Harris

Eli Lilly and Company, Health, Environmental, and Safety Department, Lilly Corporate Center, Indianapolis, IN 46285, jharris@lilly.com

In this chapter I summarise the results of efforts by the Global Environmental Management Initiative (GEMI) included in a series of documents, reports and tools that help business achieve environmental, health and safety (EH&S) excellence. This GEMI effort was the first document to systematically explore the importance of EH&S issues to procurement decisions. A complete report can be obtained from the GEMI website, at www.gemi.org.

For this study, EH&S professionals of GEMI member companies worked with their procurement colleagues to better understand how EH&S performance affects the business value of strategic sourcing. It was concluded that, by working together, improvements in both EH&S and financial performance could result. Companies that integrate EH&S concerns in their strategic sourcing can create business value by:

- Reducing downtime, product life-cycle cost and time to market
- Minimising risks and liabilities
- Enhancing reputation and market share
- Reducing overall costs arising from EH&S considerations

Today, most goods and services procured have an impact on the environment and/or the health and safety of employees, customers or surrounding communities. These impacts can affect the total cost of goods and services, the quality of products, the ability to conduct business and the reputation of the company. Greater awareness of these impacts can increase the business value of procurement decisions.

In this chapter, five topics are addressed related to business value and the ways in which strategic sourcing can enhance that value by addressing the EH&S performance of products and suppliers. Topics and case studies have been chosen to help identify and pursue selected business value opportunities, with selected suppliers, using appropriate procurement tools.

Major topics have been organised to answer a series of questions, including:

- Topic 1 (Section 3.1): Is EH&S an important source of business value in my supply chain?

- Topic 2 (Section 3.2): How do I find untapped business value in my supply chain?
- Topic 3 (Section 3.3): How can I use EH&S criteria to add business value?
- Topic 4 (Section 3.4): How can I assess and improve supplier EH&S performance?
- Topic 5 (Section 3.5): How can I improve EH&S performance through outsourcing?

A prescriptive set of examples and case studies are developed, detailing how different companies from a wide range of business sectors manage these questions and derive business value. Through this approach it was possible to address several items:

- To illustrate how business value can be enhanced by adept management of EH&S issues in the supply chain
- To encourage a selective approach that is appropriate for each unique company and for different types of suppliers and supplier relationships
- To be a practical resource for procurement staff, helping them understand and pursue business value opportunities that might otherwise be missed
- To encourage dialogue and effective collaboration between procurement and EH&S departments
- To steer companies toward practices that add business value
- These issues will be outlined in this chapter.

3.1 Environment, Health and Safety as an Important Source of Business Value in the Supply Chain

Skillful management of suppliers has become increasingly important to the corporate bottom line, and many companies are adding EH&S elements to strategic sourcing initiatives. For example:

- Texas Instruments, Motorola and General Motors subcontract on-site chemicals management to expert suppliers and share the savings that result.
- Procter and Gamble involves supplier experts on planning teams designing for environment to minimise the total costs and impact of a product through its life-cycle.
- Volvo calculates the environmental impacts associated with each car and extensively rates its suppliers' efforts to reduce those impacts.
- Bristol-Myers Squibb, IBM and Xerox have encouraged their suppliers to develop environmental management systems consistent with ISO 14001.
- Firms are paying millions to clean up sites of contract manufacturers that went bankrupt. Others report business interruptions and legal expenses resulting from supplier mishandling of EH&S requirements.
- Meanwhile, many companies and their suppliers are being inundated by a burdensome number of long and fairly detailed questionnaires from

customer companies concerning their EH&S management systems and performance metrics.

These examples illustrate different types of business value and some problems to avoid as EH&S concerns grow in business importance.

3.1.1 The Procurement Function and Environmental, Health and Safety Performance Criteria

Procurement departments are adding more EH&S criteria to product or service specifications. Many procurement departments will also add EH&S performance and management criteria to their assessment of suppliers. These statements can be made with confidence because of the following trends:

- Growing pressure from customers, advocacy groups, investors and shareholders: businesses, households and governments increasingly want to buy 'green' products. Government purchasing agencies are 'raising the bar' with new EH&S specifications for products they buy. The market for environmentally friendly goods is over US$200 billion (see Narasimhan and Carter 1998: 10). Business customers want reduced hazards. In addition, company behaviour is becoming more 'transparent'—meaning that many companies' EH&S performance is public (either voluntarily or not) and they cannot hide their risks by outsourcing them. An increasingly aware public holds companies accountable for the actions of their suppliers. Social investment portfolios include supplier efforts in screening.
- The changing regulatory landscape: given the high costs of compliance, many businesses find value in proactively addressing potential regulations or by using EH&S excellence for competitive advantage.
- Expanding definitions of liabilities and risks: definitions of product liability have been expanded. European nations have taken the lead in holding manufacturers responsible for the end-of-life impacts of their products. Liability is also being pushed back up the supply chain.

These trends are summarised in Figure 3.1. Since regulations, risks and market pressure regarding EH&S issues are likely to grow indefinitely, leading-edge companies buy the best EH&S performance they can afford. They also seek supplier allies that are committed to improving their own EH&S performance.

3.1.2.1 Proactive, Strategic Procurement
Purchasing is evolving from transaction management toward more active engagement with other departments and a greater role in shaping design decisions and product specifications. Companies are using interdepartmental teams to optimise whole-system performance of the entire supply chain, integrating departments that would otherwise strive to maximise performance from their own perspective and thereby creating a suboptimal result for the entire company. Systems approaches to supply chain management have led to more detailed accounting of costs and have produced dramatic cost reductions for many companies.

3.1.2.2 Outsourcing

Companies are outsourcing the work of entire departments; a growing number outsource their manufacturing altogether. In this environment, adaptive companies tend to be those with a nimble, well-co-ordinated horizontal network of smart suppliers and subcontractors. This is particularly true when products are complex and product innovation cycles are short.

3.1.2.3 Supplier Consolidation and Strategic Sourcing

By exerting corporate influence over facility-level purchasing decisions, and by developing strategic partnerships with suppliers, companies have radically reduced the number of suppliers and gained more control over procurement costs. Subcontract operations through outsourcing make certain suppliers more critical and extend liability throughout the life-cycle, and the result is a significant shift of corporate EH&S risks and opportunities off-site and beyond the direct control of the EH&S department. As a consequence, some of those business risks and opportunities may become the responsibility of the procurement function.

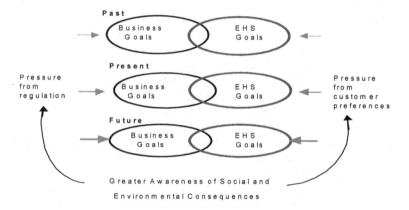

Figure 3.1. Increasing convergence of business and EHS goals

3.1.2 The Business Value of Integrating Environmental, Health and Safety Criteria into Procurement Practice

The business value that results from superior EH&S performance of inputs and suppliers can usefully be divided into six categories, shown in Figure 3.2:

- Direct costs of operations: these costs include the cost of purchasing inputs, and any expenses that are typically associated with the use of those inputs in the supply chain.
- Hidden and indirect costs: these costs are buried in other budgets and are often a very significant component of EH&S-related business value potential. Typical hidden costs include the costs of training, protective equipment, insurance, storage, waste disposal, permitting, record keeping and inspections. Hidden costs are predictable and routine but are rarely accounted for.

- Contingent costs: these costs are more visible, but they result from occasional events such as spills, accidents and lawsuits. Reducing the odds of such events is another frequent goal for managing EH&S issues in the supply chain.
- Relationship costs and benefits: these involve the perceptions of important stakeholders, including employees, stock owners, Wall Street analysts, regulators, public interest groups and customers.
- Benefits of superior products: EH&S improvements can be a product feature (*e.g.* ease of recycling, absence of toxins) or can indirectly allow for improved products.
- Marketing benefits.

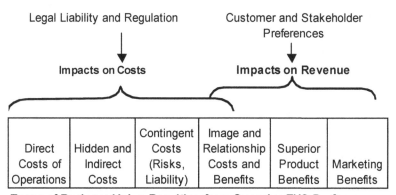

Figure 3.2. Types of business value, and their drivers

The drivers of regulation, liability and market and/or stakeholder pressure have been described in the trends listed in Section 3.1.1. Procuring better EH&S performance can have bottom-line impacts (reducing costs) and top-line impacts (increasing revenues).

The four aspects of suppliers managed by procurement (input characteristics, transportation, supplier process and supplier management) can all directly or indirectly impact business value, as shown in Table 3.1.

Once business drivers and types of business value are understood, it becomes clear that different industries have quite different levels of business value at stake in EH&S sourcing initiatives. If customers do not care about environmental impacts, the right-hand columns of Table 3.1 are empty. If processes are so benign that they encounter few EH&S risks and regulations, the business value in the left-hand columns will approach 'empty'.

Table 3.1. Procurement-managed environmental, health and safety (EH&S) impacts on business value

| | **Types of Business Value** | | | | | |
| | Impacts on Costs | | | Revenue Impacts | | |
Procurement Variable X shows a potential impact on business value.	Direct Costs of Operations	Hidden and Indirect Costs	Contingent Costs (Risks, Liability)	Image and Relationship Costs and Benefits	Superior Product Benefits	Marketing Benefits
Purchased input (product or service)						
Cost of Input	X					
EHS-related features		X	X	X	X	X
Other features		X		X	X	X
Consistency/Quality	X	X	X	X	X	X
Input "Pedigree"/origins				X	X	X
Transportation/Delivery						
On-time delivery	X	X		X		X
Transportation mode	X	X				
Supplier Process						
EHS impacts			X	X	X	X
Other qualities				X	X	X
Supplier Organization						
Financial strength			X			
Reliability of supply	X	X	X	X		X
Supplier expertise	X	X	X	X	X	X
Supplier commitment	X	X	X	X	X	X

3.2 Finding Value in the Supply Chain

Procurement departments are always looking for new opportunities to cut costs and increase total value. Procurement staff should understand which EH&S issues have business implications for their company and why, and how important they are. In this section I describe a systematic process to assess EH&S impacts (and their financial consequences) through the entire product life-cycle. The life-cycle perspective is important because customers, regulators and courts are becoming increasingly concerned about the environmental impacts before and after manufacturing occurs. Procurement has primary responsibility for the supplier management phase of the product life-cycle and sometimes a support role for manufacturing and product design, as shown in figure 3.3.

Steps in the Product Life Cycle

SUPPLY CHAIN		MANUFACTURER			CUSTOMER	
Supplier operations	Material acquisition	In-bound logistics, packaging	Value creation or manufacture	Out-bound logistics, packaging	Product use	Product retirement
Terms and Approaches for Managing the Product Life Cycle						
Life Cycle Analysis and Design for Environment						
Supply Chain / Supply Chain Management						
Supplier Management						
Materials Management						
Operations Management						

Figure 3.3. Steps in the product life-cycle: terms and approaches for managing the product life-cycle

There are three points to note about the systematic search for value opportunities in the product life-cycle:

- The value search can be focused on EH&S-related business issues alone, or it can have a broader focus on optimisation of business performance along the supply chain, considering all factors contributing to cost, quality and risk.
- The review of the product life-cycle should involve an interdepartmental team, including staff working in marketing, design and operations engineering, strategic planning, procurement, and EH&S. In the absence of an interdepartmental effort, procurement staff could work through this life-cycle review alone or with EH&S support. They could identify the value-adding opportunities within their control, such as improving supplier performance and identifying input substitutions to reduce the toxicity, allow for easier disassembly, increase recycled content and so on.
- This systematic scanning and prioritising exercise should be repeated or reviewed every year or two, because EH&S regulations, liability concepts, market preferences, competitor positioning and information are evolving rapidly.

The systematic search for value opportunities in the product life-cycle has four possible steps.

- Step 1: identify EH&S impacts along the product life-cycle of a particular product line. EH&S-related business risks and opportunities can be assessed at each point in the product life-cycle, for many different types of impacts. Table 3.2 or a similar tool can help map all EH&S impacts. Highlight the cells in Table 3.2 that indicate significant impacts on health and the physical and biological environment. The row titles on the left-hand side provide a fairly complete checklist of possible EH&S impacts.

Table 3.2. Scanning matrix for environmental, health and safety (EH&S) impacts, risks and value opportunities

Product Life Cycle Impact/Value Screening Matrix

Possible EHS Impacts	Drivers for Concern		SUPPLY CHAIN		MANUFACTURER			CUSTOMER	
	Cost/ Risk	Reputation	Supplier process	Material acquisition	In-bound logistics, packaging	Value creation or manufacture	Out-bound logistics, packaging	Product use	Product retirement
Natural Resource Use									
Energy consumption									
Depletion of water resources									
Unsustainable resource use									
Environmental Impacts									
Degradation of ecosystems									
Extinction of species									
Bio-accumulative pollutants									
Ozone depleting releases									
Global warming gasses									
Other chemicals released to air									
Water pollution (surface, ground)									
Indoor air pollution									
Hazardous solid waste									
Other solid waste									
Safety Risks									
Chemical									
Mechanical									
Electrical									
Fire and explosion									
Health Impacts									
Acute toxicity									
Carcinogencity									
Developmental/reproductive toxicity									
Irritancy, sensitization									
Ergonomics									
Noise									
Radiation									
Endocrine disruption									

Possible Coding Scheme: blank - Not relevant   ? - Impacts need more study   x - Current response adequate
highlight - Relevant   O - Clear risk or opportunity   (+) - Competitive advantage

- Step 2: identify which of these impacts have business consequences. Whether or not they do have such consequences is a function of the social, regulatory and economic environment. Screen out the EH&S impacts and issues that do not yet have financial implications, and assign a business value goal or ranking to the remaining impacts. There are three probable business value goals or rankings, with rankings and impacts varying from one industry or company to another:
  -To meet current legal requirements and prevailing standards
  -To exceed these standards by a comfortable margin (meeting voluntarily selected standards)
  - To optimise performance
- Step 3: identify value-adding opportunities to improve EH&S performance. Begin by examining impacts that are costly to manage, cause regulatory violations or are particularly salient to customers and regulators. Note any impacts where current performance is below the business goals set in step 2. Note any areas where the business goal is to optimise value and minimise risk.
- Step 4: estimate and compare the business value of current practice and likely alternatives, taking into account the six different ways EH&S performance can impact profitability by using the value dimensions defined in Table 3.1. A preliminary rough estimate should be assigned to each opportunity; only then should a more detailed calculation be considered for

the most promising options. The relative importance of business drivers will determine how much precision is sought in these estimates and how much weight is allocated to different types of costs. When liability or reputation dwarf cost as a business driver, many procurement decisions can be made without detailed calculations.[1]

3.2.1 Measuring the Costs and Benefits of Environmental, Health and Safety Improvement

Table 3.3 illustrates some approaches to measuring the costs and benefits of EH&S improvements in a procured material. Remember that costs hidden in other budgets are a major source of EH&S related savings. The benefits are ordered from hardest to softest; most would persist into future years.

Strategic advantage, reputation and public good will are often labelled as 'soft benefits' because it is difficult to estimate their financial value precisely. Companies and departments differ in terms of how much soft benefits are allowed into a calculation of the value of a given investment. At one end of the spectrum are those organisations that consider only the hardest of quantifiable benefits when evaluating programmes. The risk of this approach is that the company will miss opportunities to make strategic, quantum-leap improvements in their operations because these improvements cannot be fully justified by using only hard numbers. At this end of the spectrum, money is often 'left on the table' by disregarding intangible benefits that may have added significant, albeit difficult to quantify, value to a discarded programme.

At the other end of the 'benefit' spectrum are those organisations that allow strategic considerations to weigh heavily in their investment decisions. Because the value of soft benefits is more difficult to assess accurately, the risk here is that the company will find itself burdened with a relatively high number of unproductive and ineffective programmes that were justified largely based on fuzzy estimates of soft benefits. This is particularly a problem when no good mechanism exists for evaluating and stopping initiatives that are not realising an adequate return on investment (ROI). Assuming that most companies lie somewhere in the middle of this spectrum, Table 3.3 offers an example of how a typical company might calculate first-year benefits when evaluating a proposed strategic sourcing programme.

3.3 Adding Business Value Through Procurement

Procurement professionals usually have the primary responsibility for the supplier management phase of the product life-cycle (as shown in Table 3.2), where

---

[1] For example, Intel recently decided to join many other companies in agreeing to buy only those wood products certified to come from well-managed second-growth forests. Intel buys few wood products (e.g. pencils), so the cost consequences were small. To protect reputation, a decision was quickly made without a detailed calculation of pencil prices.

business value can result from the careful inclusion of EH&S criteria in product and supplier selection. They also play a key support role in the design and management of the rest of the product life-cycle by informing colleagues of new supply and value options and by facilitating supplier participation in collaborative planning.

Table 3.3. First year benefits of a sample sourcing initiative

Type of Benefit	How to Calculate It	Example
Reduced raw materials waste	Materials saved per unit x units of production	\$.25 less materials purchased per unit x planned volume per year of 1,000,000 units = \$250,000
Reduced transportation costs	Reduction in number of shipments or in shipping cost per load	6 fewer trailer loads received per year x average cost per load of \$3,000 = \$18,000
Reduced waste disposal costs	Disposal costs per unit of product x planned volume of production	\$.20 savings per unit of product x 1,000,000 units = \$200,000
Reduced compliance costs	Lowered consulting and legal expenses related to violations or permits	Avoided cost of new permit = \$100,000 if non-hazardous materials are used
Reduced cost of incidents	Average number of incidents per year x average cost per incident for cleanup and employee health/absence/overtime	2 reduced incidents per year x \$7,800 average cost per incident = \$15,600
Reduced risk of business interruption	Likelihood of risk x reduction of likelihood of risk x estimated cost of risk	2% likelihood x 50% reduction in chance of plant shutdown for a week due to supplier interruption x \$6,000,000 lost fixed cost and revenue = \$60,000 cost reduction
Customer retention rate increases	Percentage increase in repeat sales x profit per unit sale	5,000 units sold beyond plan because of increase in customer retention x \$100 profit per unit = \$50,000
Increased market share as a result of enhanced reputation	Number of new customers per year x sales per customer per year x profit per unit sale	20,000 new customers per year x 2 units sold per customer per year x \$100 profit per unit = \$4,000,000

In this section I cover what the procurement function can do once a value opportunity is identified. The procurement function has a role to play wherever these opportunities are found in the product life-cycle, but the phase of the life-cycle will determine how procurement can add value.

As shown in Table 3.4, proactive procurement departments may add value by participating in the evolution of design decisions and in the development of specifications, identifying superior alternative inputs, bringing knowledgeable suppliers into the planning process and co-ordinating the requirements of different departments. Such integrated proactive procurement focused on whole-chain financial optimisation typically results in cost savings of 10%–20% and a substantially greater increase in profits (Burt and Pinkerton 1996: 8-10). The procurement function should therefore seek an active role in initiatives such as Six Sigma and DfE (design for environment), which strive for whole-system optimisation and recognise the critical importance of the design phase in reducing costs and environmental impacts.

Table 3.4. Procurement roles and value opportunities in the product life cycle

Value Opportunity in Product Life Cycle	Procurement Role
Redesign of company's product or service	Describe available supply options, facilitate collaboration with expert suppliers, and help craft accurate purchasing specifications.
Redesign for process improvement	Describe available supply options and facilitate collaboration with expert suppliers, and help craft accurate purchasing specifications.
Decision to subcontract process	Conduct a make or buy analysis.
Streamline materials management (integrated procurement or inventory minimization)	Facilitate collaboration with other departments and suppliers to optimize whole-system supply chain performance.
Obtain better products and services	Identify and propose superior alternatives. Select new suppliers and/or work with current suppliers to improve supplier quality and process. Perform quality assurance checks.
Improve in-coming transportation and logistics	Assess total costs of logistics system and develop appropriate control mechanisms.
Improve supplier processes	Monitor process, provide education and suggestions, and facilitate expert-to-expert collaboration.
Improve supplier organizational strengths (financial soundness, reliability)	Evaluate supplier management systems, screen out weak organizations, and help suppliers improve management systems.

Although procurement integration at the design stage is a major source of potential business value, this tool is focused on supplier management, where procurement plays the lead role in securing business value. This role is critical. For a typical manufacturer, purchased inputs typically account for 60% of all product costs and 50% of quality problems in operations. Effective collaboration with suppliers can cut time to market by 25% (Burt and Pinkerton 1996: xi).

In looking at the management activities of core suppliers the tool addresses value opportunities through the assessment and management of four aspects of suppliers: purchased products and services, transportation, supplier processes, and supplier organisational characteristics. EH&S criteria can be smoothly integrated into all four, with potential impacts on business value being as shown in Table 3.1. Some of the specific aspects within Table 3.1 that link to EH&S characteristics are:

- EH&S-related features
- Input 'pedigree' or origins
- Transportation
- Supplier process
- Supplier organisation

EH&S-related features of purchased products and services are major contributors to the dangers and liabilities encountered in a company's operations. Hazardous inputs lead to very high indirect costs such as those relating to training, personal protection, record-keeping, insurance premiums, worker compensation, and waste disposal. Typically, the cost of managing hazardous materials far exceed their purchase cost. The relatively few companies that have analysed these costs were surprised by their results. Hazardous materials also increase the odds of non-routine contingent costs being incurred for fines, crisis management and lawsuits. In contrast, superior EH&S-related product features can be selling points. Examples include absence of toxins, increased content of recycled materials, ease of recycling and energy efficiency. For many companies, superior input specifications represent the greatest potential contribution of the procurement function to EH&S-related business value.

Input 'pedigree' refers to historical characteristics of product origin that are separate from the qualities of the product itself. Examples are carbon emitted to produce electricity, old-growth forests cut down to produce lumber, dolphins killed to catch tuna, and chemicals on-site or released during supplier manufacturing. Even though these characteristics arise in the supplier's process, they become 'attached' to the product. Because 'pedigree' characteristics are not part of the input itself they do not affect the manufacturer's direct and indirect costs. Their impact on value is entirely at the sales end of the supply chain, where it can be significant. Pedigree issues are becoming increasingly important to household and business customers. Volvo asks suppliers to remove blacklisted chemicals not only from the components it buys but also from their entire facility.

Transportation arrangements can sometimes be adjusted to improve business value by taking into account the EH&S impacts of transportation mode and frequency of delivery. For example, a value assessment could take into account carbon emissions from use of lorries compared with that from use of trains, the paperwork costs of receiving hazardous shipments and the cost of disposing of obsolete or unused materials. This may alter the calculated economic order quantity.

The supplier process can be extremely important as the source of 'pedigree' features described above. Also, whenever a buyer company ends up legally liable for a supplier's negligence or past practices, the issue is almost certain to involve environment, health or safety impacts. Newsworthy cases of egregious EH&S negligence by suppliers can also impact the buying company's reputation if the two companies are closely linked in the public's mind.

Supplier organisational strengths maintain and improve the quality of the supplier process. The procurement function already evaluates critical suppliers to ensure that they are efficient, reliable and committed to quality. Financial strength, strong leadership, clear priorities, thorough training and mechanisms for continuous improvement are also critical to EH&S performance. Specific EH&S-related aspects include the supplier's policies regarding environmental management systems, routines and certification, and special expertise and research efforts related to EH&S impacts.

3.3.1 Implementing Environmental, Health and Safety Criteria

For each EH&S impact deemed to have business value, the value it offers and the link to one or more business goals need to be determined. For example:

- When the business goal is reduction of legal liability, the procurement function must be very certain that critical products, services and suppliers meet or exceed specific legal standards.
- When the business goal is to protect reputation, a larger number of EH&S-relevant products, services and suppliers should perform adequately enough not to attract attention and should possibly excel in one or two aspects of particular interest to the public.
- When the business goal is to reduce operating costs and increase sales revenue, procurement should screen out clearly inferior products, services and suppliers and weight the selection process in favour of features that lower total operational cost and that please customers.

These goals may directly translate into the minimum requirements, ranking criteria, performance targets and/or contract incentives. The way the procurement function formulates and pursues these goals may vary, depending on the closeness of the supplier relationship.

3.3.1.1 Determining Supplier Importance
Suppliers can be classified at one of four levels, based on the intimacy and mutual dependence of the relationship, as shown in Table 3.5. This hierarchy of relationships often corresponds to the procurement activity that takes place at different levels of the company. Purchasing decisions made locally by individual plants are likely to be price-focused, whereas alliances and partnerships are often directed by corporate staff and consider many factors in addition to price.

Whereas reputation is most at stake in closer relationships, note that EH&S issues can have business importance at any level of supplier relationship. Product specifications are often the greatest single source of EH&S value and can be applied to any supplier. Surveyed GEMI companies reported instances where legal liabilities were incurred or production interrupted as a result of supplier EH&S failures. Most of these examples involved spot-purchase or short-term relationships, typically with waste handlers and construction contractors. This experience supports procurement experts' observation that the most serious purchasing mistakes can be traced to vague requirements in the hands of marginal suppliers.

Table 3.5. Four levels of supplier importance

Level I Spot Purchasing	There is little or no relationship with or knowledge of the supplier. Price is the key determinant of purchase. To the extent that quality is important, it is assessed based on predictable product characteristics or supplier reputation alone. Each transaction is its own business contract. Commodity items such as coal, sand, mops, and pencils are often purchased on the spot market. To control EHS impacts, change products or product specs. For example, Intel prohibits the purchase of pencils made from old growth forest resources.
Level II Competitively Based Incumbent Relationships	Suppliers have a longer-term business relationship, typically an annual contract against which purchase orders are issued. Contracts are renewed annually. Your business is theirs to lose. Relatively little technical cooperation is invested in these short-term relationships, because a better supplier may be located the next year. To control EHS impacts, change specs for the annual bid, and let the world know you are always looking for suppliers who can better meet these specs.
Level III Preferred Supplier	The intention is for a long-term relationship, that requires and benefits from fairly frequent communication and collaboration to improve or adjust supplier inputs over time. To control EHS impacts, include EHS issues in the periodic visits and meetings where progress and quality are discussed, and targets may be set.
Level IV Strategic Partnerships or Alliances	Relationships involve an even deeper level of commitment. Typically, there is an explicit or implicit understanding that supplier and buyer will share the business benefits of effective collaboration. To influence EHS impacts, add EHS to the agenda of problems the partnership must address. Write contracts so that the business value of better EHS performance is shared among the partners.

What differs across the four levels of supplier relationships described in Table 3.5 is not the business importance of EH&S impacts, but the procurement tools available to communicate and assure standards and seek improvement and the aspects of supplier performance that are evaluated. With spot purchases the focus is almost entirely on the input purchased. In closer, longer-term relationships the procurement function has time and incentive to assess the supplier process and organisation. There are also means and incentives to assess the supplier process when service providers are working on-site. Table 3.6 summarises the supplier aspects likely to be managed for each level of supplier relationship.

Table 3.6. Supplier assessment

What to assess?	Level of Supplier			
	1	2	3	4
Product quality	x	x	x	x
Transportation		x	x	x
Supplier process		x	x	x
Supplier management			x	x

In summary, the desired business outcome will determine the best ways to procure superior EH&S performance for products, services and suppliers. There may be little procurement value in conducting a one-size-fits-all comprehensive supplier survey, asking many suppliers questions about many aspects of EH&S performance. Such a survey is not adequate to minimise legal liability, tends toward overkill for the purposes of protecting reputation and is not particularly effective as a means to stimulate supplier commitment to continuous improvement.

3.4 Assessment and Improvement of Supplier Environmental, Health and Safety Performance

In this section I describe how EH&S criteria can be integrated into existing strategic sourcing tools and processes. I address:

- Assessment of supplier EH&S performance, going beyond product characteristics to focus on the supplier's EH&S performance and management systems (Section 3.4.1)
- Collaboration with suppliers to continuously improve EH&S and business performance along the supply chain (Section 3.4.2)

3.4.1 Assessing the Environmental, Health and Safety Performance of Suppliers

Companies that plan to assess their suppliers' processes and management systems should not reinvent the wheel. EH&S-related criteria can usually be integrated with the procurement tools and steps already in use to screen, select, negotiate with and monitor suppliers. Also, the EH&S metrics should be consistent with and derived from the buying company's own evolving set of EH&S performance metrics and priorities. In this section I will:

- Summarise typical procurement tools available at each stage of the procurement process (Sub-section 3.4.1.1)
- Describe how to select metrics to assess the EH&S performance of suppliers (Sub-section 3.4.1.2)
- Explain how to verify performance on selected metrics (Sub-section 3.4.1.3)

3.4.1.1 Typical Procurement Tools

In Table 3.7 typical procurement tools are summarised. Their applicability to the different levels of supplier relationships described in Box 3.1 is also noted. The tools can be broken down into three categories according to the stage reached in the process:

- Stage 1: pre-screening communications
- Stage 2: qualification and negotiation
- Stage 3: monitoring and continuous improvement

3.4.1.2 Selecting Metrics to Assess Supplier EHS Performance

Many of the procurement tools that could be used to improve EH&S performance in the supply chain require the use of metrics (*i.e.* standards for measuring suppliers' EH&S performance). In this section I briefly review the range of metrics options, then present guidance as to how companies can select a few metrics to assess and improve supplier performance.

For most procurement tools, EH&S criteria can easily be added as yet one more aspect of quality to be managed. However, in several challenging respects EH&S criteria differ from quality criteria. Unlike quality metrics, EH&S metrics are often focused on external impacts that the business has little incentive and few good tools to measure. Many companies are wrestling with the challenge of developing their own set of useful and workable EH&S performance metrics. Procurement departments should take advantage of lessons learned by their EH&S colleagues in this struggle and apply to suppliers a subset of the metrics the company has found workable and important to measure its own EH&S performance.

Figure 3.4 lists candidate EH&S performance indicators. Note that many of the metrics correspond to the impacts already listed in the assessment matrix of Table 3.3. For most of these indicators, five different questions can be used to assess supplier performance:

- Is the supplier aware of this impact or issue?
- Does the supplier have goals or policies regarding this impact?
- Does the supplier have detailed plans in place to measure, manage and improve this impact?
- What is the supplier's performance regarding this impact during the most recent year?
- Is the supplier's performance improving over time, and by how much?

The list of metrics from the table, multiplied by the five questions that could be asked, leads to 100 + possible metrics for assessing EHS performance. If individual chemicals are listed, and both corporate and facility-level assessment will be performed, the set of possibilities is even larger. Few supplier companies could assemble all this information, and few buyer companies could collect, track, assess, and verify it. To prioritize, one should apply the following three filters: 1) business value, 2) availability, and 3) procurement goal.

Table 3.7. Typical procurement tools by stage in process and level of supplier

LEVEL				TOOL	COMMENTS AND EXAMPLES
1	2	3	4		
Stage 1: Pre-Screening Communications					
	√	√	√	Policy Statements (EHS and Procurement)	These can communicate buyer goals and set the tone for collaboration.
	√	√	√	Code of Conduct for Suppliers	These communicate how business will be done with suppliers, and specifies standards and sanctions that may be applied if they fail.
	√	√	√	Minimum EHS Performance Standards	Most companies have contract language requiring suppliers to self-certify themselves .
	√	√	√	Product Specifications	Product constituents and performance characteristics can be specified.
	√	√	√	Lists of Chemicals to Avoid	Kodak, Canon, Sony, and Volvo are among companies circulating lists of chemicals for their suppliers to eliminate or reduce.
Stage 2: Qualifying and Negotiating					
√	√	√	√	List of Pre-Approved Materials	Many companies screen materials onto pre-approved lists to speed purchasing decisions.
	√	√	√	Requests for Proposal	Requests for Proposal can include explicit evaluation criteria for the supplied product or service, and for the supplying organization. Many ask for safety performance statistics and evidence of continuous improvement.
	√	√	√	Surveys and Questionnaires	Companies require suppliers to complete self-assessment forms that vary widely in detail.
	√	√	√	Required Standards of EHS Performance	Standards are referenced in contract documents, and may be customized for level 3 and 4 suppliers, or suppliers of EHS-sensitive services such as waste disposal, construction, and remediation.
		√	√	Supplier Selection Criteria/Ranking	Anheuser-Busch, Texas Instruments, Volvo, and Canon use various selection criteria and ranking beyond traditional measures.
√	√	√	√	Pre-Approved Supplier Lists	The EHS department may screen suppliers and prepare lists for Procurement to use.

Table 3.7. (continued)

LEVEL				TOOL	COMMENTS AND EXAMPLES
1	2	3	4		
Stage 3: Monitoring and Continuous Improvement					
		√	√	Audits	On-site audits are typically conducted for toll manufacturers, critical suppliers, and suppliers that dispose of waste.
		√	√	Regular Supplier Visits	Anheuser-Busch regularly visits packaging suppliers to review continuous improvement efforts and environmental management systems.
	√	√	√	Performance Reviews	These typically involve quarterly, six month, or annual progress and performance reports in formats developed by procurement, possibly with supplier participation.
		√	√	Collaboration to Solve EHS Problems	To increase recycling, Anheuser-Busch worked with a packaging supplier to develop standards for plastic binding on shipments. Motorola safety staff worked with a chair manufacturer to redesign chairs for better ergonomic performance. Collins & Aikman (carpet manufacturers) reduced volume of volatile organic compounds (VOCs) by collaborating with a supplier to reformulate products and modify manufacturing processes.
		√	√	Supplier Training and Seminars	Herman Miller holds semiannual conferences for all employees and suppliers on waste minimization, pollution prevention, lifecycle analysis and environmental design
			√	Collaboration on R&D and New Product Development	Intel works with suppliers and cross-functional teams to design new semiconductor manufacturing tools that will operate with minimum EHS impacts

Filter 1: Prioritize indicators based on business value

The first filtering criterion is importance or business value. Rate each metric based on its probable business value or strategic importance to your company. Assign one of four importance levels to each possible EHS metric:

- Unimportant - not on our radar screen of risks regulators, courts, or customers care about – ignore these aspects of performance for now
- Relevant - low liability, low cost, low marketing benefit – an appropriate goal is to have suppliers meet minimum standards, comply with laws

Common Measures of Supplier Impacts
Natural Resource Use
      Amount of energy consumed
      Depletion of water resources
      Unsustainable resource use

Environmental Impacts
      Amount. of bio-accumulative pollutants released
      Amount of ozone depleting releases
      Amount of global warming gasses released
      Amount of reportable chemicals released
      Water pollution, ground and surface
      Hazardous solid waste generated
      Amount of solid waste to landfill
      # and amount of reportable effluent spills, threshold

Safety Performance
      Does facility fall under Process Safety Management?
      Accidents/year per 100 employees
      Worker Comp costs, other insurance claims

Health Impacts
      Exposure levels for toxic chemicals
      Noise levels

Fines paid for EHS-related violations,
      Notices of Violation (NOVs)

Other measures for Supplier Impacts
      from buyer organization's EHS performance metrics

Supplier Purchasing Decisions
      Based on best practices?
      Are upstream suppliers held to EHS standards?

Supplier Process Characteristics
      Amount of specific chemicals used in process
      Amount of specific chemicals used on site
      Amount of toxic chemicals purchased
      Amount of ozone-depleting chemicals used

Supplier EHS Management Systems
      Are there comprehensive goals and policies?
      Is there adequate implementation effort?
      Rate of recent improvements
      Annual public reporting of EHS performance
      Suggestion, incentive, education programs
      Self-certified consistent with ISO14001
      Third-party certification for EHS MS
      Baldridge-style scoring of environmental quality
      Describe recent challenges and accomplishments

Normalization Variables
      # of full time equivalent employees
Revenues and value added
Production units or mass units

Figure 3.4. Exemplary supplier EHS performance metrics

- Important - area of significant risk or potential benefit – an appropriate goal is to optimize performance in the supply chain

- Critical - high risk issue – the appropriate goal is near-zero probability for major disruption, liability, or public relations crisis. Bear in mind that these metrics may be critical for only a few suppliers, and important or relevant for others.

Now consider only the metrics judged critical and important for at least some suppliers. Each critical issue should be actively monitored with major suppliers, and verified in a less costly way for relevant minor suppliers. If the total set of critical and important issues is small in number, you may be able to assess them all. However, if there are more than a few important issues, you may simplify in one of several ways:

- Focus on one or two important metrics as proxies for the larger set, with the assumption that performance on a few key indicators is a fairly reliable indicator of company EHS capability and commitment.
- Focus on the quality of the supplier's Environment and Safety Management System, with the assumption that a good management system will translate into better performance on a myriad of metrics that are not directly assessed.
- In work with current first-tier suppliers, focus on one or two important metrics at a time, then after a year or two shift focus to the next important issues.

For all but the most critical suppliers, focusing on a few metrics as proxies for others is a practical and adequate strategy. Using one or two selected metrics may be a reasonably effective means to identify the most problematic suppliers and encourage attention to EHS performance.

Filter 2: Prioritize performance indicators based on their availability
If possible, use metrics that are easy to obtain and verify. Try to limit them to:

- Quantitative information suppliers can easily assemble.
- Qualitative information that can be conveyed in short written answers and assessed in face-to-face meetings.
- Information that can easily be verified by checking against public records, accepting third party certification, or conducting on-site audits.

Sometimes cutting-edge indicators can be determined despite probable difficulty obtaining them. Some possible ways to acquire them may include these four alternatives:

- Work with industry groups to develop common tools, standards and certification methods.
- Ask suppliers only for indicators that a buying company is already calculating internally.
- Consider using another more available indicator as a proxy for the preferred indicator.
- Work with groups of suppliers to collaboratively plan ways to measure the new metric.

Filter 3: Choose indicators based on procurement goal

Select metrics appropriate to the stage in the procurement process and the level of relationship. For an initial screening of new suppliers, use a few readily available metrics. For contract provisions, use easily verified quantitative metrics. In discussions and audits with key first-tier suppliers, you can afford to touch upon a greater range of metrics, and more qualitative ones.

3.4.1.3 Verifying Performance on Selected Metrics

Assessment involves both obtaining and verifying information. Table 3.8 shows methods to accomplish these goals that are appropriate to the importance of the metric and the closeness of the supplier relationship. Table 3.8 also indicates where supplier questionnaires or surveys may be used productively.

Table 3.8. Choosing tools to collect and verify EHS performance metrics

Importance of metric	Many less important suppliers (Level II)	Relatively few, very important suppliers (Level III and IV)
Relevant metrics: suppliers should meet minimum standards	Publish standards; no active verification. Don't ask. Deal with substandard behavior if and when it occurs and is noticed, by dropping supplier.	Publish standards, ask for self-certification of compliance, spot check during audit.
Important metrics; suppliers should optimize performance	For *a small set* of metrics, request self-certification and annual reports. Verify by spot check against public records.	For *all* important metrics get progress reports, discuss in periodic meetings, check during audits.
Critical risk or goal	Use credible third party to measure or verify performance.	Verify by periodic audit, discuss in periodic meetings

Strategies may include:

- Self-certification: self-certification by suppliers is a control strategy that works on the low-cost deterrent principle. Buyers have a published standard to protect their reputation, and suppliers know the buyer can cancel their contract if they fail in a glaring way to meet standards.
- Third-party certification: third-party certification can provide information and verification at the same time. The more supplier performance can be certified by reliable third parties, the less work the procurement department has collecting and verifying information.
- Site audits: third parties can be used to verify compliance for peripheral suppliers, but for key suppliers site audits are advised for additional protection.
- Use of approved lists of suppliers: the procurement department can cut its workload by asking the EH&S department to create a pre-approved list of

suppliers, using members of industry associations when possible. For example, several chemical companies ask suppliers to complete a half-page survey if they are already members of the chemical trade organisation and a three-page detailed survey if they are not. Suppliers who have already reported to the chemical trade organisation are spared the redundant reporting to a particular buyer.

3.4.2 Collaborating for Continuous Improvement

Assessment does not necessarily build commitment. In addition to (and often separate from) supplier assessment activities, many companies sponsor development activities for their network of suppliers. For example, since 1993 Intel has held annual 'supplier days' at which more than 700 equipment suppliers come together to discuss Intel's directions and expectations, including those for EH&S. Motorola, General Motors and Ford sponsor universities open to their own and supplier employees. Most companies try to cultivate long-term relationships with a reduced number of suppliers, with one goal being the frank and fruitful communication that mutual trust allows. Supplier and customer companies stand to gain from collaborative effort and information sharing with the goal of improving their environmental performance and the environmental profile of their products.

Improving EH&S performance is a topic well suited for communication within supplier networks and between supplier and customer. Like quality, it is relevant to virtually all players and can be discussed in useful detail without forcing potential competitors to reveal proprietary information. Thus EH&S issues are easily integrated with the varied communications and relationship management tools that companies use to cultivate continuous improvement in their supply chain.

3.5 Improving Environmental, Health and Safety Performance Through Subcontracting

In this section I focus on and describe how companies can outsource EH&S-intensive functions to their suppliers to better focus on core competences and improve their own bottom line. Often, suppliers have valuable specialised knowledge and expertise that enables them to identify cost and risk reduction opportunities that their customers may not realise on their own. Suppliers sometimes provide project funding in return for receiving a share of the savings they achieve. In many instances, outsourcing of non-core business elements can result in immediate efficiency gains and increased profits, with minimal risk to the customer company.

Outsourcing is more than subcontracting in that outsourcing is more of a partnership between the service provider and the customer. The closer the collaboration, the greater the opportunities for improved EH&S performance and cost savings for all parties. The following actions can contribute to an effective partnership:

- Foster communication: communication needs to be open, two-way and frequent between the supplier and the customer. The expectations for the partnership need to be clearly established and communicated to both parties.
- Negotiate contracts that allow for flexibility and creativity: it is essential to set up the contractual relationship in a way that rewards proactive, result-oriented innovation and that drives continued cost reduction. A contractual relationship that addresses how those savings will be sharedis essential (*i.e.* there should be provision for 'gain sharing').
- Protect intellectual property: the agreement or relationship should be structured so that the supplier retains the ability to develop partnerships with other customers, which may be competitors of one another. Likewise, it is critical to protect the customer's privacy because the supplier is intimately familiar with aspects of the customer's day-to-day operations.

Outsourcing shifts the procurement focus beyond products to a service-based relationship. In Section 3.5.1, an example from the chemical industry shows that the outsourcing of EH&S-related services that traditionally have been handled in-house can be completed for the betterment of EH&S performance.

3.5.1 Outsourcing Chemical Management Services

In a traditional model, manufacturers purchase chemical products only from their chemical suppliers, and the products are sold by volume. The customer retains responsibility for managing the use and handling of the chemicals. The supplier's profit is based on selling as much product as possible; the customer's profit is based on buying as little product as possible. Thus, the aims of the supplier and customer are at odds. The net result is that suppliers have no incentive to help their customers use their products more safely or efficiently—in fact, the case is just the opposite. The supplier's market differentiation strategy is price based—the lower it can drive its unit price, the more it stands out. There is no partnership relationship here—just purchase orders. There is little collaboration and little incentive to collaborate.

As companies search for innovative ways to reduce costs and limit EH&S liabilities they are restructuring their relationship with suppliers and contracting out more and more of the chemical services and chemical management responsibilities. Table 3.9 describes alternatives for the structuring of contracts and incentives.

Several leading manufacturers have hired chemical companies to manage their chemicals through the entire procurement and production phases of the supply chain (see Resikin *et al.* 1999: 1931) This makes EH&S risk management the role of the supply chain.

Table 3.9. Chemical services industry models

Model	What is for sale? How is it priced?	Net Result for Cost Savings and Environmental Opportunities
$/pound	Chemical is sold by volume	Suppliers have no incentive to help customers use their products efficiently - in fact, just the opposite.
$/pound + Services	Chemical is sold by volume Higher price includes some consulting services	Services associated with the proper use/handling of the chemicals are a more prominent component of the relationship. These services might involve logistics, EHS/compliance, and applications. This strategy is an initial market differentiator for the supplier.
Chemical Management	Chemical is sold by volume Management services sold on itemized basis	Supplier brings greater expertise to performing chemical management activities previously handled by customer. Management fee reduces incentive to increase chemical sales for higher revenues. This model is good first step towards increased collaboration.
Shared Savings	Supplier is paid a fixed fee to meet the "chemical performance needs" of the customer.	Supplier and customer's goals to reduce waste and save money are financially aligned Both parties make money by reducing chemical use over time.

3.6 Broader Issues

In this final section, two major emerging trends with implications for EH&S supply chain management are presented. The first focuses on globalisation issues, the second on e-commerce issues.

3.6.1 International Challenges

One of the major challenges in the globalisation of companies is how to maintain high EH&S standards in activities and simultaneously keep a competitive edge. There are many challenges, including:

- The logistics of visiting distant suppliers and communicating in different languages
- Adapting operations to different climates and cultures
- Managing varying levels of regulation
- Managing the dilemmas of different standards around the world
- Implementing EH&S performance standards for suppliers internationally

Regarding managing varying levels of regulation, US companies, for example, may face different levels of regulatory requirements in Europe. Also, in developing countries EH&S regulations may be non-existent or less restrictive than those found in more developed countries. Company global standards may be needed where there are inadequate local standards to reference in self-certification of compliance.

With respect to managing the dilemmas of different standards around the world, multinational corporations (MNCs) operate facilities in countries with often quite different wage rates and different standards for practices and products. The weight of the evidence suggests that MNCs typically raise the standards of EH&S performance in developing countries. The issue is how much to raise them, and what leverage there is to accomplish this when it involves asking suppliers to do 'extra' things that they are not accustomed to doing.

Finally, a customer may face difficulties in implementing EH&S performance standards for suppliers internationally, particularly when there is only a single supplier and customer leverage is not great.

3.6.2 Using E-commerce

Business-to-business relationships are entering a new era of e-commerce, and businesses are becoming increasingly interlinked through electronic media. Procurement is being fundamentally affected as Internet-based procurement systems reduce the average fulfillment cycle, lower material and service costs and significantly lower the administrative costs associated with supplier searches, product feature assessment, process comparison, order entry, status tracking and payment processing.

The Internet and electronic communication tools can also be used to expedite and improve assessment of and communication with suppliers. Companies that have developed web-accessible reporting databases for collecting EH&S performance metrics from their own sites could quickly expand those systems to collect the same information from suppliers.

At this point in time, the following developments seem likely:

- Buying organisations will collaborate to share information about suppliers, to reduce supplier evaluation costs and to obtain quantity discounts. In such collaborations, a key issue will be the fate of criteria other than price: will higher EH&S or social standards prevail if they are needed by some but not all of the collaborating companies? Also, the need for third-party certification of EH&S quality for products and suppliers is likely to increase, to allow for non-price qualities to be obtained through net-enabled purchases.

- An increased ability to buy on the spot market, through buying consortia and through strategic partners, will allow major companies to greatly reduce the number of level-2 suppliers they manage. First-tier suppliers with technical expertise are unlikely to lose ground to spot purchases and will be even more closely integrated with their buying companies through use of e-mail and intranets. However, where the principal advantage of a first-tier supplier is to provide a 'middle-man' service of managing information or smaller suppliers, the 'make-or-buy?' decision may be reversed to favour in-house management with use of the Internet.
- Within three years, fairly standard EH&S performance information for companies will be easily retrieved from web-based central databases maintained by states, industry groups, advocacy groups or companies (e.g. Dun and Bradstreet). Some will republish numbers already in the public domain;[2] others will accept and then carry out spot checks on data supplied by companies.[3] A questionnaire will no longer be the most efficient means to learn about a supplier's EH&S performance.

3.7 Conclusions

In this chapter I have summarised the efforts of GEMI, in the USA, related to EH&S supply chain management. This chapter has also included a brief overview of the many tools, cases and lessons learned that are available from GEMI. Five major topics relating to determining the business value and implementation of green supply chain practices, metrics and evaluation have been presented with examples of descriptive and prescriptive tools and measures that can be, and that are, used by organisations. Much of the work reported has taken years of development and is still evolving. The importance of this topic should not go unnoticed by organisations, and GEMI is at the forefront of making sure that such organisational awareness of EH&S issues is at the utmost highest level.

References

Burt, D., and R. Pinkerton (1996) *A Purchasing Manager's Tool to Strategic Proactive Procurement* (New York: American Management Association).

Fletcher, A.C. (1999) 'Building QCI International World-Class Performance with the Baldrige Criteria', www.qualitydigest.com/ aug99/html/body_baldrige.html.

---

[2] For example, Environmental Defense draws on data from two US federal agencies to publish a 'chemical scorecard' website, ranking some 17,000 sites by cancer risk according to releases covered by the Toxic Release Inventory.

[3] For example, E&Q Rating AB, a new subsidiary of Scandia Insurance Co., plans to roll out a web-based EH&S quality rating based on a 250-question on-line survey, validated by public declaration and random spot checks.

GEMI (Global Environmental Management Initiative)(1999) *MNC Study* (Washington, DC: GEMI).

Narasimhan, R., and J. Carter (1998) *Environmental Supply Chain Management* (Tempe, AZ: Centre for Advanced Purchasing studies).

Reiskin, White, Johnson and Vota (1999) 'Servicizing the Chemical Supply Chain', *Journal of Industrial Ecology* 3.2 (http://mitpress.mit.edu/journals/JIEC/sample-article.htm): 1931.

US NAE (US National Academy of Engineering) (1999) 'Industrial Environmental Performance Metrics: Challenges and Opportunities', http://books.nap.edu/catalog/9458.html.

4

Integrating Quality, Environmental and Supply Chain Management Systems into the Learning Organisation

Toufic Mezher and Maher Ajam

American University of Beirut, Faculty of Engineering and Architecture, Engineering Management Program, PO Box 11-0236, Riad El Solh Beirut 1107 2020, Lebanon, mezher@aub.edu.lb, mrajam@dgjones.com

4.1 Introduction

Customer satisfaction and the quality of products have become major concerns globally to firms. Initially, Japanese organisations were able to capitalise on these issues throughout the 1970s and 1980s. The USA and Europe followed suit in the late 1980s and 1990s. Today, no firm can survive without good quality products that match customer satisfaction.

In order to meet quality requirements, many companies implemented quality management systems (QMSs) of some sort. In addition, environmental concerns have become more visible, especially within the past two decades. This increased concern has put pressure on companies to become more socially responsible and even to take a leadership role in corporate citizenship. Companies have adopted varying environmental management systems (EMSs) in order to face these environmental challenges.

The implementation of QMSs and EMSs has contributed to companies becoming more efficient and effective in delivering their products. This has led to the implementation of different tools and techniques, such as just-in-time (JIT) procurement systems, and, eventually, to the management of the supply chain. In a similar way to which organisations utilise QMSs and EMSs organisations use supply chain management (SCM) systems to deal with and collaborate with different stakeholders outside their organisations, such as suppliers, distributors, customers, government agencies, society and so on. These different management systems are very much connected and must be managed simultaneously and in harmony in order to meet customer and stakeholder needs.

However, this is not enough to ensure the survival of an organisation. The dynamic change of global markets and the speed of technological development are having a significant impact on firms. In today's world market, it is not enough to have a good quality product, be environmentally friendly and able to manage the supply chain; firms must learn how to manage change. This can only be done by continuously monitoring the external and internal environment of firms and by managing the knowledge gained in order eventually to lead to organisational learning.

Our objective in this chapter is primarily to link together QMSs, EMSs and SCM systems, considered separately in Sections 4.2 to 4.4, respectively. In addition, we will try to model how knowledge gained from these systems can be used in order to contribute to the creation of a learning organisation (Section 4.5). In Section 4.6 we integrate our analysis of the three systems, presenting conclusions and recommendations to complete the chapter. First, in the next section, we present a perspective on pressures for organisational learning.

4.2 The Internal and External Environment of Business

The global business environment as we know it is changing fast. Decision-makers in organisations cannot make the right decision without continuously scanning the external and internal environment. The external environment includes all elements that exist outside the organisational boundaries, and the internal environment includes all elements within the organisation (Daft 2000). Organisational learning must occur within both these environments.

4.2.1 The External Business Environment and Organisational Learning

In terms of the external environment, a company cannot change without understanding future consumer behaviour (*i.e.* the market). Therefore, there is a need to forecast future consumer needs and not just the present needs. Tracking government regulations and recognising the leading role of government in giving the private sector the correct signals is crucial for strategic planning. This process is linked to the concept of the 'technology triangle', which refers to strategic interaction and co-operation among (a) the scientific and research community; (b) business and industry and (c) institutions of governance, both national and international. It focuses on business development, and emphasises building up the infrastructure, wealth-generating capacity and competitiveness of a country (Mezher 2000). Such interaction encourages all the elements of the triangle to adopt quality systems in order to communicate effectively and to satisfy their own needs. In addition, use of the concept of the technology triangle helps firms to improve their position in trade treaties. Technological development through knowledge networking is important because knowledge is a critical driver of social and organisational change. Globalisation imposes new demands and new opportunities in access to knowledge and its applications.

Gains from knowledge networking include, identifying 'best practice', eliminating technology barriers, facilitating 'leapfrogging' in information technology (IT), protecting quality and environmental control, retaining access to knowledge frontiers and obtaining practical experience. Organisational adaptation and change is not possible without well trained and developed human capital.

4.2.2 The Internal Business Environment and Organisational Learning

The internal environment includes:

- Management commitment to resources

- Keeping the change 'magic' alive (culture)
- Education and training, including new knowledge and skills needed
- Human resource policies
- Performance gaps (the disparity between existing and desired performance levels)
- Integration of QMSs, EMSs and SCM systems in organisational functions

Top management must always be committed to resources, and probably their most difficult job is to change the existing culture to fit the new management paradigm and keep the change 'magic' alive over time. Education and training is a continuous process, especially during the adoption of new technology where knowledge and skills are needed. Human resource strategies should concentrate on recruiting the best professionals and retaining them. Skilled employees are like good customers, when you lose them it is hard to attract them back. Performance measures are dynamic indicators for improvement. They show variations in customer satisfaction, products, processes and so on. Finally, management systems (QMS, EMS, SCM) must be integrated in core business functions of the organisation, keeping track of and integrating the learning that derives from these various internal environmental sources. Later in this chapter, in Section 4.5, we will flesh out such learning organisations even further. First, we provide an overview of the various management systems.

4.2 Quality Management Systems: Total Quality Management

Total quality management (TQM) is an enhancement to the traditional way of doing business and has been part of the lexicon of business management practices for over two decades. It is, basically, the art of managing the whole to achieve excellence (Besterfield *et al*. 2003). TQM requires six basic concepts:

A committed and involved management team to provide long-term top-to-bottom organisational support

- An unwavering focus on the customer, both internally and externally
- Effective involvement and utilisation of the entire workforce
- Continuous improvement of business and production processes
- The treatment of suppliers as partners
- The establishment of performance measures for the processes considered

The TQM philosophy stresses a systematic, integrated, consistent and organisation-wide perspective involving everyone and everything. It focuses primarily on total satisfaction for internal and external customers, within a management environment that seeks continuous improvement of all systems and processes. The TQM philosophy also emphasises use of all people, usually in multi-functional teams, to bring about improvements from within the organisation. It stresses the need for optimal life-cycle costs and uses measurement within a disciplined methodology to target improvements. Additional key elements of the philosophy are the prevention of defects and an emphasis on quality in design.

Important aims include the elimination of losses and the reduction of variability (Ho 1994).

Many companies in the USA, Europe, and Asia have adopted TQM systems. In third world countries, many companies have started adopting the ISO 9000 series of standards (Besterfield *et al.* 2003) , and many of them in the late 1990s, but few are moving from ISO 9000 to TQM. However, the ISO 9000 series is nothing but a first step towards excellence and TQM, yet many companies are adopting this certification as their only QMS (Mroz 1998; Seddon 1999)

The ISO 9000 series brings many benefits to a company but, still, it suffers a number of weaknesses. By itself, it is not enough to achieve a high standard of quality for customers in the long term. According to 'quality legend' Joseph Juran (quoted in Patton 1999), the ISO 9000 series criteria fail to include some of the essentials needed to attain world-class quality, such as quality goals in the business plan, quality improvement at a revolutionary rate, training in managing for quality, and participation of the workforce. In addition, he stated that although the basic concept of the ISO 9000 series has some merits to it, there has been no empirical evidence that certified companies have products that are superior to products of those that are not certified.

The contents and sectional organisation of ISO 9001 (Besterfield *et al.* 2003), the actual specification for the QMS, have been completely revised. Quality system requirements are now organised into four sections:

- Management responsibility
- Resource management
- Product and/or service realisation
- Measurement, analysis and improvement

This new organisation makes ISO 9001 more compatible with the ISO 14001 (environmental) standard, and is consistent with the ISO 9004 plan–do–check–act (PDCA) cycle for continuous improvement (Besterfield *et al.* 2003).

Even though the new ISO 9001 is a much more complete standard than those before, it still has major weaknesses. The new standards of the ISO 9000 series still make it a minimalist quality system and a first step on the quality journey. Its requirements and scope should be correctly understood along with TQM principles and practices, and it should present a foundation for future improvement (Bradley 1994; Fletcher 1999; Ho 1994; Kirchner 1996; Puri 1995). The objective of the ISO 9000 series should be understood as being primarily of achieving customer satisfaction by preventing non-conformity at all stages, from design to servicing. All other aspects such as leadership, human resource management, strategy, customer and market focus, impact on society, continuous improvement and business results are barely or not mentioned at all (Yung 1997). In brief, the ISO 9000 series by itself is a minimalist quality system that can provide a solid foundation for future improvement and eventually business excellence, but it is not the end of the quality journey, only the beginning. Yet, the ISO 9000 series may have a significant influence on and relationships to the ISO 14000 series relating to EMSs (Besterfield *et al.* 2003).

4.3 Environmental Management Systems

A well-tuned EMS would incorporate a continual cycle of planning, implementing, reviewing and improving the processes and actions that an organisation undertakes to meet its business and environmental goals. An EMS can also be viewed as an organisational approach to environmental management in which the goal is to incorporate environmental considerations into day-to-day operations. An EMS is a management structure that provides facilities and parent organisations with a framework to minimise their environmental impacts, to ensure compliance with environmental laws and regulations and to reduce wasteful use of natural resources. (Darnall *et al.* 2001)

An effective EMS is built on TQM concepts. To improve environmental management practices, organisations need to focus not only on what went wrong but also on why problems occurred. Over time, the systematic identification and correction of system deficiencies leads to better environmental and improve performance. Most EMS models (including the ISO 14001 standard, described in Section 4.2) are built on the PDCA model introduced by Shewart and Deming (Besterfield *et al.* 2003) (see Hortensius and Barthel 1997). This model endorses the concept of continual improvement. The keys to a successful EMS, that are similar to those for successful TQM, are discussed next.

4.3.1 The Keys to a Successful Environmental Management System

In the following seven sub-sections we discuss seven keys to a successful environmental management system:

- Top-management commitment
- A focus on continual improvement
- Fleixbility
- Compatibility with organisational culture
- Employee awareness and ivolvement
- Preformance measures
- Partnering with suppliers

4.3.1.1 Top-management Commitment
Social responsibility and environmental stewardship can occur only if top management is committed to business excellence. Top management cannot be committed to quality without taking the environment into consideration. Both quality and environment must be integrated into all functions of the business. Environmental integration should be an organisational priority, and top management must be committed to its environmental policy and planning and to its continuous improvement process. Therefore, resources must be allocated to help in implementation and control and to identify and solve problems (which should be viewed as opportunities) by carrying out checks and taking corrective actions. Management commitment also means having employee involvement and

partnering with suppliers (discussed in Sub-sections 4.3.1.5 and 4.3.1.7, respectively).

4.3.1.2 A Focus on Continual Improvement
This is a never-ending process. The business environment is always changing, and new improvements are needed. Business excellence requires an organisation to learn from its mistakes and to use environmental and managerial best practices.

4.3.1.3 Flexibility
An effective EMS must be dynamic to allow the organisation to adapt quickly to change. For this reason, an EMS should be kept flexible and simple for managers and employees to understand and implement. This must also be complemented with a more flexible (flat) organisational structure and excellent human resource development practices.

4.3.1.4 Compatibility with Organisational Culture
The EMS approach and an organisation's culture should be compatible. For some organisations, this involves a choice: (1) tailoring the EMS to the culture or (2) changing the culture to be compatible with the EMS approach. Bear in mind that changing an organisation's culture can be a long-term process. Keeping this compatibility issue in mind will help to ensure that the EMS meets the organisation's needs.

4.1.3.5 Employee Awareness and Involvement
Employee awareness and involvement in the implementation of an EMS can help reduce many obstacles. Resistance to change could occur. To overcome potential 'roadblocks' in implementation, it is necessary for employees to understand the need for change, their role in that change and the environmental and cost impacts of EMS on the organisation. Employees must be involved throughout, from early decisions to adopt EMS in the organisation, to participation in the continuous improvement process. Employees must be involved in developing the environmental policy of the organisation to ensure the implementation of that policy and in the planning process.

4.3.1.6 Performance Measures
Performance measures are collected for a number of reasons (Wever 1996):

- To compare where the organisations is and where it wants to be
- To evaluate how well the organisation has satisfied its customers and all stakeholders
- To evaluate where improvements are needed in the way the organisation manages, operates and designs products, processes and services
- To inform decision-making and communications processes

Some of the measures include:

- Financial, technical and human resources
- Consumption of physical resources (e.g. energy, water, materials)

- Emissions and waste
- Efficiency (the ratio of useful output to input)
- Risks
- Environmental impact of business activities
- Customer-related issues

These measures are helpful in management planning, control and checking and in taking corrective actions.

4.3.1.7 Partnering With Suppliers

Consumers and society are putting pressure on firms to improve their environmental practices not only by producing a friendly product but also by selecting environmentally friendly inputs (materials, energy and so on). Therefore, firms must establish a procedure to carefully select their suppliers and establish long-term partnerships with them. This is clearly related to SCM, which will be discussed in more detail in the next section.

4.4 Supply Chain Management

Various and numerous definitions of SCM have been offered in the past few years as the concept has gained momentum and popularity. The Supply-Chain Council (SCC 1997) defines 'supply chain', a term used by logistics professionals, as a process that encompasses every effort involved in producing and delivering a final product, from the supplier's supplier to the customer's customer. Four basic processes—plan, source, make, deliver—broadly define these efforts, which include managing supply and demand, sourcing raw materials and parts, manufacturing and assembly, warehousing and inventory tracking, order entry and order management, distribution across all channels and delivery to the customer. In addition to defining the supply chain, many authors have tried also to define the concept of SCM. Monczka and Morgan (1997) stated that integrated supply chain management is about going from the external customer and then managing all the processes that are needed to provide the customer with value in a horizontal direction. Lummus and Vokurka (1999) defined 'supply chain' as all activities involved in delivering a product, from raw material through to the customer, including the sourcing of raw materials and parts, manufacturing and assembly, warehousing and inventory tracking, order entry and order management, distribution across all channels, delivery to the customer and information systems necessary to monitory all these activities. All these definitions are valid and provide different insights into and perspectives on SCM.

4.4.1 The Benefits of Supply Chain Management

The benefits of SCM can be better visualised in line with the immediate, short-term and long-term impact they may provide. Table 4.1 summarises some of these benefits.

Table 4.1. The Benefits of Supply Chain Management

Immediate Benefits	Medium-term Benefits	Long-term Benefits
Lowered inventory risk	Sustainable cost savings through increased productivity and streamlined business processes in procurement and purchasing, order fulfillment, accounts receivable and payable, and exception management	Increased flexibility for changing market conditions
Lowered inventory costs	Accelerated product delivery times	Dramatically improved customer responsiveness
Reductions in warehousing costs	More efficient product development efforts	Improved customer service and satisfaction
Reductions in distribution costs	Lower product manufacturing costs	Increased customer retention
Reductions in transportation costs		More effective marketing

Businesses can use these SCM capabilities to manage demand as well as supply. Even though one may argue that there is a large difference between the supply chain and the demand chain, in reality there is only one type of chain—the value chain. What the company sees as demand its customers see as supply. In other words, 'demand' and 'supply' are functions of a firm's position relative to other participants on the same value chain (Horvath 2001).

4.4.2 Success Factors for Effective Supply Chain Management

The driving force of effective SCM is collaboration. Strategic SCM demands collaboration among all participants in the value chain, whatever their size, function or relative position. The increase in levels of collaboration and information-sharing associated with strategic SCM will demand a new form of collaborative technology infrastructure. The exact infrastructure capacities required will vary depending on the role and size of the participant, but there are certain fundamental attributes of an SCM network that will remain constant, and these include (Horvath 2001):

- Open and low-cost connectivity
- Very large, flexible, multimedia data-storage capabilities
- Systems and channel integration
- Higher-level self-service capabilities
- Intelligence-gathering and analysis
- Supply chain collaboration exchanges
- Sophisticated security capabilities
- New electronic commerce capabilities

4.4.3 Supply Chain Management and E-commerce

E-commerce capabilities are essential attributes in the SCM network. These new capabilities can reduce costs and speed settlements, and hence the collaborative SCM infrastructure will incorporate innovative financial arrangements, such as electronic billing and payment, automatic progress payments and expensive engineered products (Horvath 2001). Moreover, the evolution of modern e-commerce practices in SCM may be traced back to the 1970s, when the incipient use of electronic data interchange (EDI), or computer-to-computer digital communication, began to displace the traditional forms of data and information interchange (Murillo 2001).

Table 4.2 shows a 3×3 matrix depicting the interaction between the economic agents of business, consumer and government. In modern SCM applications, the infrastructure and advantages provided by the Internet and e-commerce are continuously being incorporated in such technologies as enterprise resource planning systems (Latamore 2000).

Table 4.2. Supply chain management: interaction between business, consumers and government

	Business	Consumer	Government
Business	B2B Supply Chain, Wholesalers	B2C Retailer (good or services)	B2G Contract bidding, Privatization
Consumer	C2B Public bidding Marketplace, Auctioneers	C2C Public flea markets	C2G Public government auctions
Government	G2B Tax and fees collection	G2C Tax and fees collection	G2G Budget allocation

In short, e-commerce can act as a backbone infrastructure for SCM, helping to integrate all the players through the virtual space and thus minimising time and cost while providing a dynamic interactive media in which all communications and data needed for an efficient and effective SCM are stored. This can be used to develop a knowledge management model which in turn is a vital resource for the learning organisation (as described in Section 4.5).

4.4.4 Green Supply Chain Management

SCM usually takes into consideration issues of minimising end costs (market costs), creating efficient logistical aspects and ensuring timely delivery of goods (Cox 1999). Yet shifts are occurring where many business chain partners are formed to participate in implementing environmentally friendly practices that reduce waste and pollution (Melnyk and Handfield 1996; see also many chapters in this book).

Environmental supply chains (green supply chains) feature three issues that are important to the theory of ecological modernisation and thus important to environmental management and sustainability in general: first, the inclusion of environmental aspects in integrated chain management of industrial chains for the manufacturing of goods; second, the integration of technological innovations for environmentally beneficial outcomes throughout the industrial supply chain; and, last, the participation of a broader range of industrial actors for the environmental management of industrial production to strengthen the capacity-building of environmental governance (Berger *et al.* 2001).

Since this field is new, much of the work done until now has concentrated on defining environmental supply chain management and its importance to businesses as well as on investigating case studies and various practices adopted by companies (Green *et al.* 1998; Lamming and Hampson 1996; McIntyre *et al.* 1998; Sarkis 1999; Wycherly 1999). One may argue that looking at environmental management from a more sophisticated approach also means including the whole system of production and consumption in the analysis. Waste and emissions caused by the whole supply chain have become a major source of environmental problems caused by industrial market economies. Yet definitions of environmental supply chain management differ widely with regard to focus and implications (as has been discussed in this book).

Environmental SCM has to make business sense to be carried out by companies, and potential savings or even profits are obviously a driving force. In other words, if we consider the economy–ecology relationship from a policy-making perspective, economic issues still dominate governmental and business thinking.

Time and speed are crucial in today's fast-paced competitive markets. Enterprises that are not keeping pace with fluid marketing demands and changes may lose out to competitors that have the advantage of faster and speedier deliveries arising from well-planned plant locations and good marketplace selections (Khoo *et al.* 2001). Therefore computer simulation is a useful tool offering a wide range of decision scenarios, saving time, energy and cost. The proper management of the supply chain can lead the company only to TQM and the incorporation of all environmental considerations that are an essential part of the environmental process. None of which can be achieved unless information is shared and unless knowledge is gained and properly managed by firms. This is where the learning organisation comes in.

4.5 The Learning Organisation

In this section we link the previously discussed management concepts into one model, relating them through their contribution to the learning organisation (see Fig. 4.1). Through a continuously improving process, knowledge of best practice and lessons learned from the external and internal business environment of the organisation can be added to the organisational knowledge base. If established and maintained correctly, this knowledge base will contribute to overall organisational learning.

The decision to only adopt the ISO 9000 and/or ISO 14000 series is usually made for marketing purposes (many organisations argue that these systems are baseline systems for quality and/or environmental purposes and that their systems are actually more advanced). The adoption and implementation of TQM systems, EMSs and SCM systems is essential in a fast-changing business environment. But these systems cannot guarantee that organisations will prosper and stay in the market. These systems must be accompanied by the building of a knowledge base of best practices and lessons learned. To be effective, TQM systems, EMSs and SCM systems must be integrated at all levels of the organisation. One of the most critical principles—continuous process improvement—should not only be applied to the different management systems but also include reviewing the internal and external environment of the organisation in order to improve the strategic planning process and building up organisational knowledge. This, in turn, will contribute to the learning organisation and will help give the firm a competitive edge. This new management paradigm will help organisations to respond quickly, efficiently and effectively to the changing business environment.

4.5.1 Knowledge Management

The most important benefit from adopting TQM, EMSs and SCM is the capturing of previous experience and knowledge through continuous process improvement. This capture could be completed through knowledge management, which is the effort to systematically gather knowledge, make it widely available the organisation and foster a culture of learning. Existing techniques that support knowledge management are information systems, artificial intelligence systems (expert systems, case-based reasoning and so on), the Internet, groupware, planning tools, data warehousing and data mining (Daft 2000; Turban and Aronson 2001). These should be used to retain and learn from previously gained knowledge in order to improve future strategic planning processes and improving the competitive edge of organisations. This in turn will ultimately contribute to the learning organisation in which everyone is engaged in identifying and solving problems, enabling the organisation to continuously experiment, improve and increase its capability. The captured knowledge should include explicit and tacit types of knowledge (see Table 4.3). Explicit knowledge is objective, rational and technical (*e.g.* policies, decisions, strategies, information systems and white papers). It is easily documented, transferred, taught and learned. In contrast, tacit knowledge is subjective and cognitive (*e.g.* expertise, know-how, ideas, organisational culture and values). It is hard to document, transfer, teach and learn. It involves a lot of human interpretation.

The knowledge captured in building a knowledge management system (KMS) should include the best practices and lessons learned not only from within the organisation but also from similar industries all over the globe. In addition, an intelligent KMS should be very innovative in order to be able to adopt best practices from different industrial sectors. Knowledge management is a process of elicitation, transformation and diffusion of knowledge throughout an enterprise so

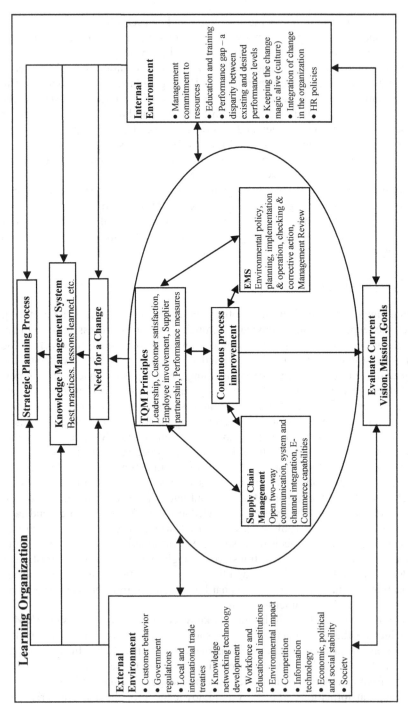

Figure 4.1. An environmental and business integrative process model of a learning organization

that it can: be shared and thus re-used; help organisations find, select, organise, disseminate and transfer important information and expertise; and transform data and information into actionable knowledge to be used effectively anywhere in the organisation by anyone. The objective of knowledge management is to create knowledge repositories, improve knowledge access, enhance the knowledge environment and manage knowledge as an asset.

KMS manages knowledge creation through: learning; knowledge capture and explication; knowledge sharing and communication through collaboration; knowledge access; knowledge use and re-use; and knowledge archiving. The cyclic model of knowledge management includes (Turban and Aronson 2001): the creation, capturing, refining, storing, management and dissemination of knowledge.

There are many other issues related to building an effective KMS, but these will not be discussed here (Awad and Ghaziri, 2004).

4.5.2 Definition of a Learning Organisation

The learning organisation can be defined as one in which everyone is engaged in identifying and solving problems, enabling the organisation to continuously experiment, change and improve, thus increasing its capacity to grow, learn and achieve its purpose. Developing a learning organisation means making specific changes in the areas of leadership, team-based structure, employee empowerment, communication and information sharing, participative strategies and an adaptive culture (Daft 2000). Each of these changes is in line with the adoption of TQM systems, EMSs and SCM systems.

From this perspective, learning organisations are closely linked to the concepts of knowledge management (Hedlund 1994) and knowledge creation (Nonaka and Takeuchi 1995), both of which use theories of organisational learning as a platform for providing insight into how organisations can acquire, interpret, distribute and 'enculturate' knowledge to facilitate and create a competitive edge. Top managers categorise and interpret the information and knowledge they accumulate in order to improve organisational performance (Thomas *et al.* 1993). Many learning models exist in the literature, such as those relating to searching and noticing (Huber 1990). Other models focus on active learning from outcome creativity and exploration (March 1995) as opposed to operative (Nelson and Winter 1982), adaptive (Senge 1990) and history-dependent types of learning that rely on exploiting emergent routines and processes (Levitt and March 1988). Some models are based on the ability to mine events as they occur for expertise and new knowledge (Henderson *et al.* 1998) and on becoming better informed about possible future states of the world (Mosakowski and Zaheer 1996).

Thomas *et al.* (2001) focus on strategic learning capability and the need to learn dynamically how ongoing discovery will affect future events and, ultimately, firm performance. In addition, there are four apparent strategic learning characteristics. First, data-collection efforts are targeted. Second, the firm times these efforts to coincide with its strategic action horizon. Third, the data collected leverages the firm's ability to generate, store and transport rich 'de-embedding' knowledge across multiple levels for the purpose of enhancing firm performance. Last, the firm will have institutionally based sense-making mechanisms in place, with associated well-defined validation processes. Based on these characteristics, a number of propositions have been introduced to build a strategic learning model through knowledge management. These propositions are important guidelines in capturing, managing and disseminating knowledge throughout an organisation. The conversion of information into knowledge and learning will be most beneficial for those organisations in a turbulent business environment.

In general, Table 4.3 summarised the different types of knowledge that could be captured from inside and outside the organisation. Figure 4.1 illustrated an environmental and business integrative process model of the learning organisations. It shows the links between TQM, EMS, and SCM. Through continuous process improvement, an organisation evaluates it current vision, mission, goals, and strategies taken into consideration the internal and external environment of the organisation. If there is a need for a change (*i.e.*, new customer requirement, new government regulations, new technology, etc.), then the knowledge management system gives the organisation, through its updated knowledge base, the ability and the confidence to make strategic decisions to stay competitive. All this eventually will contribute to the learning organisation.

4.6 Conclusions

Good implementation of TQM systems, EMSs and SCM systems will improve the quality of products and improve the ability to meet customers' needs. These systems help organisations to recognise the need for organisational change, for continuous evaluation of the current vision, mission, goals and strategies and to improve the strategic planning process. The continuous-improvement process can help organisations in building and formulating the knowledge they gain as an important driver for change. Therefore, there is a need for knowledge networking in accessing basic data, for robust information and for coherent knowledge about the issues at hand. All this will eventually contribute to developing a learning organisation.

Technology offers good opportunities for enhancing strategic learning. The use of videos, images and rich text can be developed to organise the knowledge-acquisition process and be used to leverage the organisation by preparing it for future strategic actions. This process acknowledges that the tacit–explicit distinction is not a dichotomy but part of a continuum (Cook and Brown 1999). When strategic learning is over a distance, collaborative technologies such as electronic bulletin boards and listservs can be used, especially in the knowledge-validation process (Boland and Tenkasi 1995). In addition, technology offers firms the capability of developing their 'lessons-learned' databases and organisational doctrines digitally in a structure allowing the availability of increasing complexity through hierarchically linked levels of contents (Thomas *et al.* 2001).

Strategic organisational learning can be strengthened through alliances with local and international firms, by using knowledge networking for knowledge-building and capacity-building and through collaboration with the elements of the

Table 4.3. Sources of knowledge for the learning organization

	TQM		SCM		EMS	
	Explicit	Tacit	Explicit	Tacit	Explicit	Tacit
Customer Satisfaction	Customer complaints and response rate, customer satisfaction reports, surveys, customer service, front-line people, market segment, consumer reports & other publicly available information, statistical reports, delivery time, etc.	Customers behavior, employees' perception of the customer, market behavior, history of customer changing needs, etc.	Customer complaints and response rate, customer satisfaction reports, surveys, customer service, front-line people, market segment, consumer reports & other publicly available information, statistical reports, delivery time, etc.	Customers behavior, employees' perception of the customer, market behavior, history of customer changing needs, etc.	Customers' expectations, local regulations and laws, customer environmental information, Environmentalist and NGOs reports and records, environmental safety records of products, etc.	Customer awareness, social norms and believes, employees' perception of the customer, skills and believes, market behavior, history of customer changing needs, etc.
Organization	Policies, procedural guides, internal reports, core competencies, best practices, annual reports, lessons learned, competition, government regulation, relations with universities, technology management, statistical reports, history of TQM adoption, history and product changes, auditors and raters reports, share value, cost of poor quality, etc.	Vision, objectives, believes, core values, culture, understanding concept of quality, lessons learned, know how, market behavior, intellectual assets, need for change, employees' need, informal communication channels, etc.	Policies, procedural guides, internal reports, core competencies, best practices, annual reports, lessons learned, competition, government regulation, relations with universities, technology management, statistical reports, history of SCM adoption, history and product changes, auditors and raters reports, share value, cost of poor supply chain, supply chain guidelines, security measures, etc.	Vision, objectives, believes, core values, culture, understanding concept of supply chain, lessons learned, know how, market behavior, intellectual assets, need for change, employees' need, informal communication channels, etc.	Policies, procedural guides, internal reports, core competencies, best practices, annual reports, lessons learned, competition, government regulation, relations with universities, clean and end-of-pipe technology management, statistical reports, history of EMS adoption, history and product and process changes, auditors and raters reports, shared value, cost of poor environmental practices, environmental guidelines, security measures, environmental reports, green manuals and procedures, pollution data, waste generation, etc.	Social responsibility and Environmental leadership, motivation and leadership skills, know-how and environmental perception sharing, etc.

Table 4.3. (continued)

	TQM		SCM		EMS	
	Explicit	Tacit	Explicit	Tacit	Explicit	Tacit
Suppliers & Inputs	Suppliers' customer satisfaction records, on-time delivery, JIT delivery targets, certifications and rating, reports, raw material data and information, SPC, market reports and trends, partnership history, etc.	Trade secrets, appreciation of TQM philosophy, experience, market insights, sharing of supplier-customer believes, objectives and expectations, etc.	Warehousing information and techniques, raw material data, consumption, suppliers' customer satisfaction records, on-time delivery, JIT delivery targets, JIT procedures and manuals, certifications and rating, reports, raw material data and information, SPC, market reports and trends, partnership history, etc.	Market experience and skill, capabilities pooling and sharing, teamwork experience, etc.	Supplier's environmental records, waste generation, energy efficiency, manuals, environmental market indicators & standards, Suppliers' customer satisfaction records, on-time delivery, JIT delivery targets, certifications and rating, source of raw material, raw material data and information, SPC, market reports and trends, partnership history, etc.	Experience record, supplier commitment to environmental policy, and teamwork, etc.
Technology	Information management & knowledge management systems, stores, web and networks connectivity, history of process and technology changes due to quality, R&D, etc.	Technology forecasting, competition future technology, innovation culture, etc.	Information management & knowledge management systems, stores, web and networks connectivity, history of process and technology changes due to supply chain, R&D, etc.	Technology forecasting, competition future technology, innovation culture, etc.	Information management & knowledge management systems, stores, web and networks connectivity, history of process and technology changes due to environmental concerns, R&D, etc.	Technology forecasting, competition future technology, innovation culture, etc.
Employees	Labor market, selection and recruiting, quality training & development, know-how & skills, recognition & rewards, layoffs and turnovers, grievances, accidents, suggestions and improvements, etc.	Labor market changes, motivation, empowerment, quality perception, needs are satisfied, etc.	Labor market, selection and recruiting, supply chain training & development, know-how & skills, recognition & rewards, layoffs and turnovers, grievances, accidents, suggestions and improvements, etc.	Labor market changes, motivation, empowerment, supply chain perception, needs are satisfied, etc.	Labor market, selection and recruiting, environmental training & development, know-how & skills, recognition & rewards, layoffs and turnovers, grievances, accidents, suggestions and improvements, etc.	Labor market changes, motivation, empowerment, environmental perception, needs are satisfied, etc.

technology triangle—government institutions, the business sector and educational institutions. Finally, use of the learning organisation model will have positive environmental implications for the organisation. It will ensure that all products are environmentally friendly and produced with a minimum of pollution. Best environmental management practices will ensure environmental stewardship, the selection clean technologies in preference to cleaning (end-of-pipe) technologies, highly trained employees and an improved the public image for the organisation. In addition, it will encourage partnership with suppliers, distributors and customers, and this will contribute to a 'cradle-to-grave' relationship with products.

References

Awad, E.M., and Ghaziri, H.M. (2004), Knowledge Management, Upper Saddle River, New Jersy, Prentice Hall.

Berger, G., A. Flynn, F. Hines and R. Johns (2001) 'Ecological Modernisation as a Basis for Environmental Policy: Current Environmental Discourse and Policy and Implications on Environmental Supply Chain Management', Innovation 14.1: 55-72.

Besterfield, D.H., C. Besterfield, G. Besterfield and M. Besterfield-Sacre (2003) Total Quality Management. 3nd edition, Upper Saddle River, New Jersy, Prentice Hall.

Boland, R.J., and R.V. Tenkasi (1995) 'Perspective Making and Perspective Taking in Communities of Knowing', Organisational Science 6.4: 350–72.

Bradley, M. (1994) 'Starting Total Quality Management from ISO 9000', The TQM Magazine 6.1: 50-54.

Cook, S.D.N., and J.S. Brown (1999) 'Bridging Epistemologies: The Generative Dance Between Organisational Knowledge and Organisational Knowing', Organisational Science 10.4: 381–400.

Cox, A. (1999) 'Power, Value and Supply Chain Management', International Journal of Supply Chain Management 4.4: 167-75.

Daft, R.L. (2000) Management (Orlando, FL: Dryden Press, Harcout College Publishers, 5th edn).

Darnall, N., D.R. Gallagher and R. Andrews (2001) 'ISO 14001: Greening Management Systems', in J. Sarkis (ed.), Greener Manufacturing and Operations: From Design to Delivery and Back. Sheffield, UK: Greenleaf Publishing: 178-190.

Fletcher, A.C. (1999) 'Building QCI International World-Class Performance with the Baldrige Criteria', www.qualitydigest.com/ aug99/html/body_baldrige.html.

Green, K., B. Morton and S. New (1998) 'Green Purchasing and Supply Policies: Do They Improve Companies' Environmental Performance?' Supply Chain Management 3.2: 89-95.

Hedlund, G. (1994) 'A Model of Knowledge Management and the N-Form Corporation', Strategic Management Journal 15: 73–90.

Henderson, J., S.S. Watts and J. Thomas (1998) 'Creating and Exploiting Knowledge for Fast-cycle Organisational Response', in D. Ketchen (ed.), Advances in Applied Business Strategy 5 (Stamford, CT: JAI Press): 103–128.

Ho, S.K. (1994) 'Is the ISO 9000 Series for Total Quality Management?', International Journal of Quality and Reliability Management 11.1: 74-89.

Hortensius D., and M. Barthel (1997) 'Beyond 14001: An Introduction to the ISO 14000 series', in C. Sheldon (ed.), ISO 14001 and Beyond: Environmental Management Systems in the Real World. Sheffield, UK: Greenleaf Publishing: 19-44.

Horvath, L. (2001) 'Collaboration: The Key to Value Creation in Supply Chain Management', International Journal of Supply Chain Management 6.5: 205-207.

Huber, G. P. (1990) 'A Theory of the Effects of Advanced Information Technologies on Organisational Design, Intelligence and Decision Making', Academic Management Review 15: 47–71.

Khoo, H.H., T.A. Spedding, L. Tobin and D. Taplin (2001) 'Integrated Simulation and Modeling Approach to Decision Making and Environmental Protection', Environment, Development and Sustainability 3.2: 93-108.

Kirchner, R.M. (1996) 'What's Beyond ISO 9000?', www.qualitydigest.com /nov96/iso9000.html.

Lamming, R., and J. Hampson (1996) 'The Environment as a Supply Chain Management Issue', British Journal of Management 7: 45-62.

Latamore, G.B. (2000) APICS: The Performance Advantage, Production and Operation Management, 2nd edition. McGraw Hill Ohio.

Levitt, B.J., and G. March (1988) 'Organisational Learning', Annual Review of Sociology 14: 319-40.

Lummus, R.R., and R.J. Vokurka (1999) 'Defining Supply Chain Management: A Historical Perspective and Practical Guidelines', Industrial Management and Data Systems 99.1: 11-17.

McIntyre, K., H. Smith, A. Henham and J. Pretlove (1998) 'Environmental Performance Indicators for Integrated Supply Chains: The Case of Xerox Ltd.', Supply Chain Management 3.3: 149-56.

March, J.G. (1995) Exploration and Exploitation in Organisational Learning', in M.D. Cohen and L. Sproull (eds.), Organisational Learning (Thousand Oaks, CA: Sage Publications): 101–23.

Melnyk, S., and R. Handfield (1996) 'Greenspeak', Purchasing Today (July 1996): 32-36.

Mezher, T. (2000) 'Strategies for Industrial and Technological Development of Middle Eastern Countries', International Journal of Environmental Studies 57.5: 543-62.

Monczka, R.M., and J. Morgan (1997) 'What's Wrong with Supply Chain Management?' Purchasing 122.1: 69-73.

Mosakowski, E., and S. Zaheer (1996) The Global Configuration of Speculative Trading Operation: An Empirical Study of Foreign Exchange Trading (working paper; Minneapolis, MN: Strategic Management Research Centre, University of Minnesota).

Mroz, J. (1998) 'ISO 9000 in 2000', www.qualitydigest.com/ oct98/html/cover.html.

Murillo, L. (2001) 'Supply Chain Management and the International Dissemination of E-commerce', Industrial Management and Data Systems 101.7: 370-77.

Nelson, R.R., and S.G. Winter (1982) An Evolutionary Theory of Economic Change (Cambridge, MA: Harvard University Press).

Nonaka, I., and H. Takeuchi (1995) The Knowledge-creating Company (New York: Oxford University Press).

Patton, S.M. (1999) 'A Century of Quality: An Interview with Quality Legend Joseph M. Juran', www.qualitydigest.com/feb99/html/body_jural.html.

Puri, S.C. (1995) ISO 9000 Certification and Total Quality Management (Ontario, Canada: Standards–Quality Management Group).

Sarkis, J. (1999) How Green Is the Supply Chain? Practice and Research (working paper; Worcester, NY: Graduate School of Management, Clark University).

SCC (Supply-Chain Council) (1997) 'What is the Supply Chain', 20 November 1997, www.supply-chain.org/info/faq,html.

Seddon, J. (1999) 'ISO 9000:2000: ISO 9000 Digs its own Grave', www.vanguardconsultant.co.uk/grave.htm.

Senge, P.M. (1990) The Fifth Discipline: The Art and Practice of the Learning Organisation (New York: Doubleday/Currency_.

Thomas, J.B., S.M. Clark and D.A. Gioia (1993) 'Strategic Sensemaking and Organisational Performance: Linkages among Scanning, Interpretation, Action and Outcomes', Academic Management Journal 36.2: 239–70.

Thomas, J.B., S.S. Watts and J. Henderson (2001) 'Understanding Strategic Learning: Linking Organisational Learning, Knowledge Management and Sensemaking', Organisation Science 12.3: 331-45.

Turban, E., and J.E. Aronson (2001) Decision Support Systems and Intelligent Systems (Upper Saddle River, NJ: Prentice Hall, 6th edn).

Wever, G. (1996) Strategic Environmental Management: Using TQEM and ISO 14000 for Competitive advantage (New York: John Wiley).

Wycherly, I. (1999) 'Greening Supply Chains: The Case of the Body Shop International', Business Strategy and the Environment 8: 120-27.

Yung, W.K.C. (1997) 'The Values of TQM in the Revised ISO 9000 Quality System', International Journal of Operations and Production Management 17.2: 221-30.

5

Network Dynamics of an Energy Supply Chain: Applicability of the Network Approach to Analysing the Industrial Ecology Practices of Companies

Kirsi Hämäläinen

University of Jyväskylä, School of Business and Economics, PO Box 35, FIN-40014
University of Jyväskylä, Finland, kirsi.hamalainen@tampere.fi

Currently firms face environmental challenges, which require more emphasis on inter-organisational environmental improvement efforts. A field of study of industrial ecology offers some novel approaches for inter-organisational environmental management. It emphasises symbiotic linkages between firms or other organisations in which waste of one actor is used as raw material by others. The industrial network approach has been suggested as a possible framework for developing managerial implications for industrial ecology. This study analyses an example of an industrial ecosystem with notions of the network theory, and concludes that the industrial network approach is useful in analysing and describing the development of eco-industrial relationships even though it is not highly suitable for explaining the reasons behind the development. On the basis of the network approach, some implications for the management of industrial ecosystems are also proposed in this study.

5.1 Introduction

Currently, firms face a wide and constantly changing range of environmental challenges. Some of these challenges render environmental management efforts that concentrate solely on one company's internal performance less effective and require more emphasis on inter-organisational efforts (Sinding 2000). Common requirements for extended producer responsibility and the management of the environmental impacts of products throughout their lifecycle are examples of such challenges. As a consequence, inter-organisational environmental management is needed.

Industrial ecology offers some novel solutions for inter-organisational environmental management. It proposes joint action with companies and organisations that exceeds traditional supplier or customer relationships. It aims at co-operation among various actors of society, which would produce system-wide environmental innovations that go beyond specific process modifications or improvements to single products (Wallner 1999).

Joint efforts at environmental improvement among various actors of society are needed, as organisations are open systems that operate in uncertain and ever-changing environments. A higher degree of internal differentiation within an organisation or within a network of organisations makes that organisation or network quick and effective in dealing with changing circumstances (Morgan 1986). Thus, co-operation between different types of companies and organisations enhances the possibilities for companies to respond to various and constantly changing environmental challenges.

There is probably no best way to organise efforts aimed at adapting to changing environmental circumstances. However, according to Morgan (1986), the management of organisations and operations should be concerned, above all else, with achieving "good fits" with other organisations. Hence, appropriate fits with other organisations would also help companies to become adjusted to environmental constraints imposed on them by public environmental regulation and by stakeholders and society at large.

Industrial ecology thinking also emphasises the need for good fits. One of its basic concerns is to form symbiotic linkages between firms or other organisations in which the waste of one actor may serve as raw materials for another (Ayres and Ayres 1996). Unfortunately, the field of industrial ecology has concentrated mostly on the ideological and technical aspects of inter-firm co-operation. Consequently, it has failed to offer practical managerial implications. However, Pesonen (2001) has suggested that the industrial network approach could be used as a possible framework for understanding the managerial aspects of industrial ecology.

In this chapter, the industrial network approach is integrated with notions of industrial ecology. The research questions addressed in this paper are:

- How do concepts of the network approach relate to those of industrial ecology?
- Is it meaningful to apply the industrial network approach to analyse the development of industrial ecology practices in companies?

In order to answer the first research question, secondary data analysis is used, and in order to find an answer to the second question, a case study is examined. In the case study, an example of an industrial ecosystem—the energy supply system of Jyväskylä, Finland—is analysed within the framework of the network approach. In order to gather the information needed for this analysis, interviews were carried out with people who have been responsible for the development of the energy supply system in question.

In the next section, basic notions of the network approach as well as industrial ecology are presented and compared. Thereafter, notions of the network approach are used in analysing an example of an industrial ecosystem. Finally, the usefulness of the network approach in connection with industrial ecology is discussed.

5.2 From Industrial Networks to Industrial Ecosystems

The network approach has been developed as an answer to criticism toward the traditional marketing concept based on the idea of marketing mix. The traditional marketing mix concept has been perceived as unrealistic in real-life business settings (Turnbull *et al*. 1996). The central purpose of the network approach is to understand the totality of network relationships between companies and other organisations (Easton 1992).

A rapidly growing field of industrial ecology has also been developed as an alternative to the traditional model. According to the seminal developers of this new field of study, the traditional, linear model of industrial activity needs to be transformed into a cyclical model, into a so-called industrial ecosystem (Frosch and Gallopoulos 1989). One major goal of this new field of study is to utilise the knowledge of natural ecosystem structure and behaviour to reduce material demands and waste production in industrial systems (Allenby and Cooper 1994). This analogy between natural and industrial systems requires that a systems view is taken towards the activities of industrial society and to environmental problems. Thus, industrial ecology employs a holistic view to study, assess and improve the flow of natural resources in an industrial society (Van Berkel *et al*. 1997).

Networks are regarded as arrangements for inter-organisational interactions of exchange and as arrangements for solving inter-organisational problems that cannot be solved by a single organisation alone (Agranoff and McGuire 1998). Industrial ecosystems can similarly be considered as networks. The industrial analogy to an ecosystem is an industrial park, or some larger region, that exchanges physical materials among the network partners so that each partner's waste products serve as raw materials for another partner (Ayres and Ayres 1996). This kind of material exchange is an arrangement for striving collectively towards lower emissions, as the possibilities of a single organisation alone to reduce its negative environmental impacts are very limited, even with substantial efforts (Ehrenfeld and Gertler 1997).

The emergence of an industrial network does not proceed according to a single organisation's master plan or the like (Håkansson and Johanson 1988). Every single actor establishes his or her own relationships in the network. Each relationship must still be mutual, and an interaction between network actors is needed. Organisations outside the network can participate by interacting with organisations already engaged in the network. Hence, the structure of the network may change over time as organisations enter or leave. The biological analogy used in industrial ecology also highlights the evolutionary nature of change in natural and industrial systems. Various components of these systems are organised according to an evolutionary process. This means that different components come into existence at different times and are therefore in different stages of evolution. Hence they do not come into being as a result of centralised planning or any systematic design process (Levine 1999).

A firm's network of relationships is a source of opportunities and constraints. Networks might provide a firm with access to information, resources, markets and technologies (Gulati *et al*. 2000). In addition, a firm might learn know-how and capabilities from its partners (Kale *et al*. 2000). Further, inter-firm linkages might help a firm to withstand changes and improve survival prospects and financial performance (Ahuja 2000). However, networks also have a potential dark side. They may lock a firm into unproductive relationships or restrict it from partnering with other viable firms (Gulati *et al*. 2000). Further, by joining in a network a firm may lose some of its freedom to act independently (Van de Ven 1976). In addition, a firm must invest its scarce resources and energy in developing and maintaining relationships with other organisations, when the potential returns on this investment are often unclear or intangible (Van de Ven 1976). Conflict is also inherent in all inter-firm linkages because of the risk of opportunistic behaviour by partner organisations (Kale *et al*. 2000).

Industrial ecology thinking in companies also offers opportunities and potential risks. The economic gains from industrial ecosystems would be obtained either as a positive flow, from selling of by-products or waste, or from obtaining raw materials at prices below those for virgin materials. Economic gains would also be obtained as savings in waste management or those relative to environmental legislation or so-called 'green' taxes. (Ehrenfeld and Gertler 1997) According to Esty and Porter (1998), industrial ecology thinking might also broaden the perspective of corporate decision-makers and encourage innovation.

The buyer of by-products or waste takes some risk by tying itself to a single, outside supplier and by exposing itself to the uncertainty of supply continuity. The seller also takes some risk given that the buying of the by-products may be interrupted. If this were to happen, the by-products would instantly become waste to the seller and would need to be disposed of according to the relevant regulatory requirements. (Ehrenfeld and Gertler 1997) Further, the costs of closing some loops might exceed the benefits gained (Esty and Porter 1998), and discovery costs (*i.e.*, the costs required to learn of the existence of an opportunity for exchange) may be too high (Ehrenfeld and Gertler 1997).

As illustrated by the above discussion, the concepts of industrial networks and industrial ecosystems have a fair amount in common. The similarities between these concepts are summarised in Table 5.1.

The similarities between the concepts of the network approach and industrial ecology are obvious when we consider industrial ecology at its meso level. At the meso level, industrial ecology investigates regional material flows and interdependence between firms. However, there are two other scales at which industrial ecology can be analysed: the micro level, concerning operations within a firm (such as unit processes), and the macro level, concerning grand cycles of materials and national and international resource flows. When industrial ecology is analysed at its micro or macro levels, the concepts of the network approach do not fit that well with those of industrial ecology.

The concept of industrial ecology is relatively new and undeveloped. Many aspects of this concept are still in need of further research. The network approach, instead, is more developed. Hence, the managerial implications derived from the industrial network approach could well contribute to the development of eco-industrial relationships between firms.

5.3 Network Co-operation: A Case of a Finnish Energy Supply Network

In this section, the theory of industrial networks will be used to analyse co-operation in the energy supply network of the city of Jyväskylä in Finland. The nature and operation of this network resemble the characteristics of an industrial ecosystem (Korhonen *et al.* 1999). The boundaries of the network are defined from the perspective of an energy-generation company called Jyväskylän Energiantuotanto Oy. The interests of this company determine which organisations and activities are included in the study. Thus, the analysed network captures all those relationships that the focal firm considers relevant in terms of the study in question.

Table 5.1. Similarities between the concepts of the network approach and industrial ecology

Similarity	Network approach	Industrial ecology
They take an alternative approach	It is an alternative to the traditional marketing concept of a marketing mix	It is an alternative to the traditional linear model of industrial activity
They take a holistic view	It is concerned with understanding the totality of network relationships	It is concerned with taking a holistic and systems view to activities of an industrial society and to environmental problems
They consider inter-organisational exchanges	It underlines inter-organisational exchange interactions	It underlines exchanges of materials among companies in the sense that each company's waste products are used as raw materials by another company
They promote inter-organisational problem solving	It provides an arrangement for solving inter-organisational problems	It provides an arrangement for solving collectively those environmental problems caused by industrial production
Change is a central feature	Change is considered as a central factor in industrial networks	Evolutionary change is a substantial characteristic of natural (and industrial) systems
They emerge spontaneously	The emergence of an industrial network does not proceed according to a master plan	Different components of natural (and industrial) systems do not come into being as a result of centralised planning
They are two-edged swords	Network of relationships is a source of opportunities and constraints	Industrial ecology thinking offers opportunities and potential risks

The regional energy-generation system of Jyväskylä utilises mainly local energy resources: peat and wood fuel. The primary fuel has been milled peat because of the large number of peat bogs reserved for peat fuel production in the surroundings of Jyväskylä. During recent years, the use as fuel of wood residuals from neighbouring industrial plants and from forest harvesting has increased substantially. The annual primary energy consumption of the system is about 2100 gigawatthours (GWh).

Annually, about 900 GWh district heat is produced in the system. District heating means the distribution of hot water or steam from one or more sources to multiple buildings (Lehtilä *et al.* 1997). The main user, the city of Jyväskylä, uses the district heat for the purposes of space heating. The total process steam production of the system is about 400 GWh per year, and the annual electricity production is slightly more than 400 GWh. The whole power generation system is based on the co-production of heat and electricity. The heat generated in electricity production is used in the district heating system. The alternative would be to discharge this so-called waste heat into lakes or into the air, as is done by conventional condensing power plants.

The major power plant of the region is owned and governed by the focal firm of this study, Jyväskylän Energiantuotanto Oy. The power plant produces all the electricity generated in the energy supply system of Jyväskylä. In addition, the power plant produces most of the district heat produced and consumed in the energy system in question. Furthermore, the power plant produces all process steam consumed by a local paper mill, Kangas Paper Mill, which is owned by well-known Finnish pulp and paper company M-real. Previously this paper mill had its own steam production facilities but, in connection with plans to construct one major power plant in the region, it decided to drive down its own energy-generation capacity and rely on the established power-generation company as its steam supplier.

Two competing energy-generation companies jointly own the focal firm of this study. The local energy company, Jyväskylän Energia, is owned by the city of Jyväskylä and is thus governed by local politicians. The other owner company is Fortum Oyj, which is a leading energy company in the Nordic countries, and is partly owned by the state of Finland. These two companies operate in the same energy-generation and energy-transmission fields and are thus rivals to each other. However, they have joined forces in the case of the energy supply system of Jyväskylä and have invested in a common power plant.

In addition to the local paper mill, a paper machine producer, the Rautpohja plant, which is owned by Metso Paper, Inc., the world's leading manufacturer of machinery for the paper industry, relies on the local energy generators for its process steam supply. The local energy company Jyväskylän Energia provides the Rautpohja plant with process steam, as a consequence of which the plant does not need to construct its own energy-generation facility. The fewer small energy-generation facilities there are, the better it is for the state of the environment in Jyväskylä.

A park and garden area called Viherlaakso, situated near the power plant, utilises the ash produced by the power plant for landscaping. This saves the need for thousands of tonnes of gravel that would otherwise be extracted for such landscaping purposes. The utilisation of ash from the nearby power plant decreases remarkably the need for transportation, which would be considerably higher were the gravel needed by the park and garden area to be transported from further off. The need for transportation would also be greater were the ash produced by the power plant to be transported further to a disposal site.

The heat for a centre that cultivates and sells plants, called Viherlandia, is extracted from the return steam condensate of the local paper mill. In addition, a pedestrian precinct in the Jyväskylä city centre is kept clear of snow and ice in winter by using district heat extracted from the district heating waters returning from the city. These ways of utilising lower-quality heat in the energy supply system are excellent examples of energy cascading, which is a typical feature of industrial ecosystems.

5.3.1 Analysing the Co-operation Through the Network Approach

Håkansson and Johanson (1992) have proposed a network model, according to which networks actually involve three overlapping networks: networks of actors, networks of activities and networks of resources. The nature and dynamics of the whole network can be analysed in terms of actor bonds, activity links and resource ties. Next, the energy supply network of Jyväskylä is analysed with use of this network model.

5.3.1.1 Actor Bonds
The following actors are engaged in the energy supply network of Jyväskylä:

- The energy-generation company, Jyväskylän Energiantuotanto Oy, which is the focal firm of this study
- Two owners of the focal firm:
  A local energy company, Jyväskylän Energia
  A national energy company, Fortum Oyj
- The city of Jyväskylä, which is an owner of the local energy company, Jyväskylän Energia, and also represents individual inhabitants (*i.e.*, energy consumers in the city of Jyväskylä)
- Wood suppliers
- Peat suppliers
- Kangas Paper Mill, owned by M-real
- The Rautpohja plant, owned by Metso Paper, Inc.
- A centre for the cultivation and sale of plants called Viherlandia
- A park and garden area called Viherlaakso.

The basic assumption of the industrial network approach is that networks are essentially heterogeneous in nature (Axelsson and Easton 1992). This is also the case when the energy supply system of Jyväskylä is concerned. In this system, commercial companies as well as municipal authorities are involved. The size of the companies vary from small-scale peat and wood-chip suppliers situated in small municipalities in sparsely populated areas around Jyväskylä, to internationally operating and well-known companies such as Metso Paper, Inc. The partners of this energy-generation system represent different stages of energy lifecycle, including raw material producers, energy producers, energy transmitters and energy consumers. Heterogeneity of actors involved in a network is said to be beneficial for innovativeness (*e.g.*, Axelsson and Easton 1992). The development of combined efforts aimed at environmental improvement is an innovative process in which technical and co-operation-based innovations are of utmost importance.

Therefore, a heterogeneous network of actors, such as that in the case studied here, is a good requisite for further developing industrial ecology practices.

Along with interaction and relationship building, different kinds of bonds arise between the network actors (Håkansson and Snehota 1995). The focal company of this study, the energy-generation company Jyväskylän Energiantuotanto Oy, has **economic bonds** (*i.e.*, financial agreements and terms of payment) with the fuel suppliers and energy consumers in the network. In addition, it has economic bonds with its owner companies in terms of economic investments. In addition to economic bonds, the focal company has **legal bonds** (*i.e.*, formal contracts) with these partner groups.

Technical bonds have arisen between the focal company and the fuel suppliers in the form of adjustments to the energy production process. The power plant of the focal company has been modified in order to enable the use of local wood and peat fuels. Technical bonds have also arisen between the focal company and Kangas Paper Mill. The paper mill has driven down its own energy-generation capacity, and the energy-generation processes of the power plant have been modified to meet the process steam needs of the paper mill. In addition, a technical bond has arisen between the local energy company Jyväskylän Energia and the Rautpohja plant. The Rautpohja plant has driven down its own process steam production capacity, and process steam pipes have been built from a power plant for a test paper machine.

Further, technical bonds have arisen between the focal company and its owner companies, which have united their energy-generation capacities and built common technology for energy generation. In addition, a form of technical bond has arisen between the focal company and the city of Jyväskylä, which represents the citizens of the region and thus the consumers of heat and energy. In particular, in the case of heat the technical bond is fairly strong because of the district heating network, which connects most of the houses to the major power plant of the region. Finally, technical bonds have arisen between the focal company of the study and the commercial garden called Viherlandia, and the park area called Viherlaakso. In order to provide heat for the commercial garden, a separate heat exchanger has been installed that extracts lower-quality heat from the return steam condensate of the paper mill. The park and garden area called Viherlaakso, utilises the ash produced by the major power plant for landscaping purposes, which requires technical equipment for to separate ash from the combustion gas of the power plant and for transportation of the ash.

Planning bonds (*i.e.*, logistic co-ordination and common modification of planning systems) have arisen between the focal company and all the network partners considered in this study. These planning bonds involve co-ordination of fuel and energy delivery, as well as modification of planning systems in general.

In addition to planning bonds, **knowledge bonds** and **social bonds** have arisen between the focal company and all its network counterparts. An exception to this is the city of Jyväskylä, which represents all heat and power-consuming citizens of the region. The focal company does not have much knowledge about these individual consumers, nor can it have personal relations with all of them. When all other partners are considered, the focal company has a fairly good knowledge about them and personal relations with their key representatives. In particular, in the case of the commercial garden called Viherlandia, and the park area called Viherlaakso, the existence of personal relations, mutual confidence and personal liking have been most significant for the development of co-operation.

Strong bonds between networking firms tend to provide a more stable and predictable network structure and one that is more likely to be able to withstand change. Yet this does not mean that a strongly bonded network will remain stable over time. (Axelsson and Easton 1992) In the energy-generation network of the city of Jyväskylä, evolutionary changes have happened in the process for some years. The various actors of the network have participated in the network at different times and thus have enlarged the boundaries of the network. In addition, critical changes, such as the possible selling of the local energy company, which is at the moment under discussion, encounter resistance among various network actors. Actor bonds and interdependencies between organisational and personal actors are likely to slow down and direct the possible change (Easton 1992).

Most of the actors in the energy-generation system of Jyväskylä have expressed a mutual orientation (*i.e.*, in their preparedness to interact and their willingness to generate mutual knowledge) by joining the common project, aimed at establishing the environmental and economic benefits gained from the development of regional co-operation in the energy supply system. However, there was one actor, the local paper mill, which did not want to invest time and effort to the work of the project group. Yet the paper mill is the actor whose operations are dependent on the network in terms of the steam supplied by the network. This implies that the network that is considered as central from the focal firm's viewpoint might not be considered that important from some other actor's viewpoint. Thus, the actors have different assumptions regarding the network, depending on their role and position in the network.

5.3.1.2 Activity Links and Resource Ties
The actors of the energy supply network of Jyväskylä control the network activities by choosing which activities will be performed and which resources will be used. The fuel-production, energy-generation and energy-consumption activities performed by these network actors are linked together, as a result of which various technical, administrative, commercial and other activity links are formed.

Resources in the network can be controlled directly or indirectly. Direct control usually consists of ownership, and indirect control stems from relationships with and dependence on other actors in the network. All other relationships in the energy supply network of Jyväskylä are based on indirect control over resources, except in the case of the relationship between Jyväskylän Energiantuotanto Oy (the focal firm of the study) and its owner companies. In addition, the relationship between the city of Jyväskylä and the local energy company, Jyväskylän Energia, is based on direct control.

Inducements for companies to form relationships with other companies can be related to their need for resources (Ahuja 2000). The focal firm of this study, the energy-generation company Jyväskylän Energiantuotanto Oy, has formed relationships with wood suppliers and peat suppliers in order to gain access to resources that are, to some extent, necessary for its survival and growth. Naturally, the focal firm could find other fuels for its processes, which would mean, mainly, the purchasing of imported fossil fuels. But because of the investments made in the power plant that enable the use of local wood and peat fuels it is by no means beneficial for the focal firm to change its raw materials to foreign fossil fuels. In addition to the focal firm, Kangas Paper Mill, the Rautpohja plant, the commercial garden called Viherlandia, and the park and garden area called Viherlaakso, have formed relationships in order to obtain resources that are of vital importance to their survival. The paper mill could not continue producing paper without process steam; the Rautpohja plant could not use its test paper machine without steam; the commercial garden could not cultivate and sell its plants without heat; and the park and garden area could not enlarge its area without access to raw material suitable for its landscaping purposes.

Inducement of companies to form relationships with other companies can also be related to their search for competitive advantage (Galaskiewicz and Zaheer 1999). Firms seek to obtain control over those resources that are inimitable and not readily substitutable. The energy-generation companies of the focal network have gained unique intangible resources by developing material and energy cascading features in their operations. The use of wood residuals as well as lower-quality energy and the utilisation of power-plant ash provide the companies with a positive environmental reputation that might be beneficial for their competitive advantage.

As the relationships between actors develop, the actors make adaptations to their resource features and to the use of resource combinations. As a consequence, the companies become mutually and increasingly interdependent. (Håkansson and Snehota 1995) The focal firm has made adaptations to its power plant processes as a result of its relationship with wood and peat suppliers. This has been a conscious strategy for the focal firm and it will also be one of the main development strategies for the firm in the future. In addition to these fuel adaptations, the focal firm continuously makes adaptations to its power plant processes according to the process steam demands of the local paper mill. It changes the amount and temperature of the generated steam in accordance with the demands of the paper mill.

Adaptations to resources can be connected to investments. An investment may be a completely traditional hard investment, such as the purchase of a new machine. Still more often, the resources in an investment process are people, their know-how and time. When the actors of the focal net are considered, all of them have invested more to the relationships with the focal firm than what is needed in order to execute basic exchanges. For example, the energy companies with common ownership of the focal firm of this study have made remarkable hard investments by constructing the major power plant of the region. In addition, they have invested in the modification of the plant, which enabled the use of wood and peat fuels. As an another example, the individuals responsible for the park and garden area as well as those at the plant cultivation and sales centre have made significant soft investments in order to initiate the utilisation of power-plant ash and lower-quality heat. Some hard investments have naturally been needed as well, but the most significant investments have been intangible in nature.

The above-mentioned adaptations and investments create dependence in these inter-organisational relations. The focal energy-generation company can be considered to be somewhat dependent on wood and peat suppliers, even though the size of these suppliers and thus the bargaining power, related to their size, is not remarkable. In general, though, a large firm with considerable resources has a greater possibility of dominating its partners than does a small firm.

The Kangas Paper Mill is highly dependent on the focal energy-generation firm in terms of process steam production. Steam supplied by the focal firm is an important input to the paper manufacturing processes. This implies that the focal energy-generation company has power over the paper mill. In some senses, dependence and power relations established in this relationship may be regarded as the price the paper mill has to pay for the benefits that the relationship offers.

In every network, two dialectical processes are present: competition and co-operation (Axelsson and Easton 1992). The simultaneous presence of co-operation and competition is clearly illustrated in the energy supply network of Jyväskylä, where two competing energy companies have joined forces in order to achieve common benefits. They have constructed a common power plant and have established a common company to govern the operations of the plant. Nevertheless, they are competing nationally with each other for electricity customers. This kind of co-operation between competing companies is not a very common arrangement because of the risk of opportunistic behaviour, which causes network partners to be less willing to share information and know-how with each other.

In the case of these two co-operating and competing energy companies, fears of opportunistic behaviour were clearly present at the beginning of the relationship. However, a high level of personal interaction between these actors has been able to diminish these fears. After successful co-operation, the partners have expressed willingness to proceed with joint efforts. This supports previous research findings according to which two firms with prior alliances are likely to trust each other more than they are to trust other firms with which they have had no co-operation (Ring and Van de Ven 1994).

5.4 Towards Industrial Ecosystems

In Section 5.3 the theory of industrial networks was used to analyse an example of an industrial ecosystem. The same factors influencing the dynamics of industrial networks were found to have an effect on the dynamics of an industrial ecosystem. This implies that the industrial network approach could well be used in the analysis of industrial ecosystems. However, it must be admitted that as this network model concentrates on describing the network structure and relationships, it does not give much aid in explaining why and how the structure and relationships have evolved. Therefore the analysis made according to this network model is on a rather general and shallow level.

On the basis of the understanding evolved during the analysis, it could be argued that the network approach may also be used for highlighting the managerial implications of the development of industrial ecology practices. The following list gives examples of concepts within the network approach that may be regarded as useful in relation to industrial ecosystem development.

- Relationship formation: relationships between actors are formed through various exchange processes, giving actors access to the resources of other actors and possibly a competitive advantage.
- Adaptation: actors make adaptations to their resources, processes, routines or administrative procedures, creating mutual interdependence between actors. Adaptations are investments in specific relationships and they should be regarded as a conscious strategy.
- Bond building: bonds between network actors are likely to arise. The conscious building of bonds requires investments and prioritisation of existing and potential relationships.
- Mutual orientation: mutual orientation is needed in attempts to build long-life and strong relationships. In order to achieve mutual orientation, partners need to develop their relationship based on the needs of both parties and provide the partner organisation with additional value.
- Trade-off between co-operation and competition: partners need to decide on a trade-off between co-operation and competition regarding the control of resources.
- Generation of trust and interaction: in order to avoid opportunistic behaviour, partners need to generate a high degree of trust and interaction.
- Heterogeneity of actors: heterogeneity of actors is valuable in innovation processes.
- Co-ordination processes: network processes can be considered as a form of co-ordination between organisations in addition to traditional planning and control.

Interesting dimensions for further analysis would be a more precise examination of changes in industrial relationships and in operations of industrial ecosystems. An analysis of such changes could help to explain how industrial ecosystems come into being and how they can be transformed in the preferred direction.

According to the network approach, change occurs mainly through co-ordination and mobilisation of activities and resources (Hertz 1992). According to Lundgren (1992), the co-ordination of activities refers to continuous changes in the network, whereas the mobilisation of resources refers to discontinuous, more radical, network changes. Network processes and bonds between network actors can be considered as a form of co-ordination, directing network changes. Thus, changes in networks are evolutionary in nature, as network inertia and interdependencies slow down and shape changes (Lundgren 1992).

Mobilisation of resources and actors is a prerequisite for radical changes to occur in networks. Mobilisation processes are aimed at expanding or extending the network. Alternatively, mobilisation processes are aimed at establishing new activities, breaking old activities or combining previously unrelated activities together (Lundgren 1992). According to Axelsson and Easton (1992), technical knowledge is not enough for new operational solutions to occur, as these require changes within the network structure as well as within the firms involved. Hence, firms must adapt their old relationships and internal activities and develop new relationships before radically new forms of interaction can arise.

Radical changes or innovations often occur for network partners as a result of attempts to solve mutual problems together (Axelsson and Easton 1992). Thus, social relations and social exchanges between network partners that reveal common problems may facilitate the creation of new operational solutions. Social exchanges between different actors in society may then be needed in order to facilitate the development of industrial ecosystems.

A careful examination of how continuous co-ordination processes occur in the energy supply system of Jyväskylä and direct changes in that would be a worthy extension to this research. In addition, it would be appealing to examine more precisely how so-called radical changes have happened in the energy network in question: that is, how the operational solutions currently involved in the network have been initiated and put into practice. Such empirical studies might provide some insight into the essential question of how industrial systems can be transformed toward more environmentally benign industrial ecosystems.

5.5 Conclusions

It is not always enough that environmental management of a company focus solely on the firm's internal operations. Often, firms face environmental challenges that require inter-organisational responses. A newly emerged field of study, industrial ecology, offers some novel approaches to inter-organisational environmental management. It emphasises the need for "good fits" between organisations. These fits are formed through the formation of symbiotic linkages between firms or other organisations, in which the waste products of one actor are used as raw materials by others. So far, the field of industrial ecology has concentrated mostly on the ideological and technical aspects of inter-firm linkages. Consequently, it has failed to offer practical managerial implications for the development of such linkages.

Industrial network theory has been suggested as a possible framework for understanding the managerial aspects of industrial ecology. On the basis of this, in this chapter, industrial network theory was integrated with notions of industrial ecology. Some common features between these concepts were found and illustrated. The suitability of the network approach in relation to industrial ecosystems was then tested. An example of an industrial ecosystem was analysed with use of the network approach.

The results of this chapter suggest that the industrial network approach is useful in analysing and describing eco-industrial development, even though it is not that useful in explaining why and how the development has occurred. In addition, it is useful in developing managerial implications for industrial ecosystems. As a result, some implications for the development and management of industrial ecosystems have been proposed on the basis of the network approach. These highlighted the importance of adaptations and the conscious building of bonds between partners. In addition, development of relationships based on the needs of both parties was emphasised as being important in the achievement of a mutual orientation between partners. Heterogeneity of actors was proposed to be valuable to innovation processes, and the generation of trust and interaction were suggested to be essential in order to avoid opportunistic behaviour between partners.

References

Agranoff R, McGuire M (1998) Multi-network management: collaboration and the hollow state in local economic policy. Journal of Public Administration Research and Theory 8.1 (January 1998):67–92

Ahuja G (2000) The duality of collaboration: inducements and opportunities in the formation of interfirm linkages. Strategic Management Journal 21:317–343

Allenby B, Cooper WE (1994) Understanding industrial ecology from a biological systems perspective. Total Quality Environmental Management (Spring 1994):343–354

Axelsson B, Easton G (eds.) (1992) Industrial networks: a new view of reality. Routledge, London

Ayres RU, Ayres L (1996) Industrial ecology: towards closing the materials cycle. Edward Elgar, Cheltenham, UK

Easton G (1992) Industrial networks: a review. In: Axelsson B, Easton G (eds.) Industrial networks: a new view of reality. Routledge, London, pp 3–27

Ehrenfeld J, Gertler N (1997) The evolution of interdependence at Kalundborg. Journal of Industrial Ecology 1.1:67–80

Esty DC, Porter ME (1998) Industrial ecology and competitiveness: strategic implications for the firm. Journal of Industrial Ecology 2.1:35–44

Frosch D, Gallopoulos N (1989) Strategies for manufacturing. Scientific American 261.3:94–102

Galaskiewicz J, Zaheer A (1999) Networks of competitive advantage. In: Andrews S, Knoke D (eds.) Research in the sociology of organisations. JAI Press, Greenwich, CT, pp 237–261

Gulati R, Nohria N, Zaheer A (2000) Strategic networks. Strategic Management Journal 21:203–215

Håkansson H, Johanson J (1988) Formal and informal co-operation strategies in international industrial networks. In: Contractor FJ, Lorange P (eds.) Co-operative strategies in international business. Lexington Books, Lexington, MA, pp 369–379

Håkansson H, Johanson J (1992) A model of industrial networks. In: Axelsson B, Easton G (eds.) Industrial networks: a new view of reality. Routledge, London, pp 28–34

Håkansson H, Snehota I (eds.) (1995) Developing relationships in business networks. Routledge, London

Hertz S (1992) Towards more integrated industrial systems. In: Axelsson B, Easton G (eds.) Industrial networks: a new view of reality. Routledge, London, pp 105–128

Kale P, Singh H, Perlmutter H (2000) Learning and protection of proprietary assets in strategic alliances: building relational capital. Strategic Management Journal 21:217–237

Korhonen J, Wihersaari M, Savolainen I (1999) Industrial ecology of a regional energy supply system: a case of Jyväskylä region. Journal of Greener Management International 26:57–67

Lehtilä A, Savolainen I, Tuhkanen S (1997) Indicators of CO_2 emissions and energy efficiency: comparison of Finland with other countries. Technical Research Centre of Finland VTT, Espoo, Finland

Levine SH (1999) Products and ecological models: a population ecology perspective. Journal of Industrial Ecology 3.2–3:47–62

Lundgren A (1992) Co-ordination and mobilisation processes in industrial networks. In: Axelsson B, Easton G (eds.) Industrial networks: a new view of reality. Routledge, London, pp 144–165

Morgan G (1986) Images of organisation. Sage, Beverly Hills, CA

Pesonen HL (2001) Environmental management of value chains: promoting life-cycle thinking in industrial networks. Greener Management International 33:45–58

Ring PS, Van de Ven AH (1994) Developmental processes of co-operative interorganisational relationships. Academy of Management Review 19.1:90–118

Sinding K (2000) Environmental management beyond the boundaries of the firm: definitions and constraints. Business Strategy and the Environment 9.2:79–91

Turnbull P, Ford D, Cunningham M (1996) Interaction, relationships and networks in business markets: an evolving perspective. Journal of Business and Industrial Marketing 11.3:44–62

Van Berkel R, Willems E, Lafleur M (1997) The relationship between cleaner production and industrial ecology. Journal of Industrial Ecology 1.1:51–66

Van de Ven AH (1976) On the nature, formation and maintenance of relations among organisations. Academy of Management Review 1.4:24–36

Wallner HP (1999) Towards sustainable development of industry: networking, complexity and eco-clusters. Journal of Cleaner Production 7:49–58

6

Strategic Business Operations, Freight Transport and Eco-efficiency: A Conceptual Model

Booi Hon Kam[1], Geoff Christopherson[1], Kosmas X. Smyrnios[1]
and Rhett H. Walker[2]

[1]School of Management, Business Portfolio, RMIT University, Melbourne, Australia,
booi.kam@rmit.edu.au; kosmas.smyrnios@rmit.edu.au; geoff.christopherson@rmit.edu.au
[2]School of Business and Technology, La Trobe University, Bendigo, PO Box 199, Bendigo,
Victoria 3552, Australia, rhett.walker@rmit.edu.au

In an editorial in the *International Journal of Physical Distribution and Logistics Management* for a special issue devoted to the environmental aspects of logistics, McKinnon (1995: 3) described the logistics literature concerning environmental issues as 'small but expanding'. Since then, the literature on the environmental sustainability of freight logistics has flourished; as Murphy and Poist (2000: 5) posited, 'environmentalism has "come of age" as a major topic in logistics and no longer can be regarded as a peripheral concern or fad'.

Referred to commonly as green logistics, this topic of inquiry has also been widely acknowledged as one of the key issues facing logistic management in Europe in the new millennium (Skjoett-Larsen 2000). Despite the growing attention given to topics relating to green logistics, most research in the area seems to be preoccupied either with exploring ways to achieve environmentally sustainable logistics or with determining strategies considered as most cost-effective for managing and responding to environmental issues in logistics. There are a number of studies directed at exploring solutions for environmentally sustainable logistics. Examples include Beamon's (1999) paper on designing the green supply chain; Wu and Dunn's (1995) recommended steps toward achieving a proactive environmental management focus in logistics; and Lin *et al.*'s (2001) process perspective on environmental practices and assessment.

In the case of research aimed at identifying green logistics strategies, a survey by Murphy and Poist (2000) of companies in the USA, Canada and the European Union (EU) on the usage patterns of green logistics strategies is an illustration. The work by Prendergast and Pitt (1996), who surveyed marketing executives of UK companies on their perceptions of trade-offs between traditional marketing and logistical functions and environmental pressures to reduce, recycle and re-use packaging, is another. A further example involves Fernie *et al.*'s (2000) analysis of UK grocery retailers' assessment of factors that affect grocery supply chain as a result of a need to accommodate e-commerce initiatives and environmental pressures. Unfortunately, scant attention has been directed at developing theories that   can assist senior management in making decisions on supply chain strategy that meet the necessary requirements of environmental sustainability.

Building on the notion of sustainable development—that is, meeting the needs of the present without jeopardising the needs of future generations—contained in the report of the Brundtland Commission (WCED 1987), De Simone and Popoff (1997: 21) defined eco-efficiency as 'the business response to the challenge of sustainable development'. These investigators argued that eco-efficiency is more than the reduction of pollution, energy, and material throughput, as contended by Schmidt-Bleek (1994). The principle is seen to transcend even pollution prevention, product stewardship and other concepts addressed in industrial ecology (Ayres 1989; Ayres *et al.* 1994). Eco-efficiency encompasses four key ideas: creating value for business as well as society; building long-term targets for improvements; linking business excellence with environmental excellence; and ensuring production as well as consumption is sustainable. To understand how businesses can attain these eco-efficiency aspirations in their continual striving to achieve and sustain competitive advantage, it is clear that links between strategic operations and business decisions need to be established on empirics.

In this chapter we argue that understanding behavioral responses of firms to the changing nature of business environments is a prerequisite to dealing with issues arising from impacts that freight transport has on the environment. In Section 6.1, drawing on the transport system model of Gudmundsson and Höjer (1996) we develop a conceptual framework linking three sets of factors regarded as having a direct effect on the manner and extent to which companies devise strategies to deliver products to customers. In Section 6.2, the strategic product delivery actions adopted by three organisations in different business activities, to optimise business growth and profitability, are presented and examined (Sections 6.2.1–6.2.3). Ways in which these actions have led to dissimilar patterns of product delivery regimes, creating different levels of impact on the environment, is discussed in Section 6.2.4. We conclude the chapter, in Section 6.3, by highlighting the applicability of the proposed conceptual model. This framework provides a useful platform for conceptualising future green logistic research aimed at understanding impacts of changing logistical demands and concomitant business responses on the environment.

6.1 Freight Transport and Environmental Impact: A Conceptual Causal Framework

Kordi *et al.* (1979) suggested that the physical components of any transport system include the vehicle (*e.g.*, car, train, vessel, aircraft), the energy source (*e.g.*, petrol, liquefied petroleum gas, diesel oil, electricity) and infrastructure (*e.g.*, road, railway, airport, harbour). Moreover, Gudmundsson and Höjer (1996) contended that a functioning transport system also encompasses a social system. Individuals, who dictate the structure of the social system through their choices and institutional make-up, also operate within the system and are affected by it. The imposition of the social system gives rise to two other important elements: vehicle operators and organisations. Gudmundsson and Höjer (1996) asserted that these five elements—

vehicle, infrastructure, energy source, vehicle operator, and organization-concurrently, and through their interactions, determine the type of environmental impact generated by a given transport system. The size of the impact, in turn, is dependent on the type of impact and is also a function of the use imposed on the system arising from the transport services demanded.

A careful examination of Gudmundsson and Höjer (1996), and Kordi *et al.*'s (1979) models indicates that these paradigms provide only a partial view of the causal chain that links transport to its impact on the environment. Interestingly, factors determining the nature of the five elements that constitute the transport system and those governing its use have been largely ignored. In passenger transport, the significance of behavioral traits on trip generation, mode, and route choice was recognised over 30 years ago, beginning with the travel-demand system estimated by Domencich and McFadden (1975). In freight transport, in contrast, there is a dearth of literature concentrating on the behavioral underpinnings of product delivery patterns relating to satisfying customer needs, to meeting production deadlines (such as just-in-time [JIT] delivery) and to optimising value in the management of the supply chain. These underpinnings have a substantial effect on nature and use of transport systems.

Conceptually, the choice of delivery vehicles, the scheduling of deliveries, and decisions concerning frequency and mode of product delivery, and type of fuel to use represent company responses to the joint actions of three main sets of factors. The first of these relates to external influences, covering economic, political, social, and physical issues. These influences set the stage and the context within which business entities operate. The second set of factors pertains to company demographics, such as size and nature of business and internal policy environments. An organisation's internal environment encompasses elements regarding a company's mission, long-term goals, competitive position, and resource constraints along with other characteristics commonly examined under a SWOT (strengths–weaknesses–opportunities–threats) analysis. The final compendium of factors involves the state of available technology, enabling firms to select cost-efficient solutions from a technically feasible set. Figure 6.1 shows this conceptualisation, depicting causal chains linking the joint effects of these three sets of factors on product delivery responses and their subsequent environmental impact.

The conceptualisation proposed in Figure 6.1 is consistent with that of Skjoett-Larsen (2000). Outlining the emerging trends in international logistics, and describing how European companies are facing up to the new challenges, Skjoett-Larsen (2000: 379) pointed out that many companies are moving 'towards a reduction in their supplier base and differentiated supplier co-operation'. This shift suggests that companies are differentiating their strategies for sourcing supplies in relation to suppliers, according to the relative importance of the purchase and problems associated with managing purchases. This shift is the result of three major factors:

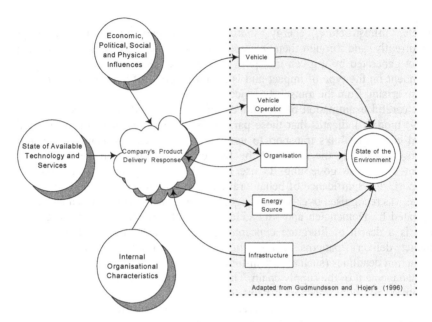

Figure 6.1. A conceptual framework for analysing the environmental impact of freight transport

- Realisation that searching, selecting, and evaluating suppliers on an ongoing basis is not as cost-efficient as it seems, because long-run costs associated with training, quality problems, and delivery delays might more than offset short-run reductions in costs achieved by playing one supplier against another
- The move from in-house production of nearly all components to outsourcing production processes that are not part of a company's core competences
- The rapid pace of technological development that makes it more important to use suppliers at the leading edge of new technology, rather than suppliers who are competitive in price

The first of these factors is evidently a response to changing circumstances occurring in the external business environment. The second signifies a corollary of the business strategy companies adopt to remain competitive. The third factor represents solutions adopted by businesses in their attempt to capitalise on the state of available technology.

To further illustrate the theoretical underpinnings of our proposed conceptual model, the product delivery strategies of three different companies located in disparate industry sectors are presented below. Secondary information from a number of different sources (*i.e.*, SCR 2000; Sullivan 2001b; Simon and Ashton 1998) was used to develop a profile of the logistics systems put in place by these three companies. We then examine the manner in which these three companies respond to the three sets of factors: external influences, internal organisational characteristics and business goals, and the state of available technology and services. Relationships between responses employed by these three companies and the five transport elements are discussed to highlight links between company responses and their impact on environmental integrity, which underpins ecologically sustainable logistic operations.

6.2 Case Studies

6.2.1 Case Study 1: Herron Pharmaceuticals (Supply Chain Review, 2000)

Herron Pharmaceuticals (hereafter referred to as Herron), a leading name in analgesic products in Australia, manufactures approximately 900 products besides its flagship Herron Paracetamol. Herron uses an indicator of 'delivery in full and on time' (DIFOT), which it acquired through a benchmarking group. This indicator is employed as a yardstick for gauging product delivery performance. The indicator comprises two scores—an internal rating of delivery 'in full and on time', and a measure of external carrier's delivery performance. Prior to joining the peer benchmarking group, Herron employed a measure in terms of stock-keeping units (SKUs) to gauge fulfilment performance. Since joining the benchmarking group, Herron has adopted measure in terms of fulfilment by-order—a more stringent yardstick.

For Herron, a critical success factor in freight transport is the delivery of goods to large supermarket distribution centres within one-hour delivery-windows. A delivery made outside the allocated time-slot could mean a delay of two-to-three days before the carrier receives another time-slot. Adoption of the DIFOT customer service measure enabled Herron to deal with inefficiencies associated with the original delivery company. Owing to costs associated with lower performance ratings, Herron changed its transport contractor to one with a larger interstate volume to enable it to consolidate and manage deliveries much more accurately. This new contractor also assisted Herron to track its goods while in transit.

In addition, Herron instituted a 'reason code' for deliveries. This coding system provided Herron with the capability of monitoring shipment of goods, including reasons for late delivery. Before the introduction of the reason code, Herron assumed that dispatched products were delivered as scheduled.

As Herron's key account representatives were in Sydney and Melbourne, whereas the company's manufacturing headquarters was in Brisbane, long lead-times were built into the company's delivery system. As a result, Herron increased its stock holdings from an average of two weeks to six-to-eight weeks as a means of limiting the number of stock-outs and to maintain its delivery performance at a consistently high level. These changes culminated in significant customer service benefits and reduced the quantity of back-orders and re-servicing fees. Part of the rationale behind this strategy came from the 80/20 rule (*i.e.*, 80% of sales came from 20% of SKUs).

Although this policy tied up a considerable amount of capital in finished goods and heightens the costs of warehouse storage, part of these expenses were offset by efficiencies gained from larger production runs and decisions to remain out of stock for longer periods with slow-moving and obsolete lines. To Herron, the competitive advantages created by increasing customer satisfaction had far outweighed the costs associated with holding extra inventory, even in the face of increasing interest changes. Possibly, and not surprisingly, maintaining high-stock levels helped Herron to engage in rapid growth. It appears that Herron's success, in sum, lied in making optimal trade-offs: one that needed to be made in achieving targeted levels of delivery performance.

Put in the context of the conceptual model outlined in Figure 6.1, the principal external environmental factors that have a particular influence on Herron's responsiveness concern the critical advantage of delivering pharmaceutical products within one-hour delivery-windows and the interstate locations of their largest accounts. The principle of delivering in full and on time constitutes the main mission of the company: satisfying customers—in particular the large interstate accounts. The availability of a transport contractor with large interstate volume represents a third factor, prompting Herron to change its transport operator. The interactions of these three sets of factors led to two important decisions: to engage a transport contractor (vehicle operator) whose interstate consignments necessitated the utilisation of large vehicles, and to increase levels of stock-holdings from an average of two weeks to six-to-eight weeks (*i.e.*, representing a change in organisation). This conceptualisation is shown in Figure 6.2.

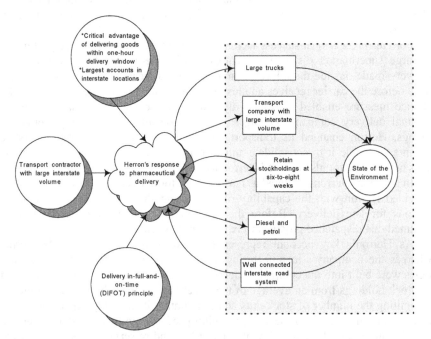

Figure 6.2. Relationships between Herron's product delivery strategy and the environment

6.2.2 Case Study 2: Solomon Distribution (Supply Chain Review, 2000)

Once a family milk-run, Solomon Distribution (hereafter referred to as Solomon) has emerged as one of Brisbane's major wholesale food distributors. Solomon's foremost business goal is to cultivate long-term customer relationships, and its business principle is to provide fast, responsive delivery services to customers.

In 1993, in response to an Australian Federal Government announcement that the dairy industry was to be deregulated by 1999 (Australian Bureau of Statistics [ABS], 2004), Solomon set about extending its geographical customer base by acquiring four milk-runs. Deregulation meant that customers were no longer tied to a single milk distributor. Factoring economic trends and contingencies into its business plans, Solomon adopted a strategy of controlled diversification as its motto for growth. Milk was seen as an appropriate base to expand into the food-services industry. As a first step in its diversification program, in 1997 Solomon acquired Classic Fruit Juice and upgraded its premises to incorporate a bottling facility on-site. Diversifying into food services also enabled the company to increase the utilisation of its vehicular fleet. Solomon vehicles operated on two shifts: milk distribution from midnight to 7 am, and food service from 5 am to 5 pm. Although the customer base also expanded geographically, business growth had been planned strategically.

Solomon's expansion also concentrated in servicing city take-aways, restaurants, and coffee shops. With a modern warehouse located close to the city centre, Solomon was capable of delivering to customers within 10 minutes. As a fast, responsive service, Solomon's trucks, at times, operated below capacity on daily runs, as against competitors who delivered to certain areas only two-to-three times a week. This solution helped Solomon to develop sound levels of customer loyalty and to expand into new markets.

Achieving a fast and responsive service has also meant minimising vehicle downtime. In order to maximise uptime for its vehicular fleet, Solomon courted Brisbane Isuzu to form a strategic partnership. Within the context of this strategic alliance, Isuzu provided on-site vehicle servicing on Solomon's fleet of Isuzu trucks to ensure fast, reliable service delivery.

Within the framework of the proposed conceptual model, the announcement by the Federal Government to deregulate the dairy industry triggered Solomon into adopting an expansion, then diversification, path. Coupled with Solomon's overarching goal to provide fast, responsive delivery services to customers, and the agreement of Brisbane Isuzu to perform on-site servicing of vehicles, Solomon opted to operate its medium-sized trucks at below capacity on daily runs when necessary. In the process, the company strategically sited its new warehouses in close proximity to the city centre so that urgent deliveries could be made to city customers within minutes. This interpretation is illustrated in Figure 6.3.

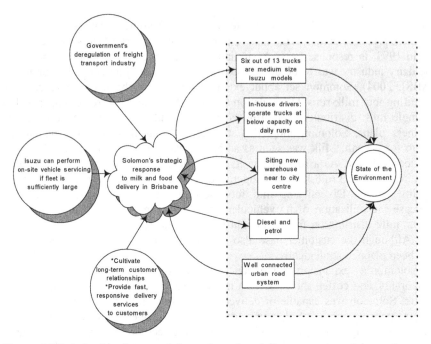

Figure 6.3. Relationships between Solomon's product delivery strategy and the environment

6.2.3 Case Study 3: Wall's Ice Cream (Simon and Ashton, 1998)

In the early 1990s Wall's Ice Cream (hereafter referred to as Wall's), one of the most well-known brand names in China, was confronted with a distribution problem when it commenced operation in Beijing. Noted for extreme levels of traffic congestion, Beijing's road system is a challenge for any foreign organisation involved in distribution, let alone in the distribution of perishables requiring refrigeration. Finding that its trucks were unable to deliver to the many stores located around the city of Beijing, Wall's began in 1994 to commission bicycle and motorcycle owners to deal with this logistical problem. Distributors, on bicycles and motorcycles equipped with small freezers, picked up ice cream at Wall's main Beijing distribution centre, delivering to assigned stores, many of which were outfitted with dedicated freezers provided by Wall's. The condition attached was that freezers were to be used only to store and sell Wall's products. This successful solution to a logistical and environmental problem had since been applied by Wall's to Shanghai and several other major cities throughout China.

Obviously, Wall's ingenious arrangement of using bicycles and motorcycles for its delivery grew out of the unique transport environment in Beijing. An inadequate road system plagued with extreme traffic congestion necessitated an uncommon solution. Capitalising on the omnipresence of bicycles and motorcycles in the road network of Beijing, Wall's outfitted these vehicles with small freezers to enable them to deliver a highly perishable product. Figure 6.4 depicts the interaction of these factors in the context of our proposed framework.

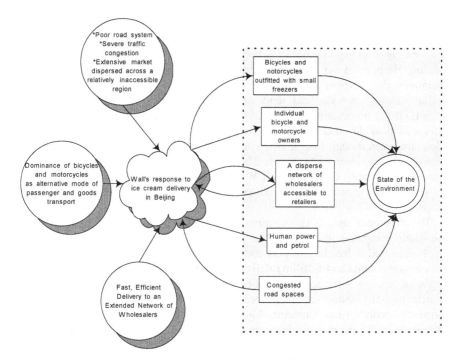

Figure 6.4. Relationships between Wall's product delivery strategy and the environment

6.2.4 Discussion

These three case studies highlight that the search for cost-effective modes of product delivery requires creative methods, operating within the context of three sets of factors: external influences, internal policy directives, and the state of available technology. These factors should be considered within the context of existing transport systems. In the case of Herron, the consignment of large, rather than small, trucks seemed to offer the most cost-efficient solution to this company's product delivery problem. In comparison, Solomon benefited through the use of small, frequent consignments, even when at times the runs were below capacity. Wall's adopted a novel approach to overcome a hostile traffic environment. However, in eco-efficiency terms, outcomes associated with the various strategies are somewhat different.

Wall's solution can be regarded as the best environmentally, as cycling is one of the least polluting transport modes available. It is a solution to a particular distribution problem and, from a freight logistics viewpoint, it is not a solution that can have a more general application. Solomon's decision to position warehouses close to its customer locations and to deliver at times when the road traffic was relatively light (at night or in the early morning) reduced trip distances and congestion and so can be classed as an ecologically sustainable operation. In contrast, Herron's decision to supply customers from a central point over quite long distances with use of heavy vehicles does have adverse environmental effect. Bamford (2002) commented that such operations are 'bad news' from an environmental viewpoint because they can involve up to 30% vehicle-empty running. He pointed out further that there is current criticism in the EUof logistic operations that involve the sourcing of products from distant locations   when similar, although not identical, items are available locally. This criticism is based, in part, on the observation that larger vehicles contribute more to greenhouse gas emissions than do smaller vehicles in terms of grammes of carbon dioxide per kilometre, even though freight efficiency in terms of tonnes of carbon dioxide per kilometre is improving (Austroads 2001). On this point, in Australia, the Road Transport Authority of New South Wales (NSWRTA 2001) pointed out that road transport alone contributes approximately 14.5% to greenhouse gas emissions, nationally.

The divergent and, at times, innovative logistical adaptations made in these case-study companies suggest that the principle of delivery in full and on time, so highly regarded in freight transport operations, is not a simple matter of optimising vehicle movements and fulfilling delivery schedules based on least-cost solutions. Logistics requires a fundamental understanding of plausible behavioral responses in order to assist businesses to survive, compete, and grow in fiercely competitive, constantly evolving environments. More importantly, green logistics demands an understanding of interactions between eco-efficiency of businesses, their bottom lines, and other financial considerations. The largely unrecognised downside of risk and upside of potential linked to business operations seeking to meet targets of eco-efficiencies is a key area requiring further exploration. The proposed conceptual framework is intended to facilitate the development and testing of grounded theories that explain complex causal relationships between strategic positioning, product delivery, freight transport logistics, and the environment. Application and testing of these theories is an important first step for future research aimed at broadening eco-efficiency agendas of logistics operations.

6.3 Conclusions

Compared with passenger travel planning, freight transport planning is far more complex. As far as designing a conceptual framework to link freight transport with environmental sustainability is concerned, one of the many intractable issues is to establish the functional and physical 'distances' between locations of enterprises and sites of environmental damage that result from logistical structures of freight operations. As our three case studies illustrate, decisions concerning stocking, vehicle sizes and vehicle scheduling are made at locations far away from affected areas. A challenge in the context of achieving eco-efficiency is 'whether and how the logistical chains can be influenced "backwards" from that location' (Hesse 1995: 39). Clearly, it would be hard to confront this challenge without an understanding of factors underpinning strategic business decision-making processes.

In this chapter we have provided a theoretical framework within which to explore the impact that freight transport has on the environment. The present conceptualisation is based on the view that companies' attempts to meet customer needs are predicated on the concurrent actions of three important factors: external environmental conditions; company internal characteristics, policies and strategic mission; and the state of available technology and services. The interactions of these elements shape company responses, influencing the extent of a delivery that is in full and on time. Systematically, a company's level of response has significant ramifications for the environment. That is, responses affect type of vehicle, vehicle operators, and logistics organisation, which, together with the energy used and the condition of the transport infrastructure, determine the extent of eco-efficiencies.

The Australian Trucking Association has developed an environmental management system (EMS) covering a range of environmental sustainability issues, which it aims to implement and audit. The extent to which government agencies are prepared to accept such industry self-regulation as an alternative to government regulation appears to be determined by the degree to which industry can provide genuine signs of commitment to making the EMS work (Sullivan 2001a). The proposed model offers an analytical platform to examine company responses to comply with requirements of whatever EMS is in force.

The significance of the framework proposed in this chapter lies in its emphasis on understanding behavioral underpinnings of companies in their attempts not only to compete but also to sustain their competitive advantages (Porter 1996), as a prerequisite to assessing the impact that freight transport has on the environment. More importantly, perhaps, is the growing popularity of e-commerce, which has warranted a re-engineering process in supply chain management and freight transport (Wagner and Frankel 2000; Lewis 2001; van Hoek and Chong 2001). Businesses have high expectations about the speed of product delivery that e-commerce offers, and these expectations significantly heighten demands for freight transport. In comparative terms, changes heralded by rising demands for e-commerce in the freight transportation and logistics industry is reckoned to pale the impact generated by the JIT revolution of the 1980s and 1990s (Golob and Regan 2001).

Exploring the potential effects of information technology on transportation, Golob and Regan (2001: 104) stated that 'e-commerce may lead to smaller, more frequent shipments and significant freight flows from points where neither shipper nor recipient are present'. In other words, the use of light, small, rather than large, vehicles for freight transport is expected to dominate. In Australia, this trend has been emerging in the freight usage pattern over a three year period. Figures released by the Australian Bureau of Statistics (ABS 2001) show that between 1998 and 2000 the total laden business kilometres travelled by freight vehicles in Australia rose from 18,967 million to 20,997 million. Over 70% of this differential is attributable to the surge in laden kilometres travelled by light commercial vehicles. Interestingly, small freight vehicles contributed a mere 6.5% of the increase in tonnes-kilometres travelled. The increase in tonnes-kilometres travelled by light commercial vehicles over the 1998–2000 period represents a rise of 23%, compared with 14% for articulated trucks and 11% for rigid trucks (ABS 2001).

The increasing usage of light commercial vehicles has far-reaching implications, particularly in terms of fuel consumption and load carrying efficiency. In a review of freight activity and energy use in 10 industrialised countries over the period 1973–92, Schipper *et al.* (1997) found that heavy trucks use three-to-five times more fuel per kilometre than do light trucks, and about 30% more fuel per kilometre than do medium-sized trucks. Schipper *et al.* also reported that as the number of small trucks entering the national stock to carry smaller loads of freight increased, the modal fuel intensity, defined in terms of megajoules per tonne-kilometre, rose. This phenomenon was evident in Denmark and Sweden during the 1980s (Schipper *et al.* 1997). In the Australian context, light commercial vehicles are also the least fuel efficient when measured in terms of fuel consumption per 100 tonne-kilometre travelled. Light commercial vehicles consume some 65.8 litres per 100 tonne-kilometres travelled, compared with only 7.4 litres for rigid trucks and 2.8 litres for articulated trucks (ABS 2001). As light commercial vehicles continue to gain prominence, however, the wide disparity between their rate and that achievable by rigid and by articulated lorries is a cause for concern from the point of view of environmental sustainability. With Internet 'going mobile' (*The Economist* 2001), the use of small trucks is expected to escalate (Golob and Regan 2001). What is even more disturbing is that the form and nature of e-commerce itself is in a constant state of flux.

The above trends merely point to the need for environmentally responsive research agendas in freight transport logistics geared to meeting rising demands for e-commerce and JIT deliveries. More importantly, these trends indicate that solutions associated with problems in freight transport lie in understanding evolving complexities emerging from responses by companies to fulfil growing demands of e-commerce. Within this context, the proposed framework serves as a useful platform on which to base future research.

References

ABS (Australian Bureau of Statistics) (2004) *Australian Year 2004*, (Canberra, Australia:ABS).

ABS (Australian Bureau of Statistics) (2001) *Survey of Motor Vehicle Use* (Canberra, Australia: ABS).

Austroads (2001) 'Environmental Performance Indicators', *Australian and New Zealand Road System and Road Authorities National Performance Indicators 2000* (Sydney, Australia: Austroads Incorporated): 37.

Ayres, R. (1989) 'Industrial Metabolism', in J. Ausubel and H.E. Sladovich (ed.), *Technology and Environment* (Washington DC: National Academy Press.): 23-49.

Ayres, R., W. Schlesinger and R. Socolow (1994) 'Human Impacts on Carbon and Nitrogen Cycles', in R. Socolow, C. Andrews, F. Berkhout and V. Thomas (eds) *Industrial Ecology and Global Change* (Cambridge, UK: Cambridge University Press): 121-55.

Bamford, C. (2002) 'Sustainable Transport: Can it be a Reality? ,' *Huddersfield Logistician*, (Huddersfield, UK: University of Huddersfield)

January 2002: 24-27.

Beamon, B.M. (1999) 'Designing the Green Supply Chain', *Logistics Information Management* 12.4: 332-42.

De Simone, L., and F. Popoff (1997) *Eco-efficiency: The Business Link to Sustainable Development* (Cambridge, MA: MIT Press).

Fernie, J., F. Pfab and C. Marchant (2000) 'Retail Grocery Logistics in the UK', *International Journal of Logistics Management* 11.2: 83-90

Golob, T.F., and A.C. Regan (2001) 'Impacts of Information Technology on Personal Travel and Commercial Vehicle Operations: Research Challenges and Opportunities', *Transportation Research Part C* 9: 87-121.

Gudmundsson, H., and M. Höjer (1996) 'Sustainable Development Principles and Their Implications for Transport', *Ecological Economics* 19: 269-82.

Hesse, M. (1995) 'Urban Space and Logistics: On the Road to Sustainability?', *World Transport Policy and Practice* 1.4: 39-45.

Kordi, I., B. Jonsson, H. Scholander and R. Westerlund (1979) 'Energy Efficiency in Passenger and Freight Transportation: A Comparative Study of the Swedish Situation', *Transportforskningsdelegationen* 6 (Solna, Sweden: Tryckindustri AB,): 106.

Lewis, I. (2001) 'Logistics and Electronic Commerce: An Interorganisational Perspective', *Transportation Journal* 40.4: 5-13.

Lin, B., C.A. Jones and C. Hsieh (2001) 'Environmental Practices and Assessment: A Process Perspective', *Industrial Management and Data Systems* 101.2: 71-79.

Domencich, T.A. and D. McFadden (1975) Urban Travel Demand : A Behavioral Analysis : A Charles River Associates Research Study. (New York, USA: North-Holland Pub. Co.)

McKinnon, A. (1995) 'Editorial', *International Journal of Physical Distribution and Logistics Management* 25.2 (Special Issues on Environmental Aspects of Logistics): 3-4.

Murphy, P. R. and Poist, R. F. (2000) 'Green logistics strategies: An analysis of usage patterns', *Transportation Journal*, 40.2:5-16.

NSWRTA (New South Wales Roads and Traffic Authority) (2001) ' Greenhouse 10', in *RTA Environment Report* (Sydney, NSW : RTA): 18.

Porter, M. (1996) 'What is Strategy?', *Harvard Business Review* (November–December 1996): 61-78.

Prendergast, G., and L. Pitt (1996) 'Packaging, Marketing, Logistics and the Environment: Are there trade-offs?' International Journal of Physical Distribution & Logistics Management, 26.6: 60-72.

Schipper, L., L. Scholl and L. Price (1997) 'Energy Use and Carbon Emissions from Freight in 10 Industrialised Countries: An Analysis of Trends from 1973 to 1992', *Transportation Research Part D*, 2.1: 57-76.

Schmidt-Bleek, F. (1994) *Wievtel Umwelt, braucht der Mensch? MIPS: Das Mass für ÖOkologisches Wirtschaten* (Berlin, Germany: Birkhaeuser).

SCR (Supply Chain Review) (2000) 'More Inventory Equals More Sales', *Supply Chain Review* 25 August 2000: 12-15, 18.

Simon, D., and D. Ashton (1998) 'China's Supply Chain Challenge: Creativity in a Giant Marketplace', in J. Gattorna (ed.), *Strategic Supply Chain Alignment: Best Practice in Supply Chain Management* (London, UK: Gower): 588-602.

Skjoett-Larsen, T. (2000) 'European Logistics Beyond 2000', *International Journal of Physical Distribution and Logistics Management* 30.5: 377-87.

Sullivan, P. (2001a) 'Quid Pro Quo', *Supply Chain Review*, 27 April 2001: 82.

Sullivan, P. (2001b) 'Going for Growth', *Australasian Transport News*, 27 July 2001: 20-23.

The Economist (2001) 'Survey: The Mobile Internet', *The Economist* 11 October 2001.

van Hoek, R.I., and I. Chong (2001) 'Epilogue: UPS Logistics: Practical Approaches to the E-supply Chain', *International Journal of Physical Distribution and Logistics Management* 31.6: 463-68.

Wagner, W.B., and R. Frankel (2000) 'The Convergent Carrier: A Critical Factor in Supply Chain Compression', *International Journal of Logistics Management* 11.1: 99-110.

WCED (World Commission on Environment and Development) (1987) *Our Common Future*. (Brundtland Report; Oxford, UK: Oxford University Press).

Wu, H., and S.C. Dunn (1994) 'Environmentally Responsible Logistics Systems', International Journal of Physical Distribution and Logistics Management 25.2: 20-38.

7

Reverse Logistics for Recycling: Challenges Facing the Carpet Industry

Marilyn M. Helms[1] and Aref A. Hervani[2]

[1]Sesquicentennial Endowed Chair and Professor of Management, Division of Business Administration, Dalton State College, 213 N. College Drive, Dalton, GA 30720, mhelms@daltonstate.edu
[2]Chicago State University, Dept. of Geography, Anthropology, Sociology, & Economics SCI-321, 9501 S. King Dr. Chicago, IL 60626-2186, ahervani@csu.edu

In this chapter we explore the issues of reverse logistics for recycling within the carpet industry, including an economic analysis of the success of carpet recycling. In particular, we look at:

- Carpet recycling (Section 7.2);
- The future of reverse logistics for the carpet industry (Section 7.3);
- A framework for understanding recycling (Section 7.4);
- Policy options and implications (Section 7.5); and
- Areas for future research (Section 7.6).

First, in Section 7.1, we present a general discussion of reverse logistics.

7.1 Reverse Logistics for Recycling

Manufacturers spend much of their time and energy coordinating their complex supply chains from raw material suppliers to producers, wholesalers, distributors, retailers and customer. With all the attention to the forward action of the supply chain, few manufacturers have considered how this supply chain can or should work in reverse to reclaim products at the end of their life-cycle and return them through the supply chain for decomposition, disposal or re-use of key components. Believing that once products are delivered the firm's responsibilities end is one of the deadly sins of logistics (Stock 2001). Taking a life-cycle approach to product distribution is vital, along with implementing educational programmes for customers, suppliers, vendors and others in the supply chain.

Strategic factors to consider in reverse logistics include costs, overall quality, customer service, environmental concerns and legislative concerns. On the operational side, factors to consider are cost–benefit analysis, transportation, warehousing, supply management, remanufacturing and recycling, and packaging. According to Dowlatshahi (2000), insights about these factors form the state-of-the-art knowledge on the keys to the successful design and use of reverse logistics systems. Other issues to consider in reverse logistics are the desires of the customers. For example, do customers feel a responsibility to recycle and return used products and do they demand recycled content in their new products? Often, incentive systems or no-cost return systems must be in place to make reverse logistics work without external governmental regulation. Because the quality of inputs for re-use is important in many situations, clean, safe return methods must be in place as well. All supply channel members must be committed to the process, and it needs to be financially attractive to participate in the process. Economies of scale must be sufficient to make environmental reverse logistics viable (Blumberg 1999).

The recycling of old materials requires collection, sorting and processing, and the profitability is influenced by the efficiency achieved through co-ordination and integration. The profits made at each stage are determined by the state of the competition and the nature of markets. The implementation of internal reverse logistics programmes often involves significant allocations of capital and/or resources for the construction of reclamation and/or redistribution facilities and the purchasing of recycling equipment. Sustainable economic growth is achieved when firms choose the production technology process that will reduce the amount of pollution by-products and allow the final product to be used or reprocessed in further production operations. The usability and reprocessing characteristics of products requires initial planning and product design to allow future re-usability.

A firm's incentive to design a more usable product will depend on whether such a change will require costly production technology (Nagel 2001). Sustainable economic growth and reverse logistics merge when both emphasise the need for changing production technologies to reduce the by-products of a final good. Reverse logistics for recycling is growing, for two reasons: (1) to reclaim value through returned products that are further re-used for recycling and (2) the environmental concerns arising from a lack of future landfill availability for disposal options.

7.2 Carpet Recycling

The Carpet and Rug Institute (CRI) based in Dalton, GA, a proactive advocacy organisation for the carpet industry, signed the voluntary National Carpet Recycling Agreement in 2002, with various carpet and fibre manufacturers, state governments, non-governmental organisations (NGOs), and the US Environmental Protection Agency (EPA), with the goal of reducing the amount of carpet and carpet fibre going to landfill by 40%, to conserve oil and to reduce greenhouse gas emissions. The carpet industry is relying on this initial informal agreement to encourage product stewardship because voluntary environmental agreements (VEAs) do not yet exist (Whaley 2002).

The CRI's *Sustainability Report* (2003) outlines the industry's progress in lessening the environmental impact of carpet, including the reduction of water use by 46% since 1991, the reduction of energy consumption per unit by 70% and the recycling of 6.4 billion pounds of waste per year. A third-party organisation, the

Carpet American Recovery Effort (CARE)[1] has established a collection network for used carpet. In addition, the industry is reviewing advances in logistics management for recycling that have found success in Europe.

Carpet can rarely be refurbished. Although some manufacturers can reapply dye to extend the life of carpets or replace or clean carpet tiles in high-traffic areas, remanufacturing has not been an important activity for the industry. Product design at the early stages is essential in the success of the recycling of old materials into the manufacture of new products after reclamation and recovery. The technology must exist in order to accelerate the recycling of used carpet back into raw materials in a closed-loop recycling system. This would lead to efficient use of the product throughout its entire life-cycle. Mills for recycling carpet are and will continue to be the largest firms operating the largest production facilities in a vertically integrated process.

7.2.1 Carpet Recycling Success Factors

For reverse carpet logistics to operate successfully a number of factors need to be present. It requires:

- Initiatives by government and private industry to launch such programmes
- Changes to the firm's structure
- Capital and long-term commitment for product redesign and resource recovery
- Collection infrastructures developed by the industry or integrated with those of independent collectors or public collection agencies
- Markets for old carpet
- Relatively high prices help to recover the collection cost and provide an incentive for resource reclamation
- Integration with the existing collection agents in the private and public sectors
- The development of new collection infrastructures
- Use of special coding to identify the product, its manufacturer and its components to ease sorting
- Development of collection centres to collect old carpets, with specified coding or markings on the back
- Changes to product design, to ease reprocessing

The industry must undertake efforts to make recycling less costly and more efficient. One way to achieve this goal could be for industry to engage in activities such as:

- Giving producers access to quantities of secondary fibre
- Providing incentives to users to dispose of old carpets appropriately
- Devising a means of collecting the used carpets to provide the producers with recovered material that is not contaminated or damaged

---

[1]Note: this should not be confused with the CARE project (Computer Aided Resource Efficiency Accounting for Medium-Sized Enterprises), described in Chapter 20.

- Establishing drop-off centres or the necessary transportation to pick up the used, dry carpet and fibres

In some regions, collection agencies collect old carpets for a fee. This fee may not provide an incentive to households to recycle, since lower or no-cost disposal fees may divert the material from recycling bins to landfill. This consumer fee does not affect the producers and thus will not provide the producer with an incentive to manufacture new products that can more easily be recycled. Some state regulations hamper carpet recycling efforts and create obstacles to the success of recycling.

The collection process for used carpet involves the final consumers or end-user, collectors, the carpet retailers and the carpet installer. The reclaimed value of old carpets and recycling is distributed among these agents. The integration of these agents can increase the efficiency of collection efforts and lead to lower costs of reclamation and to profitability in recycling. An alternative strategy for carpet collection would be for producers to take the initiative to develop a collection infrastructure to reclaim carpets. A co-ordinated collection process would ensure production lines have adequate and accessible raw materials without relying on carpet collectors for the distribution of the raw materials and supplies.

Carpet manufacturers produce in-plant carpet waste; such pre-consumer carpet waste can be easily managed through internal channels whereas the reclamation of post-consumer carpet waste is a more challenging objective. Reverse logistics operations in an economical sense take into account the opportunity cost of resources that are not reclaimed as well as a reliable (and possibly cheaper) source of secondary fibre in the manufacture of products with recycled content. A successful reverse logistic and reclamation operation in the carpet industry will require a national infrastructure capable of collecting used carpet, recycling that carpet into an alternative product and the production of carpets with a greater recycled content.

The cost of carpet recycling can include the cost of:

- Obtaining the carpet from customers
- Transportation to a storage area
- Sorting and identification of carpet waste
- The reprocessing of carpet waste
- The design and implementation of policies to address environmental issues
- The promotion of environmental awareness and concerns
- Consumer incentives (monetary incentives or an ethical awareness of recycling) and a 'greening' of the environment

Environmental concerns regarding carpet disposal relate to land degradation (in the form of landfill sites) and soil degradation (because of the nature of the materials and chemicals utilised in carpet manufacturing). The best alternative for old carpets is recycling, to divert the carpet from landfill and provide the industry with easy access to secondary fibre sources. The cost of recycling can be reduced by greater co-operation among fibre and carpet mills in the collection, sorting and transportation of carpet and fibre. An example of a reverse logistics operation in a closed-loop system for recycling old carpet is presented in Figure 7.1. This figure suggests a possible infrastructure for successful recycling. The revenue obtained

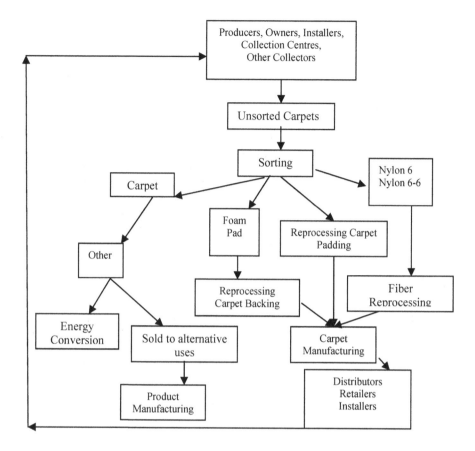

Figure 7.1. Reverse logistic operation in a closed loop system for recycling old carpets

from recycling is the value of the material sold at different stages of its life-cycle, and the cost includes the sum of collection, sorting and reprocessing costs.

The demand for old carpets as a source of secondary fibre among the carpet industry (and its use in alternative ways) has been limited. The lagging demand for and excess levels of supply of recovered carpet combined with the lack of landfill space and rising tipping fees makes recycling the best possible disposal option.

7.2.2 Re-use and Reprocessing

The re-use and reprocessing of used carpets depends on the available technologies and the promotion of recycling activities. The number of US patents aimed at the reprocessing and re-use of old carpet is increasing (see Table 7.1). The reprocessing of old carpet will allow reclamation of carpet fibres such as nylon 6, nylon 6-6, and polyester as substitutes for raw material.

7.2.3 Current state of Reverse Logistics and Recycling

There are over 292 carpet manufacturing plants in the USA, and the top 25 companies in the industry produce 98% of the total carpet produced (the top 10 producing 69%). According to the CRI, the South, and the state of Georgia in particular, holds the largest concentration of mills, accounting for 70% of the total number of manufacturing mills in the industry.

The fibre consumed in manufacturing includes nylon (66% of the total fibres used), polypropylene (23.5%), polyester (10%) and wool (0.5%). The types of carpets produced differ among mills, but the two major types are broadloom carpet and tufted broadloom carpet. The carpet is consumed in residential markets (61.4% of the total market) and in the contract market (government, institutional, educational; the remaining 38.6% of the market). The export market makes up 9% of the total carpet produced and has become more competitive in recent years.

Industry leaders in reverse logistics in the carpet industry include firms that recycle the fibre as well as firms that manufacture carpet with recycled content. Fiber producers have been working on carpet sustainability since the early 1990s (Helm, 2005). Honeywell, Inc. (www.honeywell.com), for example, has been working on green technologies since the 1990's when the firm patented the industry's first closed loop recycling system. All their Zeftron yarns have recycled content and their facility in Amprior, Ontario is the only depolymerization plant in North America. They have diverted some 200 million pounds of carpet from landfills. Solutia, Inc. (www.solutia.com) uses the face fibre of the carpet (which is 60% nylon) and recycles it to nylon or use it in the production of new carpets; alternatively, they use the fibre in the production of other products or in waste-to-energy generation. Their latest process is a patented dissolution technique to dissolve the carpet into a solvent and extract from it a nylon solution that can be re-extruded into new nylon. They are searching for partners to assist with the used carpet reclamation. Invista's Antron Reclamation Program processes some 15 million pounds of used carpet a year. More than 80% is recycled into a product and the remaining 20% is used as fuel. The company has also doubled its outlets for collecting used carpet. Their facility in Calhoun, Georgia has been operational since 1991 (Helm, 2005).

Carpet Manufacturers including Interface, Inc. (www.interfaceinc.com), Milliken (www.milikencarpet.com), Collins & Aikman Floorcoverings, Inc. (www.cafloorcoverings.com), Mohawk Industries, Inc. (www.mohawkind.com) Mannington Mills, Beaulieu of America, J&J Industries, and Shaw Industries Inc. (www.shawfloors.com) use the backing and filler to produce recycled backing and carpets. C&A Floorcoverings even earned the Recycling Works Recognition Award from the National Recycling Coalition in recognition of its closed-loop recycling of old carpet (*Floor Focus* 2002). Honeywell had developed a collection and recovery infrastructure through co-operation and integration with retailers, dealers, waste haulers and recyclers prior to suspending its operations because of lack of demand and prohibitive costs (See the 2001 Green Seal Report at http://www.greenseal.org/cgrs/Carpet_CGR.pdf).

Table 7.1. Technology and U.S. patents aimed at the reprocessing and re-use of used carpet

Year	Type of Technology	Number of US patents granted
1996	depolymerisation	4
	shredding	2
	separation	1
	recycling process	3
	total	10
1997	depolymerisation	2
	shredding	1
	separation	2
	recycling process	2
	total	7
1998	depolymerisation	5
	shredding	2
	separation	1
	recycling process	1
	total	9
1999	depolymerisation	5
	shredding	1
	separation	1
	recycling process	4
	other	1
	total	12
2000	depolymerisation	1
	shredding	1
	separation	2
	recycling process	5
	other	2
	total	11
2001	depolymerisation	1
	shredding	0
	separation	1
	recycling process	0
	other	4
	total	6

Source: http://www.uspto.gov/ US Patent Office.

Interface Inc.'s goal is to have a 100% sustainable product line by 2020. Interface Flooring has been working with LaGrange, GA where they are headquartered to cap the landfill and use the methane from the landfill as a fuel source in the plant to replace natural gas. Shaw Industries has reduced its waste production by 75% recently and their new Siemens gasification plants will reduce that waste to zero. C & A Floorcovering has been an environmental leader since the early 1990s and they recycle about 15 million pounds of carpet per year and as the cost of oil have risen, their product ER3, developed in 1998 to use recycled content is now cheaper than virgin products. At Milliken, 27 of their 46 domestic facilities produce no landfill waste and they offer commercial customers their "NO

Carpet to Landfill" reclamation programs. Mohawk Industries reclaims three billion plastic (PET) bottles turning them into 160 million pounds of polyester fiber for carpets. Beaulieu has a new carpet tile backing with an 85% post consumer recycled called Nexterra also made from PET bottles. Finally J&J Industries has developed several products with recycled content and offers a no-landfill guarantee on their Encore SD Ultima products (Helm, 2005).

Finally backing producers are also using recycled content and bio-based materials to produce backings for the carpet industry, many at the same cost as virgin materials but with improved performance. Still other sustainable flooring substitutes are hardwood flooring, eucalyptus flooring (which grows five times faster than other hardwood species), wool carpet, cork, bamboo, rubber, and Ceres recycled rubber flooring made from recycled tires (Helm, 2005).

Most of the carpets that are recycled contain nylon fibre. Carpets with polyester and polypropylene are not currently recycled because of the low cost of virgin fibres. Carpet can be recycled for its face fibre to produce secondary fibres such as nylon 6 and nylon 6-6, which in turn are used as fibres in the production of new carpet. Carpet tiles and padding can be recycled to produce recycled-content backing used in the manufacture of new carpet and carpet padding.

Carpet recycling initiatives in the USA have not been driven by state or federal laws but have been implemented for profitability, marketing, to meet the increased demand for recycled-content parts from alternative uses of old carpets or to avoid possible future legislation. Face fibre is the most valuable component of carpet and is used to produce recycled fibre; selvedge and trim can be recycled and used in carpet padding. Various carpet wastes (beam waste yarn) can be utilised in the manufacture of products such as stuffing and soft textiles or can be palletised for use as an injection-moulding compound to make hard plastic parts for automobiles. Backed carpet scrap is utilised to manufacture fibre pad—the layer under installed residential carpets. (For a summary of uses of carpet waste, see Table 7.2)

Table 7.2. End uses of carpet waste

Type	End-use option
Selvedge and trim	Carpet padding
Wool waste	Fibre reclamation
Beam waste yarn	Stuffing for pillows and toys Soft textiles
Backed carpet scrap	Fibre pad
Various carpet waste	Fan housing Fast-food trays Flower pots Plastic bags Parking-lot bumpers
Face fibre	Fibre reclamation

Several companies have included reverse logistic operations in their management plans and have implemented take-back programmes on a voluntarily basis at different points in the supply chain. DuPont is the largest producer of nylon fibre and, through its existing collection infrastructure, takes back its old carpet to recycle for further use in carpet manufacture and for other uses. Honeywell (formerly AlliedSignal) is the second-largest producer of nylon. It has developed its collection network for used carpet through integration with retailers, waste haulers and various recyclers. The primary motive for these firms to recycle old fibre has been economic, as the market for recycled fibre is on the rise. BASF also takes back its old carpets, recycling it back to fibre, which is used in the production of new carpets (Gavin 2001).

Interface Inc., through its closed loop, takes back old carpet to be used in the production of new carpets and separates the used carpets into their components to make products of lesser value (http://www.interfacesustainability.com/). Milliken refurbishes reclaimed carpets and sells it back to customers, therefore reclaiming value. C&A Floorcoverings recycles face fibre and backing into new carpet backing (http://www.cafloorcoverings.com/sustain01.html). Shaw Industries Inc. has developed the collection infrastructure that allows it to take back any used carpets from customers who buy a new Shaw product and have their old carpet recycled (http://www.commercial.shawinc.com/uc/recycling.htm). Not all the fibre producers or carpet manufacturers are engaged in reverse logistic operation and the recycling of old carpets and, despite their efforts, the recycling rate for carpets is very low. Realff et al. (1999, 2000) and Louwers et al. (1999) analysed shipping costs and location for carpet waste collection and found the main barriers to recycling are the high costs of collection, transportation, sorting and recycling compared with the cost of virgin fibre.

7.2.4 After-market Value

The value of items at the end of their life depends on a number of factors, including the price paid for the new recycled content products. The prices of recycled-content product depend on the quality perception of buyers for recycled-content carpet compared with virgin carpet product. The recycled carpets made through the reprocessing of old carpets can compete with recycled carpets made from other sources of reprocessed fibres such as carpet made from plastic bottles (from polyethylene terephthate [PET]). The increase in demand for products with recycled content can lead to greater prices paid for these products. The greater demand can come from a focus on the environment and increased regulation requiring recycled-content carpet in federal and state buildings.

Carpet mills recycle approximately 6.4 million pounds (weight) of carpet waste per year, or about 1% of the total carpet discarded. Carpets produced with reclaimed fibres and carpet pad and carpet backings are being produced with a 50% recycled content. In most carpet manufacturing, the average is about 20%–25% recycled material in carpet fibre and carpet production. Many carpet manufacturers are also working to reduce other types of waste within their production facilities, including office waste, containers and creels used to hold yarn, and water and energy in the manufacturing and dyeing process. Most are environmentally aware, responsible and continue to research viable uses and options that are profitable for their waste, with an emphasis on economically sound initiatives (Whaley 2002).

7.2.5 Production Methods and Recycling Challenges

Carpet is manufactured by tufting a face fibre into a primary backing and then using latex to bind this subassembly to a final or secondary backing for shape and stability. The latex holding the fibres in place is made from calcium carbonate. Both nylon and polypropylene are the most common materials used for face fibre and backing material.

Carpet waste comes from many sources, including construction and demolition, residential used carpet, government and military remodelling, and commercial and industrial used carpet. It can be returned to reclamation centres in rolls, baled, in squares, chipped, or unbundled and loose. Carpet must be sorted by type of fibre to determine whether the nylon fibres are recyclable into other carpet and/or other products.

The use of post-consumer carpet in other products is limited by the undeveloped infrastructure for reclamation. Industrial and institutional customers generate larger volumes of waste carpet, but a reverse supply chain logistics network must still be developed to efficiently and effectively recover the carpet in a clean and dry condition suitable for further reprocessing. Future legislation banning kerbside disposal of carpet in all US cities (including New York and Chicago, where it is currently legal) may help in the recovery process from individual homes and smaller users. DuPont leads the reclamation programme initiatives, with 80active collection sites for consolidation, receiving and sorting some 20 million pounds (weight) of used carpet and installation scraps each year (http://www.floortec.net/used_carpet_recycling.htm).

Polk *et al*. (1994) estimated that one billion pounds (weight) of fibre from used carpet might currently be finding its way into landfill in the USA. The carpet waste can be recycled in a number of ways:

- Depolymerisation of the nylon or face fibre
- Manufacture of recyclable single-material polyester carpets
- Chipping the carpet and combining the hard waste through extrusion and production into carpet tiles
- Use of the shredded carpet waste as a fuel source and/or additive for concrete or aggregate reinforcement

Because most of the face fibres and backing materials used in carpet production are by-products of the petroleum industry, re-use of waste carpet creates a secondary source of resource supply and conserves natural resources.

Some used carpeting has found new life in roadway construction to control soil erosion. Since many carpet manufacturers in the USA are selling their products through 'big-box' retail stores such as Home Depot, Lowes, Menards and others, it seems that reclamation at these locations might be a viable alternative.

The transportation mode used primarily to ship carpet is via lorries by road, and since the value of carpet per pound (weight, or by its value density) is low, recyclers must determine carefully where reclaimed carpets should be stocked geographically and how they should be shipped to recycling centres or processors. If the costs of fuel or transportation exceed the value of the recycled product, it is not economical to recycle and doing so would further deplete natural resources resulting in a negative economic added benefit. A study by Louwers *et al.* (1999) as well as Realff *et al.* (1999) developed facility location models for the collection, pre-processing and redistribution of carpet waste to support the design of the logistic structure of re-use networks for Europe and the USA.

7.3 The Future of Reverse Logistics for Carpet

Figure 7.2 shows the disposition of used carpet and how it can be broken down into various components; this involves a number of complex paths, largely because of the composition of the carpet itself. Because the designs and components vary widely, it is necessary to know the chemical composition of the face fibre to determine its ease of and applicability to recycling.

7.3.1 Accounting Standards for Valuing Waste

A cost–benefit analysis for carpet recovery and recycling has not made its way to generally accepted accounting practices or to the balance sheet and income statement. Several life-cycle costing models have been proposed to aid management in such costing analysis (Kumaran, Ong, Tan, and Nee 2001; Weitz, Smith, and Warren, 1994; and Williams, 1977). These models attempt to determine all the true costs of reclaiming the goods, including those involved in transportation, labour, storage, sorting and processing.

7.3.2 Financial Incentives and Benefits

Recycling is not simply a consideration when government regulation mandates firms and industries to include recycling in product design. Most countries in the European Union (Fishbein, 2000) have been faced with environmental regulation; however, in the USA federal mandates are not yet in place for industry. In these cases, recycling must have a positive financial impact for firms and must provide a competitive edge in the marketplace. Carpet recycling initiatives are growing but, as a whole, are still in the introductory stage of the life-cycle (Dunn 2002).

RCRD Inc. uses post-consumer carpet waste to make Camel Storm Bags® and Alligator Archery Bag® for target practice. These products use 100 percent recycled carpet fibres as filler. They also use recycled nylon 6 and a polypropylene/polyester cocktail in making carpet padding, with the addition of new (http://www.wastenews.com/) nylon 6 fibres. Invista collects post-consumer carpet and sends it to its facilities in Calhoun, GA (Helm, 2005). There, they separate the carpets by face fibre type. They then use fibre to make pellets. Invista has over 100 active collection sites where waste carpet is collected, consolidated and sorted. Their reclamation programme is collecting and processing over 116 million pounds of carpet and has conserved close to half a million cubic yards of landfill space each year. Pike Companies plans to use post-consumer nylon 6-6 to produce tiles that can be used in high-moisture areas, such as marine environments (http://www.carpetrecovery.org/annual_report/04_CARE-annual-rpt.pdf).

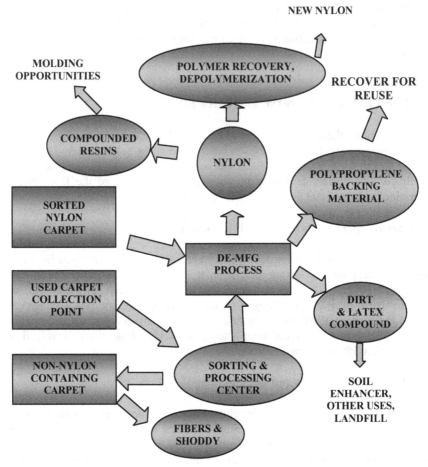

Figure 7.2. Disposition of carpet

7.3.4 Future Uses

Patent searches on products and processes for recycling carpet reveal many additional ideas for products are available, as shown in the patent search data (see Table 7.1). Other potential uses of carpet are listed below. It can be used:

- As insulation in homes or automobiles
- Inside automobile door panels and trunk voids to buffer and insulate
- As products moulded from PET
- In guardrails and guardrail stops or bumpers, park benches and swings, and pylons
- As a soil substitute to grow grass or turf that can be rolled up after use for use in indoor competitions and athletic events

- As an additive to improve soil quality and function
- As an additive in cement kilns
- As fuel (for burning) in manufacturing

These potential product ideas and uses are interesting, but the actuality of producing the goods costs too much and puts the items at a competitive and strategic disadvantage compared with virgin materials and existing substitute products. To develop the ideas requires a large amount of capital and investors, and venture capitalists are needed to invest in the most viable alternatives and to provide seed capital financing.

7.4 Framework for Understanding Recycling: The Present Business Model

The economic analysis of carpet recycling should include separate cost–benefit analysis for the four major recycling players: consumers, collectors, processors and recyclers. The interconnection of these four players makes up the recycling infrastructure. The incentives for recycling programmes lie within the constituents of and the payoffs to each player along the spectrum. For a successful recycling programme the incentives to recycle should be present for all involved and revenues must outweigh the costs at each step of the operation in order to promote continuous collection, recovery and recycling efforts. A lack of incentive in one segment will have an impact on the overall success of the recycling programme because of the strong links that exist between these segments. There is a need to develop markets for the various components of the recovered carpet in order to raise the economic incentives.

7.4.1 Incentives for Collectors

Collectors transport carpet waste and pay for the cost of transportation. The economic incentive for collection efforts is the revenue generated from the sale of waste carpet delivered to a drop-off location or to the retailer's store. In addition, collectors may also be able to charge for carpet collection (in amounts less than the landfill tipping fee) for delivery to incineration sites for energy-generation purposes. The cost of fuel, along with inflation, increases the cost of real transportation costs and will have an impact on how far the collectors will travel to collect carpet waste and the minimum price they are willing to accept.

7.4.2 Incentives for Retailers

The retailer may receive the old carpet or carpet pads that are removed by the installer and returned to the store. The economic incentive for recycling carpet resides with the retailer, who can reduce the tipping fees currently paid to dispose of the carpet. The retailer usually sells the carpet pads to the recycling processor for a face value and disposes of the rest of the waste carpet. The success of a recycling programme will depend on the retailer, who must provide an incentive to the installers to return the carpet to the store to enable the installer to cover the cost of collection and delivery.

7.4.3 Incentives for Installers

Installers are independent contractors and remove a significant volume of carpet from customer locations. The installer usually takes the waste carpet to the retailers for disposal or removes the carpet and delivers it for sale at face value. The economic incentive for the installers to deliver the old pads to the retailer is influenced by the prices paid by the retailer and whether these prices cover the cost of collection and delivery. Installers have several options when removing the carpet pads from the waste-generation source. They may:

- Have haulers deliver the old, used carpet and padding to the retailers
- Locate dumpsters for carpet waste at a designated drop-off location and possibly pay for the recycled carpet
- Sort carpet waste on-site and then deliver it to a drop-off location
- Take the waste carpet to a sorting centre to remove the pads for resale

Sorters have their own processing location for sorting carpets. If they are also performing a function as collectors then they will take the waste carpet to their own facility for sorting or they may also have contracts with other haulers to receive waste carpet for sorting. The installer may have an incentive to deliver the waste carpet to an incinerator, depending on fees received. The economic incentive for installers to return old carpet to the retailer for recycling purposes lies in the prices paid by the retailer, and these prices should reflect the highest value use for old carpets in order to attract a high level of recovery.

7.4.4 Incentives for Processors

A recycling processor with an established facility provides a drop-off location for waste carpet from stores, installers, other collectors and haulers. The recycling processor sorts and bales the carpet and disposes of the residue. The processor delivers the pre-sorted and baled old carpets to a carpet recycler. The cost of operation for the recycling processor is primarily the fixed cost of the facility and the variable costs, including labour, transportation and other operating expenses. The recycling processor must pay a fee to the suppliers of the old carpet for delivered materials in order to gain access to a reliable and continuous source of material. Landfill tipping fees may have a significant impact on prices paid for old carpet and may impose a price mechanism for the recovered material. The recycling processor reviewing the landfill tipping fees may use this information as a price mechanism to offer a relatively lower price to its suppliers. The opportunity cost of not selling old carpets is the cost of the tipping fees that suppliers of old carpets incur to avoid storage costs. The recycling processor generates revenue from the prices paid for sorted materials, such as for face fibre, carpet pads, and trims and sledges.

7.4.5 Incentives for Recyclers

The recyclers receive the old carpet materials from the processors and, using various technologies, process the old carpet into a secondary fibre or into end-products. The carpet recyclers accept and process pre-sorted post-consumer carpets into various components for reclamation. In general, the recycler will liberate face fibre from carpet backing; reclaim high-value fibre into recycled resin and process low-value fibre into marketable materials for the manufacture of other fibres. In some cases, recyclers are vertically integrated and undertake all three functions and thus benefit from the economies of scale. Non-integrated recyclers fall into three separate market segments: the resin manufacturer, the fibre producer and the carpet manufacturer. The resin manufacturer purchases nylon 6 and processes it into carpet resins and sells it to the fibre producers, who form carpet fibres. The fibre producer sells the carpet fibre to the carpet manufacturer for the creation of end-products. The economic incentive for non-integrated mills to produce resins is dependent on the price paid for the old carpet and the prices they will receive for the manufactured resins. Demand for resin is a derived demand, being dependent on the demand for the final product. Recyclers receive demand for their products (resins) mainly from (carpet) fibre producers and from other users of resins, such as the automobile industry, which uses nylon resins for the manufacture of automobile parts. The processed resin, as an input, faces competition from other inputs, such as plastic bottles (PET), that can replace resins, and demand will also depend on the prices that fibre producers or manufacturers of alternative products (such as the automobile industry) have to pay for virgin resins. The economic incentives for the carpet fibre producer is influenced by the market prices paid for the recycled resin, the recycled carpet fibre, the available substitutes for resin and the virgin fibre. The economic incentive for the carpet manufacturer lies in the price of the virgin fibre, recycled fibre and the demand for recycled-content carpets. The lower prices of recycled fibre compared with virgin fibre should make recycling profitable when the necessary infrastructure for collection and recovery of used carpets are in place.

7.4.6 Value Added from Recycling

Recycling of used carpet is less detrimental to the environment than other options largely because carpet is bulky and is not, as a rule, biodegradable, particularly when considering nylon based fibers; also, the recycling of old carpet reduces the demand on existing landfill space and poses less of a threat to future land use for landfills. The recycling of old carpet creates greater value added than the alternatives, such as incineration and landfill. Incineration and landfill ends the life-cycle of the product and, as a result, less value is reclaimed from the material. The economic value of recycling old carpet compared with the alternatives is presented in Figure 7.3.

  It is far easier to recycle and re-use post-industrial waste from internal manufacturing than it is to reclaim post-consumer waste to use as a raw material input. Although a few firms have created an infrastructure for returning used carpet, two major issues still exist. The first issue concerns how to reclaim the carpet on a large scale. The solution to this issue seems paramount to the development of new products that utilise the carpet waste and used carpet. Once sales of these products grow, there will be a greater demand for post-consumer waste carpet. It will be easier to achieve the volume needed from commercial users (large office buildings, apartments, educational facilities, government offices and military installations) than from individual homes. Working with retail outlets and installers is also paramount to gathering the carpet.

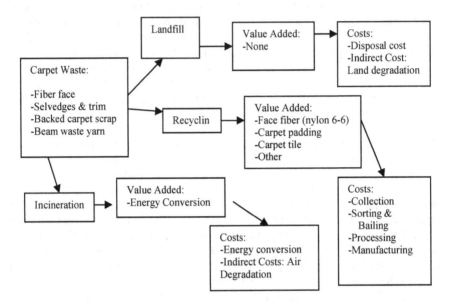

Figure 7.3. Alternative uses of old carpets: cost vs benefits

A second issue is the cost of recycling. If fuel and labour costs to reclaim and recycle carpet are too high, the resulting products created from the waste material will not be competitive or profitable in the marketplace. Education of consumers and legislative representatives on the true and total costs of recycling must be in place. There is also a need for an industry-wide initiative for recycling. Working together and developing industry solutions with a united front may be the best option in addressing the issue. Legislative changes on the kerbside disposal of carpet must be enacted to facilitate recovery.

7.5 Influencing Recycling: Policy Options and Implications

Recycling is not completely market-driven, and government intervention may be necessary to help to push the market in achieving efficient solutions to material re-use and recycling. To divert recyclable materials from landfill, many state governments have passed mandatory laws to help to increase the supply of recyclable materials. To promote demand for those recovered materials, states have implemented minimum constant standards along with various tax incentives to promote recycling. The government policy options to help to rectify the recycling bottleneck would be twofold: (1) to internalise externalities and (2) to require the producers to internalise the externality. Government can influence recycling efforts on the demand and supply side of the industry, among others, and through several policy options, as outlined in Table 7.3.

Table 7.3. Government policy options for improving recycling efforts

Policy type	Impact on recycling
Supply-oriented:	
The introduction of mandatory recycling programmes such as: Drop-off programmes Kerbside recycling	Helps to increase supply levels Encourages recycling through convenience
Imposition of a user fee in the residential sector	Helps to pay for collection efforts
Imposition of a ban on product disposal (*i.e.* carpet disposal)	Diverts old carpets from landfill
Use of volume-based pricing of waste	Reduces disposal volume (costs of disposal are incurred by consumers)
Introduction of unit pricing of waste	Encourages recycling through reduction of waste volume (costs of disposal are incurred by consumers)
The granting of tax credits to private collectors	Helps to increase supply levels by helping to pay for collection costs
The granting of tax credits for transportation costs	Helps to increase supply levels
The granting of tax credits on storage costs	Helps to increase supply levels
The introduction of a requirement for deposit programmes on recyclable items	Helps to increase supply levels
Public-sector construction of waste separation and sorting plants	Helps to increase supply levels
The granting of loans to municipalities for the creation of recycling programmes	Helps to increase supply levels
Demand-oriented:	
The creation of a standard for minimum recycled content	Helps to increase the demand: responsibility of producers (EPR)
The granting of tax credits and exemptions for recycling programmes owned by private businesses	Encourages recycling
The granting of more patents	Encourages recycling by providing technical assistance for recycling programmes

Table 7.3. (continued)

Policy type	Impact on recycling
The introduction of a requirement for producers to use coding or labelling	Eases recycling
The introduction of a requirement for take-back recycling programmes	Placing responsibility on producers (EPR)
The granting of tax credits for investment in product redesign and improvement (*e.g.* tax credits for R&D)	Eases recycling: responsibility on producers (EPR)
The creation of investment tax brackets	Encourages recycling by placing responsibility on producers (EPR)
The imposition of a tax on use of virgin materials	Encourages recycling
The introduction of a requirement for state or local agencies to purchase recycled-content products	Encourages recycling
Initiation of a tradable permit system in the recycling market	Encourages recycling

EPR = extended producer responsibility
R&D = research and development

7.6 Areas for Further Research

Understanding and identifying the variables affecting the demand and supply of used carpets is essential for successful reclamation of resources. Future research should take an empirical approach to analysing the carpet industry in identifying the factors beyond the lack of demand, including estimation of the price elasticities of demand and supply. The presence of market imperfections in the carpet industry may hamper collection efforts and recycling rates. Furthermore, given the current market structure, government policies aimed at helping the industry will be ineffective. One possible reason for the lack of demand for used carpets is the lack of buyers, discouraging collection efforts. Future study should focus on an analysis of market structures within the carpet industry.

References

Blumberg, D.F. (1999) 'Strategic Examination of Reverse Logistics and Repair Service Requirements, Needs, Market Size, and Opportunities', Journal of Business Logistics 20.2: 141-59.

CRI (Carpet and Rug Institute) (2003) The Carpet Industry's Sustainability Report 2003 (Dalton, GA: Carpet and Rug Institute (Available at: http://www.carpet-rug.com/pdf_word_docs/03_CRI-Sustainability-Report.pdf).

Dowlatshahi, S. (2000), 'Developing a Theory of Reverse Logistics,' Interfaces 30.3: 143-55.

Dunn, R.S. (2002) 'Carpet Industry, Government Unite to Increase Recycling' Textile World, February, p. 26.

Fishbein, B. K. (2000). "Carpet Take-Back: EPR American Style," Environmental Quality Management, 10(1), autumn, 25-36 and Adapted from "Extended Producer Responsibility: Materials Policy for the 21st Century," INFORM, Inc.

Gavin, K. (2001) 'Fibre Report: Business Falloff Affects Outlook,' Floor Covering Weekly (November 1, 2001): 50(29), p. 6-7.

Louwers, D., B.J. Kip, E. Peters, F. Souren and S.D.P. Flapper (1999), Reuse of carpets. A location-allocation model. Computers & Industrial Engineering 36(4), 855-869.

Floor Focus (2002) 'Newsfront', Floor Focus (August–September 2002): 14.

Helm, Darius "The Green Revolution" (2005) Focus, August/ September, p. 31-51.

Kumaran, D. S.; Ong, S.K.; Tan, R. B. H., and Nee, A. Y.C. (2001) 'Environmental Life Cycle Cost Analysis of Products,' Environmental Management and Health, 12(2/3), 260-277.

Nagel, M.H. (2001) Environmental Quality in the Supply Chain of an Original Equipment Manufacturer in the Electronics Industry (PhD Dissertation, September 2001; Delft, The Netherlands: Faculty of Mechanical Engineering, Laboratory for Production Technology and Organisation, Delft University of Technology).

Polk, M., S. Kumar and Y. Wang (1994) 'Fundamental Studies for the Utilisation of Carpet Waste', in National Textile Centre Annual Report (September 1994; Spring House: PA, National Textile Centre (http://www.ntcresearch.org/): 167-74.

Realff, M.J., J.C. Ammons and D. Newton (2000) 'Strategic Design of Reverse Production Systems', Computers and Chemical Engineering 24.2: 991-96.

Realff, M.J., J.C. Ammons and D. Newton (1999) 'Carpet Recycling: Determining the Reverse Production System Design', Polymer-Plastics Technology and Engineering 38.3: 547-67.

Stock, J.R. (2001) 'The 7 Deadly Sins of Reverse Logistics', Material Handling Management 56.3: 5-11.

Weitz, K. A., Smith, J. K. And Warren, J. L. (1994) , "Developing a Decision Support Tool for Life-Cycle Cost Assessments," Total Quality Environmental Management, autumn, 4(1), 23-37.

Whaley, P. (2002) 'Covering New Ground', Textile World (February 2002): 32-35.

Williams, J. E. (1977). "Data Requirements for Life Cycle Costing," American Association of Cost Engineers. AACE Bulletin, November/December, 19(6), 225.

Part II

Empirical Studies

8

Customer and Supplier Relations for Environmental Performance

Gregory Theyel

College of Business and Economics, California State University, 25800 Carlos Bee Boulevard, Hayward, CA 94542 USA, gregory.theyel@csueastbay.edu

This paper examines how relations between customers and suppliers affect environmental performance as indicated by waste reduction in chemical firms. Using a national survey, firm visits and phone interviews with U.S. chemical firms, this analysis suggests that a reciprocal learning process between customers and suppliers occurs as firms exchange information to set and meet environmental requirements. Firms that collaborate with customers tend to collaborate with their suppliers similarly, with the greatest successes in waste reduction occurring in firms that meet their customers' environmental standards and in turn set standards for their suppliers.

8.1 Introduction

Environmental performance is a concern of managers for regulatory and contractual compliance, public perception, and competitive advantage. Recent operations and strategic management literature suggests that firms are improving their environmental performance by setting standards, sharing resources and changing products and processes with the help of their customers and suppliers (Florida and Davison, 2001; Handfield, et al. 2001; Geffen and Rothenberg, 2000; Florida, 1996; Green et al., 1996; Sarkis, 1995). The literature offers insight on potential patterns in supply chain relations for improving environmental performance. However, more understanding is needed on the ways firms work with their customers and suppliers for environmental purposes and whether supply chain interaction helps firms improve their environmental performance.

By studying firms in the U.S. chemical industry, this research explores the adoption of environmental supply chain relations and the connection between supply chains and environmental performance. This article begins by summarizing recent contributions on environmental supply chains, focusing on the nature and extent of supply chain relations and their effect on environmental performance. The article then presents the results of a mailed questionnaire, plant visits, and phone interviews with managers of firms in the U.S. plastics and ink manufacturing sectors. The data show the levels of adoption of different types of environmental supply chain relations and describe firms' strategic use of these relations. After describing the nature and extent of environmental supply chain relations, the article pursues whether adopting these relations is correlated with leadership in chemical waste reduction. Finally, the article suggests implications of this research for environmental strategy.

8.2 Environmental Supply Chain Relations

Research by Cramer, et al. (1991) assesses the exchange of environmental information in the wood preservatives, printing inks, cosmetics, and carpeting industries in the Netherlands. By conducting interviews with managers in nearly 50 firms at various points in the production process, this research shows that the exchange of environmental information is largely informal and sporadic across and within the industries. The exchange of environmental information is most developed between manufacturers and their suppliers, but less so with their customers. Furthermore, large firms with greater resources and more market power are more successful at requiring environmental information from suppliers. They raise questions about the nature and extent of supply chain relations for environmental purposes as suppliers and customers and firms in different industries practice them.

While Cramer, et al (1991) focuses on which firms are working with their customers and suppliers, Schot, et al. (1990) researched the types of environmental relations being practiced. Schot, et al. (1990) studied the structure of the production systems surrounding 13 chemical firms in the Netherlands, including internal factors such as R&D, marketing, and quality control and external factors such as banks, insurance companies, and trade unions. Their research, which was based on in-depth interviews, concludes that environmental requirements of suppliers are infrequent and informal. However, firms are beginning to implement purchasing policies that require suppliers to assist in preventing pollution, and firms are setting restrictions on the content of goods provided by suppliers and attempting to keep stock as low as possible to limit leakage and waste. Their research suggests firms are at early stages in the development of supply chain relations for environmental purposes, but that these relations are evolving.

Cramer and Schot (1993) continue the research direction of Schot, et al. (1990) by describing two steps firms make toward what they call "environmental comakership." The first step is the exchange of information between firms on the environmental aspects of their products, and the second step is setting requirements of suppliers by manufacturers. Haveman (1995) adds to this research by suggesting a third step of supplier-customer cooperation that is even more progressive involving customers and suppliers sharing costs and resources to innovate cleaner production and products.

Green, et al. (1996) synthesize this research on types of environmental supply chain relations with case studies about the environmental practices of companies and their suppliers in six industries. They found that customers play a larger role in influencing environmental improvement than suppliers, and environmental requirements and providing information are the most common interactions.

However, they also found that opportunities for joint development of cleaner products and processes with customers and suppliers remain.

This research suggests at least three types of environmental relations between customers and suppliers: environmental requirements, sharing environmental information, and collaboration for improving environmental aspects of products and processes. Environmental requirements may be practices such as purchasing requirements, employee training, and ISO-14000 certification (Green, et al. 1996). Environmental information sharing may be in the form of new product samples, regulatory updates, and best practices. Environmental collaboration ranges from sharing personnel and equipment to co-developing recyclable products and cleaner processes by substituting materials or reducing waste and energy use (Walton, et al. 1998).

Questions remain about the level of adoption of various types of environmental relations between customers and suppliers. Firms at different points within a supply chain appear to be at different levels in their participation in inter-firm relations, and industries are at different points in the pervasiveness of the adoption of types of supply chain relations. This research seeks to understand the adoption of different types of environmental supply chain relations in the plastics and ink manufacturing industries and whether firms interacting with their customers for environmental purposes are likely to interact in the same manner with their suppliers.

8.3 Supply Chains and Environmental Performance

Frosch (1994) found that firms in the metal finishing industry are using their connections with customers and suppliers to improve environmental performance. Through extensive in-plant interviews with approximately 30 firms, he examined relationships between customers and suppliers to identify success at reducing waste between steps in production processes. Many of these inter-firm linkages were among firms in proximity to each other in regional networks, and Frosch argues that the inter-firm linkages are facilitated by this proximity and lead to improvements in environmental performance.

While Frosch (1994) focuses on proximity and inter-firm relations, Florida (1996) studies the types of relations that are likely to lead to improvements in environmental performance. Florida (1996) states that closer bonds between suppliers and customers, which can facilitate cleaner production, are the trend in manufacturing as leading firms need such close relationships with suppliers to incorporate management strategies such as just-in-time, continuous improvement, and total quality management. Closer ties to suppliers are also likely to facilitate diffusion of innovative approaches to environmental management, and environmental performance can be an added benefit of lean management initiatives as suppliers and customers share the responsibilities for improved efficiency and waste minimization (Florida 1996).

Geffen and Rothenberg (2000) build on the work of Frosch (1994) and Florida (1996) by showing the environmental benefits of collaborative relations with suppliers. They present three case studies of U.S. automobile assembly plants and their suppliers and find that relations with suppliers aid the adoption and development of innovative environmental technologies. In addition, the interaction of customer and supplier staff, partnership agreements, and joint research and development lead to improvements in environmental performance.

Dodgson (2000), Dyer and Singh (1998), von Hippel (1988), and others argue that inter-firm relations provide formal and informal mechanisms that promote trust, reduce risk, and in turn increase innovation and profitability. An important question is whether this is the case for environmental innovation and performance. This research seeks to understand whether firms with closer linkages with their customers and suppliers have better environmental performance.

8.4 Methodology

In order to answer the questions raised in the sections above, this research focuses on all of the 650 plants in two important industrial sectors of the U.S. chemical industry -- plastics and resins (365 plants) and ink manufacturing (295 plants). The plastics and resins (SIC 2821) and ink manufacturing (SIC 2893) sectors are important sectors to study because they were two of the 17 industrial sectors that the U.S. EPA (1991) identified from nearly 200 candidates as industrial sectors with great potential for improving environmental performance. The Natural Resources Defense Council (1991) also identified the three digit industrial sectors of 282 and 289 (which include plastics and resins and ink manufacturing) as two of the top ten most polluting sectors by volume and the top two by carcinogenicity.

The research design involved three tasks for data collection. The first task was the development of an extensive data set on all the plants in the plastics and resins and ink manufacturing sectors of the U.S. chemical industry. The second task involved administering a mailed questionnaire to plant managers of chemical plants to identify each plant's environmental practices and performance. The questionnaire had 60 closed-ended questions about environmental performance and about environmental management practices drawn from case studies and the technical and academic literature. Six plant managers and six environmental consultants pre-tested the questionnaire. They suggested industry-specific language for the questionnaire, and they modified the list of chemicals used for questions about waste reduction. Of the 650 surveys mailed, 188 were returned for a response rate of 28.9 percent. Based on chi-square tests, the distribution of the returned surveys by employment, organizational status, and industrial sector is statistically similar at the level of 0.05 to the distribution of the survey population. The third task involved follow-up telephone and in-person interviews with a sample of 20 survey respondents. These managers verified and elaborated on their questionnaire responses.

This research uses waste reduction as an environmental performance measure. Waste reduction is represented by firms' effectiveness at reducing waste generation of chemicals commonly used by plants in the plastics and resins and ink manufacturing sectors. Plant managers estimated their plant's percent reduction of waste generation during the past three years of the chemicals - Toluene/Xylene, Ethylene/Propylene, Styrene, and Vinyl Chloride Monomer. Managers chose from

(1) increase or no change (0%), (2) slight reduction (1%-10%), (3) intermediate reduction (11%-50%), and (4) great reduction (51%-100%). Firms' environmental performance was scored by calculating their average level of reduction (scores ranged from 1.0 to 4.0) for the chemicals relevant to their production processes.

The following results first address the pervasiveness and nature of supply chain relations for environmental management. Next, the results show the correlation between different types of environmental relations between customers and suppliers. Descriptive statistics and Pearson correlation analysis show the levels of adoption of supply chain relations and the correlation between the types of relations. Lastly, the results address the connection between firms' relations with customers and suppliers and their environmental performance. Pearson correlation shows the connection between different types of customer and supplier relations and waste reduction.

8.5 The Adoption of Environmental Supply Chain Relations

This section describes the adoption of environmental requirements, sharing environmental information, and collaboration for improving environmental aspects of products and processes. Over 55 percent of the sample plants must meet environmental requirements set by their customers (see Table 8.1). Firms reported having to ship their products in returnable packaging, supply exact amounts just in time for use, and reformulate their products so they are easier to recycle. This indicates that in addition to quality and price, which are standard requirements, many of the customers have added environmental quality as a requirement. This was especially the case for plastics firms where nearly two-thirds of the firms were required to meet environmental requirements set by their customers.

About 20 percent of the sample plants required their suppliers to meet environmental standards or prerequisites. This is significantly less than the customer requirements discussed above. A possible explanation for this difference is that because the sample plants are lower in the supply chain than their customers, there are fewer options for their suppliers, who provide basic chemicals, to certify the environmental quality of the products they sell to the sample plants.

Approximately one-quarter of the sample plants receive environmental information for plant management and operation from their customers. This result may be low because plants are hesitant to seek information from their customers lest they reveal their deficiencies. Half of the sample plants received environmental information on plant management and operation from their material or equipment suppliers. This result is not surprising because plant managers reported that they receive large quantities of product information from their suppliers, including information on the environmental attributes of products and equipment. Suppliers market their products and equipment to the sample plants, and part of their marketing is often information on the progress the suppliers have made at improving the environmental performance of their products and equipment.

Sample plants collaborate with customers for environmental innovation at the following rates: a) collaboration with customers in substituting non-hazardous or less hazardous materials to develop cleaner products (58.5 percent of the sample plants); b) collaboration with customers to develop recyclable products (33.5 percent of the sample plants); and c) collaboration with customers for cleaner production (37.6 percent of the sample firms). These results show evidence of the development of more sustainable products and processes emanating from supply chain relations.

Table 8.1. Adoption rates of environmental supply chain relations

	Percent Ink Plants Adopting	Percent Plastics Plants Adopting	Percent of All Firms Adopting
Customers Set Requirements	47.1%	63.8%	55.3%
Requirements for Suppliers	12.6	25.5	19.7
Information from Customers	27.6	20.2	24.5
Information from Suppliers	57.5	44.7	50.0
Material Substitution with Customers	46.0	73.4	58.5
Material Substitution with Suppliers	51.7	71.3	60.1
Recyclable Products with Customers	34.5	33.0	33.5
Recyclable Products with Suppliers	43.7	36.2	39.9
Cleaner Production with Customers	40.2	35.1	37.6
Cleaner Production with Suppliers	48.3	50.0	49.2

N=181 (94 Ink Firms and 87 Plastics Firms); Source: Mail Survey

The majority of the sample plants report that they collaborate with their suppliers for environmental innovation. The highest percentage of collaboration was plants working with suppliers to substitute non-hazardous or less hazardous materials to develop cleaner products (60.1 percent of the respondents). In addition, nearly 50 percent of the sample firms collaborated with their suppliers for cleaner production, and nearly 40 percent collaborated with their suppliers to develop recyclable products. These results are understandable because if a plant chooses to substitute cleaner materials or develop cleaner processes, examining the production inputs and equipment are logical approaches. These results, combined with the evidence of customer collaboration discussed above, show how firms are using their supply chains for environmental improvement.

While the information above describes the level of adoption of environmental supply chain relations, the results of Pearson correlation analysis show that the adoption of environmental requirements, information exchange, and collaboration are related. Table 8.2 shows significant correlation between customers setting requirements, receiving information from customers, and collaborating with customers for material substitution, recyclable products, and cleaner production.

Table 8.2. Pearson correlation analysis of environmental supply chain relations

Variable	1	2	3	4	5	6	7	8	9	10
1. Waste Reduction										
2. Requirements for Suppliers	.324**									
3. Information from Suppliers	.186	.279**								
4. Material Substitution w/ Suppliers	.153	.213*	.159							
5. Recyclable Products w/ Suppliers	.055	.181*	.133	.327**						
6. Cleaner Production w/ Suppliers	.173	.386**	.287**	.479**	.478**					
7. Customers Set Requirements	.263*	**.331****	.183*	.214**	.111	.178*				
8. Information from Customers	.146	.314**	**.372****	.047	.182*	.307**	.306**			
9. Material Substitution w/ Customers	.186	.196*	.073	**.435****	.083	.308**	.106	.250**		
10. Recyclable Products w/ Customers	.095	.252**	.177*	.227**	**.430****	.114	.198*	.216*	.370*	
11. Cleaner Production w/Customers	.136	.151	.227**	.180*	.027	**.199***	.277**	.418**	.395**	.270**

* = .05 level of significance; ** = .01 level of significance; Source: Mail survey

Table 8.2 also shows significant correlation between setting requirements for suppliers, receiving environmental information from suppliers, and collaborating with suppliers for material substitution, recyclable products, and cleaner production. These results suggest that firms use a suite of environmental relations with their customers and suppliers. The results in this paper do not confirm a progression in the adoption of these relations, but the results do indicate that firms are likely to use one type of relation with the other types. Future research is needed to understand if there is an order or sequence to the adoption of these types of environmental relations or if they are adopted simultaneously.

8.6 Reciprocal Environmental Relations

Firms that have environmental relations with their customers are likely to have similar relations with their suppliers and vise versa. The bolded results in Table 8.2 show firms with environmental relations with their customers (requirements, information exchange, and collaboration) also have similar relations with their suppliers. This suggests that firms transfer knowledge through their supply chains, and supports the notion that environmental learning occurs within the firms' supply chains. This was also confirmed during the follow-up interviews when plant managers discussed the source of their learning about environmental relations with customers and suppliers. Several of the plant managers stated they used the lessons they learned from their customers or suppliers and transferred these in the opposite direction in their supply chain. For example, one plastic resin manufacturer was required by its customer to remove a chemical from its product and in turn required its supplier to substitute for the chemical. The three parties (customer, sample firm, and supplier) worked together to find a replacement chemical with less toxic effects on the environment.

8.7 Supply Chains and Environmental Performance

This section addresses whether or not there is a connection between environmental relations with customers and suppliers and environmental performance. The results of the Pearson correlation analysis suggest that firms that are leading adopters of some types of environmental relations are leaders in chemical waste reduction. Table 8.2 shows the variables significantly correlated with environmental performance (column 1). Both forms of environmental requirements in this study, environmental specifications set by customers and standards or prerequisites for suppliers, are significantly related to environmental performance. Thus, firms that are setting environmental requirements for their suppliers and are being subjected to environmental specifications set by their customers are leaders in reducing their generation of chemical waste. Table 8.2 also shows that there is not a significant correlation for information exchange and collaboration with environmental performance. It is possible that over a longer time period, firms will see improvements in their waste reduction due to these other types of environmental relations (information exchange and collaboration). This speculation was partially supported during the follow-up interviews when plant managers commented that collaboration with customers and suppliers for environmental improvement is their long-term approach for improving environmental performance.

8.8 Implications for Environmental Supply Chain Relations

There are several implications of these findings for managers and their organizations. One clear advantage of closer relations within the supply chain is that new knowledge is created and shared. The reciprocal nature of practices allows the performance benefits to increase exponentially. Involvement of customers with the firm that is similarly involved with its suppliers creates added benefits compared to the sum of the individual benefits. For example, joint development of products that are recyclable with both customers and suppliers can ensure success on the first run rather than the second or third through the entire supply chain. Time to market and product quality achieved simultaneously will produce the highest possible success.

A second implication is that environmental supply chain management offers an important option for improving waste reduction and environmental innovation. There are other approaches and explanations for improving environmental performance such as management practices, investment in technology, and general experience with innovation. For example, Theyel (2000) shows how the implementation of management practices leads to waste reduction, and Klassen and Whybark (1999) show that the pattern of investment in environmental technologies affects environmental performance. In addition, Florida (1996) shows that firms that are more innovative in general are more likely to be leaders in environmental performance. Environmental supply chain management can be added to these strategies for improving environmental performance.

A third implication is that the supply chain relations discussed in this paper begin to address environmental sustainability of industrial production. A sustainable chemical industry would be, as described by Frosch and Gallopolous (1992), "an industrial ecosystem where the consumption of energy and materials is optimized, waste generation is minimized, and the effluents of one process serve as the raw materials for another process". The progress firms have made integrating environmental management throughout their supply chain represents progress toward the model of a sustainable industrial ecosystem. The benefits of this integrated approach are that significant and long-term improvements in environmental performance will encourage the evolution of sustainable systems of production.

Firms in the chemical industry and in other industries can learn from the leading firms in this research. Firms that do make closer supply chain relations part of their environmental strategy are likely to be leaders in waste reduction and environmental innovation. In-depth research, similar to that, which is presented in this paper, of firms across all industries, would identify which industries are best suited for benefiting from closer supply chain relations. Research of this nature would help managers identify whether or not there are similar and sufficient industry organization and culture in other industries to seek to improve their environmental performance through their supply chains.

Acknowledgements

Research funding was provided by grants from the National Science Foundation (Grant No. GSOG 10-70-21780-0410) and the U.S. Environmental Protection Agency, Office of Exploratory Research (Grant No. R82-009-01-0). David Angel, Sam Ratick, Richard Florida, Roger Kasperson, James Cummings-Saxton, Lisa

Lebduska, and two reviewers provided helpful comments and suggestions at various stages of this research.

References

Cramer, J., P. Dral, and B. Roes. (1991). "Product information exchange about environmental aspects between producers," Netherlands: Ministry of Housing, Physical Planning, and Environment.

Cramer, J. and J. Schot (1993). "Environmental Comakership Among Firms as a Cornerstone in the Striving for Sustainable Development" In Fischer, K. and J. Schot (eds.) *Environmental Strategies for Industry: International Perspectives on Research Needs and Policy Implications*. Washington, D.C.: Island Press.

Dodgson, M. (2000).*Management of Technology*. London: Routledge.

Dyer, J.H. and H. Singh, (1998). "The Relational View: Cooperative Strategy and Sources of Interorganizational Competitive Advantage," *Academy of Management Review*, 23, pp. 660-679.

Florida, Richard. (1996). "The Environment and the New Industrial Revolution" *California Management Review.* Vol. 38 (Fall), pp. 80-115.

Florida, Richard and Davison, Derek. (2001). "Gaining from Green Management: Environmental Management Systems Inside and Outside the Factory" California *Management Review*, Vol. 43, No.3, pp. 64-84.

Frosch, R. (1994). "Industrial Ecology: Minimizing the Impact of Industrial Waste" *Physics Today*. November.

Frosch, R. and N. Gallopoulos (1992). "Towards an Industrial Ecology" In A. Bradshaw, et al., *The Treatment and Handling of Wastes*. London: Chapman and Hall.

Geffen, Charlette and Rothenberg, Sandra (2000). "Suppliers and Environmental Innovation: The Automotive Paint Process" International *Journal of Operations and Production Management*, Vol. 20, No. 2, pp. 166-186.

Green, Ken, Morton, Barbara, and New, Steve. (1996). "Purchasing and Environmental Management: Interactions, Policies and Opportunities" *Business Strategy and the Environment*, Vol. 5, 188-197.

Handfield, R., S. Walton, R. Sroufe, and S. Melnyk. (2001). "Applying Environmental Criteria to Supplier Assessment: A Study of the Application of the Analytical Hierarchy Process" European Journal of Operations Research.

Haveman, M. (1995). "Leveraging Supply Chains for Pollution Prevention" *Pollution Prevention Northwest*.

Klassen, Robert and Whybark, D.C. (1999). "The Impact of Environmental Technologies on Manufacturing Performance" Academy *of Management Journal*, Vol. 42, No. 6, pp. 599-615.

Lamming, Richard and Hampson, Jon. (1996). "The Environment as a Supply Chain Management Issue" British *Journal of Management*, Vol. 7, Special Issue, pp. 45-62.

Natural Resources Defense Council. (1991), "Going to the Source: A Case Study on Source Reduction of Industrial Toxic Waste", NRDC Publications, New York, NY.

Sarkis, J. (1995). "Supply Chain Management and Environmentally Conscious Design and Manufacturing" *The International Journal of Environmentally Conscious Design and Manufacturing*, Vol. 4, No. 2, pp.43-52.

Schot, J., B. DeLaat, R. der Meijden, and H. Bosma. (1990). *Care for the Environment: Environmental Behavior of Chemical Firms.* The Hague: Netherlands Organization of Technology Assessment.

Theyel, Gregory (2000). "Management Practices for Environmental Innovation and Performance" International *Journal of Operations and Production Management*, Vol. 20, No. 2, pp. 249-266.

U.S. Environmental Protection Agency. (1991), *Industrial Pollution Prevention Opportunities for the 1990s*, U.S. Government Printing Office, Washington, D.C.

von Hippel, E. (1988). *The Sources of Innovation.* Oxford: Oxford University Press.

9

Horses for Courses: Explaining the Gap Between the Theory and Practice of Green Supply

Frances Bowen[1], Paul Cousins[2], Richard Lamming[3] and Adam Faruk[4]

[1]Haskayne School of Business, University of Calgary, 2500 University Drive NW, Calgary, AB, T2N 1N4, frances.bowen@haskayne.ucalgary.ca
[2]School of Management and Economics, Queen's University Belfast, 25 University Square, Belfast BT9 1NN, UK
[3]School of Management, University of Southampton, Southampton, SO17 1BJ, UK, R.C.Lamming@soton.ac.uk
[4]Ashridge, Berkhamsted, Hertfordshire, HP4 1NS, United Kingdom, adam.faruk@ashridge.org.uk

Researchers and policy-makers have become increasingly enthusiastic about greening purchasing and supply management activities. In theory, greening supply should both limit environmental damage from industrial activities, and deliver bottom line benefits to implementing firms. However, compared with other environmental initiatives, few firms have implemented extensive green supply programmes.

This chapter seeks to resolve the apparent paradox between the desirability of green supply in theory, and the slow implementation of green supply in practice. Using data from a recent series of interviews and a questionnaire in the UK, we examine the green supply practices adopted by particular types of firms, and their performance implications. We cluster the operating units in our sample into four archetypal groups of green supply adopters, and examine the characteristics of each group. We conclude that explaining the gap between the theory and practice of green supply requires looking beyond the aggregate pattern across firms. Firms are not ignoring the potential private benefits from green supply. On the contrary, they are rational actors playing to their own strengths, and designing appropriate packages of green supply activities within their own corporate environmental, procurement and performance contexts.

9.1 Introduction

Throughout the 1990s, researchers and policy-makers have become increasingly enthusiastic about greening supply[1]. Benefits of implementing green supply identified in theory range from straightforward cost reduction, to facilitating the development of co-operative relationships with suppliers, and even encouraging a life-cycle, holistic approach in managerial decision-making. The espoused benefits of integrating environmental criteria into the purchasing and supply process include benefits to society, to individual commercial firms and to the purchasing and supply process itself (see Table 9.1).

Despite all these alleged benefits, there is a growing concern that the implementation of green supply is not yet widespread in UK industry (Russel 1998; Bowen et al. 2001). Pockets of proactive environmental supply practice clearly exist in some high profile organisations such as BT (Lamming and Hampson 1996), Unilever (World Resources Institute 1999), and the Body Shop (Wycherley 1999). Most large companies in the UK have some reference to environmental purchasing in their environmental policies (Peattie and Ringler 1994; Business in the Environment 1996, 1997, 1998, 2000). However, environmental purchasing is not as well developed in these large companies as other aspects of environmental management such as employee environmental programmes (Business in the Environment 1996, 1997, 1998, 2000), and companies have been known to routinely ignore commitments made in their environmental policies in their daily business practice (Ketola 1997).

There appears to be a gap between the desirability of green supply activity in theory and the slow implementation of green supply at the aggregate level across firms. Given this gap, there is a research need to reorient green supply research from theory to commercial practice. Businesses will only undertake initiatives when they perceive them as broadly in their interests. Reexamining current green supply practice, and the performance outcomes of different types of green supply initiatives, would indicate whether and where the potential private benefits exist. Key questions which need to be addressed include:

- do managers see the potential of green supply to deliver the benefits described in theory?
- can patterns of green supply implementation be observed within the aggregate low level of implementation?
- how are these patterns related to organisational characteristics?
- are firms which implement green supply reaping the benefits expected by theorists?

---

[1] We use the term green supply to indicate "supply management activities that are attempts to improve the environmental performance of purchased inputs, or of the suppliers that provide them. They might include activities such as co-operative recycling and packaging waste reduction initiatives, environmental data gathering about products, processes or vendors, and joint development of new environmental products or processes. The term encompasses a wide range of activity, and is broader than previous definitions of environmental purchasing." (Bowen et al. 2001, p. 5)

Table 9.1. Potential benefits of green supply identified in the literature

Potential Benefits of Green Supply	Study
To Society	
Aid diffusion of environmentally sound practices through industry	Lamming and Hampson (1996); Green et al. (1996); Bowen et al. (2001); Russel (1998)
Facilitate legislative compliance	Green et al. (1996); Min and Galle (1997); Hampson and Johnson (1996)
Provide response to public concern	Drumwright (1994); Cramer (1996); Russel (1998); Miller and Szekely (1995)
Environmental benefits through co-operation	Cramer (1996)
Facilitate moves towards sustainability	Russel (1998); Miller and Szekely (1995)
Eliminate / reduce demand for environmentally harmful raw materials	Min and Galle (1997); Epstein and Roy (1998)
Encourage use of a life-cycle approach	White (1996)
To the Firm	
Reduce costs	Drumwright (1994); Green et al. (1996); Cramer (1996); Bowen et al. (2001);
Manage reputational risks	Drumwright (1994); Bowen et al. (1998)
Manage liability for environmental damage	Min and Galle (1997)
Avoid potential increase in cost of waste / disposal	Lamming and Hampson (1996); Min and Galle (1997)
Deliver legislative compliance at lower cost (current and future)	Green et al. (1996); Min and Galle (1997); Hampson and Johnson (1996)
Improve product or service quality	Cramer (1996); Noci (1997); Russel (1998)
Meet market expectations	Hutchinson (1998); Knight (1996)
To the Purchasing and Supply Process	
Support corporate environmental objectives	Lamming and Hampson (1996); Green et al. (1996); Noci (1997); Carter and Carter (1998); Carter et al. (1998); Hart (1995)
Develop co-operative relationships with suppliers	Noci (1997); Lamming and Hampson (1996)
Purchasing direct cost reductions	Carter et al. (1998); Stock (1992);
Maintain security of supply	Russel (1998); Lamming et al. (1996)
Improve purchasing's status / strategic importance	Bowen et al. (1998); Green et al. (1996)

This chapter will use the answers to these questions to resolve the gap between green supply theory and practice. Understanding the reasons for the mismatch between the theory and practice of green supply would benefit future researchers, current business strategists and regulatory authorities alike. Researchers could refine their performance predictions of green supply according to the business context. Business managers could focus their green supply efforts on initiatives which are most likely to yield benefits to the firm given their own particular context. Most of all, improved understanding of environmental purchasing activity and its performance implications in different contexts might allow regulators to better align incentives so that the social benefits of green supply activity can finally be reaped.

This chapter reports on a recent study of green supply in the UK designed to describe current green supply practice, and illicit the performance implications of integrating environmental considerations into purchasing and supply activity. The aim of this chapter is to generate explanations for the theory-practice gap by addressing the key questions above. We begin by outlining our research methods. We go on to describe purchasing managers' attitudes on green supply and current environmental supply practice. We later derive the characteristics of businesses which follow different green supply approaches from clustering the units in the sample into similar groups. We then describe the performance outcomes of various types of environmental purchasing approach. The chapter concludes with a discussion of the four factors which can help to resolve the theory-practice gap.

9.2 Research Method

We conducted a two phase survey of green supply activities in the UK[2]. The first phase involved interviews with senior managers in twenty-four business units within UK public limited companies. The aim of this phase was to assess the strategic threats and opportunities posed by the environmental agenda on purchasing and supply activity. The interviews were conducted in a broad cross-section of industries, in order to capture managerial interpretations of green supply and potential management strategies and tactics across a wide variety of business activity. They varied in their position in the supply chain from extraction, through pre-production, manufacture, distribution and retail. Various support activities, such as civil engineering and transport, were also included. The sample of business units included industries considered both 'clean' and 'dirty', and all the business units in the sample were part of large corporations (annual turnover greater than £300m).

Semi-structured interviews, each lasting around an hour, were conducted with at least one senior manager in each business unit. Most respondents were senior general managers in the business units; others included specialists in HSE (Health, Safety and Environment), purchasing or production/operations. All the interviews were undertaken at the respondent's premises and the opportunity was taken to

---

[2] Readers interested in the detail of our research method are referred to Bowen et al. (2001), where we provide further information on the sample, response rates and development of the questionnaire.

collect secondary material such as annual reports, environmental policies, and internal newsletters to support the interview data. We used the interviews to generate issues that needed further exploration at the operating level. We also asked respondents to complete a brief questionnaire to provide a top management perspective on questions we later asked of middle managers in the postal questionnaire.

In the second phase, questionnaires were sent to key respondents in operating units within the business units interviewed in phase 1. The aim of this phase was to follow up on themes we uncovered in the interviews, and to get a view of the issues from the operational 'front line'. Two questionnaires were mailed : one for completion by the operating unit general manager, and the other for completion by the person responsible for purchasing at the operating unit. Of the 138 sets of questionnaires sent out in the 24 business units, 95 useable general manager questionnaires were eventually returned (a response rate of 69%), as were a further 70 purchasing questionnaires (a response rate of 51%).

The results we report here are based on the questionnaires completed by the business unit interviewees, the operating unit general managers and the operating unit purchasing managers. We will refer to these three sets of informants as 'top managers', 'middle managers' and 'purchasing managers' respectively. We had complete sets of observations from all three sources for 70 operating units, so the following results are normally based on a sample size of 70.

9.3 The Potential of Green Supply

Our first priority in assessing the potential and practice of green supply was to examine purchasing personnel's attitudes on green supply. Some authors claim that the full potential of green supply is not being realised due to purchasing personnel's environmental illiteracy (Lamming and Hampson 1996; Bowen et al. 2001). Many purchasing managers apparently do not recognise the potential for green supply to realise economic benefits, appearing to believe that environmental initiatives are costly (Min and Galle 1997). Table 9.2 provides descriptive statistics of purchasing managers' environmental attitudes.

In our sample, over 60% of purchasing managers agreed strongly that their company should share the responsibility for the environmental impacts of their suppliers, and none strongly disagreed. Clearly, we asked in this question whether they should share environmental responsibility, and not whether they are actually sharing it. Such questions often suffer from a lack of validity, and the overwhelming agreement might simply reflect a tendency to provide socially desirable answers. Even if this is so, the overwhelming response suggests that purchasing managers are at the very least aware that they are expected by society to pay attention to suppliers' environmental performance, and that they have a responsibility to manage their supply chains in an environmentally sound way (Lamming et al. 1999). Implicitly, they recognise the potential benefits to society of implementing environmentally responsible supply practices.

Table 9.2. Purchasing personnel's attitudes on the potential of green supply

Questionnaire Item	Mean	Std. Dev.
2.1 We as a company should share the responsibility for the environmental impacts of our suppliers	5.53	1.20
2.2 There are more opportunities than threats for our business arising out of the environmental agenda	4.08	1.75
2.3 Implementing an environment-related supplier program might help us manage risk	3.89	1.18
2.4 We have the capabilities in purchasing to improve the environmental performance of our suppliers	3.84	1.61
2.5 Manipulating our supply chain for environmental improvement might give us a competitive advantage	3.57	1.63
2.6 We have the resources to experiment with different solutions to environmental problems in our supply activity	3.05	1.62

Note: all items are scored on a scale of 1: strongly disagree to 7: strongly agree

Attitudes on the potential benefits to the company of undertaking an environmental supply approach are more mixed. For example, despite the high mean score for whether environmental issues present a net opportunity to the business (item 2.2), the standard deviation is rather high (1.75). Thus despite the average perception that the environmental agenda presents an opportunity rather than a threat to the businesses concerned, attitudes on potential environmental opportunities remain highly varied in the UK context (see Peattie and Ringler 1994). As long as purchasing personnel perceive environmental issues as a threat, rather than an opportunity, they are less likely to undertake a green supply programme (Sharma and Nguan 1999; Sharma et al. 1999).

Respondents were in more agreement on some of the other potential benefits to the firm and to the purchasing function. The purchasing managers agreed that implementing an environment-related supplier programme might help them manage risk (item 2.3), and that manipulating their supply chain for environmental improvement might give them a competitive advantage (item 2.5). They also felt that they had the capabilities in purchasing to improve the environmental performance of their suppliers (item 2.4), supporting the view that environmental management practices might feasibly be spread through the network of industrial buying and selling (Green et al. 1996; Lamming et al. 1996; Russel 1998).

However, twice as many respondents thought that they did not have the resources to experiment with different solutions to environmental problems as thought they did (item 2.6). This is a concerning finding on purchasing managers' perceptions of the costs and effort necessary to implement green supply initiatives. Many green supply initiatives are aimed at taking waste, excess materials and cost out of the customer-supplier interface. Such initiatives may be aimed at improving purchasing efficiency and effectiveness, not at environmental benefit per se, and may be undertaken without dedicating specific 'environmental' resources. This finding adds support to fears about purchasing personnel's belief that environmental initiatives are costly (Min and Galle 1997), and represents a major barrier to the future integration of environmental concerns into standard purchasing and supply practice.

Overall, despite agreeing that the company should share the responsibility for their suppliers' environmental performance, there is concern among purchasing managers that such efforts may be costly, and that they do not have the resources to find the best solution. Purchasing managers recognize the potential for society and for their firm of greening their supply activity, but they are hesitant about the resources required for them to do so, and whether environmental issues present a net threat or opportunity.

9.4 Current Green Supply Practices

9.4.1 Types of Green Supply

We asked the purchasing managers whether they were implementing any of the green supply initiatives identified in the interviews. For each initiative, managers were asked whether they had been "implemented", "planned" or "not planned" in that operating unit. Figure 9.1 shows the total percentage of operating units implementing each type of initiative.

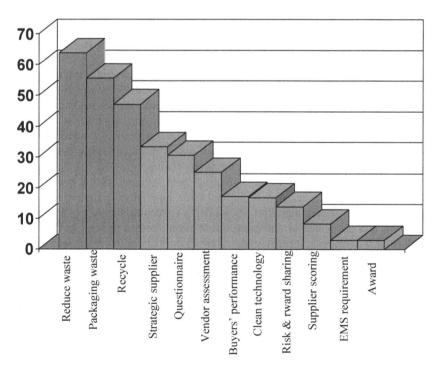

Figure 9.1. Green Supply Initiatives Implemented in the Sample Operating Units

We then conducted an exploratory analysis of implementation patterns and inductively derived three main types of green supply. A formal factor analysis could not be conducted as in similar studies (cf. Arragon-Correa 1998) due to the categorical answers received on each of the initiatives. Thus we made a trade-off between increased validity of the responses with categorical answers (i.e. being able to capture actual implementation as opposed to perceived extent of implementation as in Arragon-Correa 1998), and our ability to formally attribute the initiatives into types based on the underlying structure of variance as in factor analysis. Instead, in an attempt to gain face and content validity for the types, two expert judges were asked to allocate each of the initiatives into one of the three categories of initiatives (see Bowen 2001 for further details). The three types of initiatives also exhibited acceptable reliability (see below).

The first type, 'greening the supply process', represents adaptations made to the firm's supplier management activities aimed at incorporating environmental considerations. Green supply actions of this type are changes to the process of collecting environmental information on suppliers and assessing and ranking suppliers' environmental performance (Cronbach's $\alpha = 0.88$). The second type, 'product-based green supply', is conceptually distinct from greening the supply process in that it involves changes to the product supplied. It also includes attempts to manage the by-products of supplied inputs such as packaging. Product-based green supply includes initiatives such as recycling which requires co-operation with a supplier and efforts with suppliers to reduce waste (Cronbach's $\alpha = 0.84$). The third type of green supply, 'advanced green supply', includes more proactive measures such as introducing environmental criteria into buyers' performance or introducing joint clean technology programmes with a supplier (Cronbach's $\alpha = 0.71$).

The most popular practices are the product-based green supply initiatives. Over two-thirds of units had implemented either recycling initiatives which require co-operation with a supplier, collaboration with a supplier to reduce packaging, or had made efforts with a supplier to reduce waste. It is unsurprising that these were the most popular initiatives since they have the best direct potential to reduce waste and cost in supply activity. They are the so called 'win-win' green supply initiatives and focus on the purchase and delivery of a product.

Between a quarter and a third of operating units have introduced environmental criteria into their selection of strategic suppliers, supplier questionnaire or vendor assessment system. These activities are all part of greening the supply process (see Table 9.3), and are aimed at gathering information about suppliers' environmental performance. Examples of greening the supply process from our interviews included several retailers' introduction of scoring suppliers on their environmental performance. This enables firms to identify potential reputational problems before any negative publicity appears in the media. A further illustration of greening the supply process is provided by several companies which had already gained certified environmental management systems such as ISO 14001, or which were working towards this aim, which sent out environmental questionnaires to their suppliers. Unlike product-based green supply initiatives, both of these examples represented cost increases to the businesses concerned. Greening the supply process is often undertaken for commercial reasons which reside outside the normal purchasing and supply process - corporate risk management in the case of the retailers, and competitive positioning in the case of gaining certification for environmental management systems.

Table 9.3. Green supply activity and the percentage of implementers in each group

	G1	G2	G3	G4
Product-based Green Supply				
Recycling initiatives which require co-operation with a supplier	0	69	33	100
Collaboration with a supplier to eliminate packaging	0	81	50	100
Efforts with suppliers to reduce waste	0	81	100	100
Greening the Supply Process				
Building environmental criteria into the vendor assessment system	0	25	50	0
Scoring system to rank suppliers on their environmental performance	0	0	33	0
Environmental supplier questionnaire	0	25	83	29
Environmental criteria in the selection of strategic suppliers	11	25	67	29
Supplier environmental award	0	0	17	0
Requirement on suppliers to have an environmental management system	0	19	17	0
Advanced Green Supply				
Environmental criteria in evaluation of buyers' performance	0	6	17	100
Environmental criteria in risk and reward sharing agreements	0	0	0	29
Joint clean technology programmes with a supplier	11	13	0	57

Note: numbers are percentages of units in that group which have implemented the specific green supply initiative

We did find some evidence of advanced green supply initiatives, but examples of such behaviour were quite rare. Only the largest, most high impact companies in our sample had embarked on significant joint clean technology programmes with a supplier, and only the most proactive of retailers had environmental criteria in the performance assessment of their buyers and relationships with suppliers. One company (with two operating units in our sample) has introduced a supplier environmental award to raise the profile of environmental issues in their purchasing function and among suppliers both inside and outside the firm. It is surprising that the prevalence of Supplier Environmental Awards is so low in our sample since they can have high positive environmental visibility (Bowen 2000), but with a low cost outlay to the company.

Having examined the overall prevalence of the various types of green supply in our sample, we attempted to find groups of similar operating units in our data[3]. Clustering the operating units into similar groups allowed us to derive archetypal patterns of green supply implementation, draw lessons on the characteristics of units in the groups, and the average performance outcomes for each group based on their green supply activity. We found four archetypal groups based on the types of green supply implemented at operating units. Table 9.3 shows the proportion of units within each group which have implemented each of the green supply initiatives. The groups show striking differences in their patterns of green supply implementation.

The 18 operating units in Group 1 scored less than average on all three types of green supply - they are the least likely to implement any environmental supply initiatives, and are the least proactive on environmental supply issues. Group 2 units are characterised by undertaking product-based initiatives to a more than average extent, but have lower levels of implementation of the other two types of green supply. This is the largest group in our sample made up of nearly half the operating units (n = 32). The Group 3 units focus on greening the supply process. These 12 units also implement some advanced initiatives, though they also undertake product-based green supply to an average extent. The seven Group 4 units are distinguished by their extensive implementation of product-based green supply, but also their high implementation of advanced green supply initiatives. These are the most proactive green supply implementers.

The unit of analysis for this clustering exercise was the operating unit. A legitimate concern, therefore, could be that the clustering exercise captures corporate policy more than supply activity. However, analysis of the clusters suggested that there was considerable variation in green supply activity within business units. For example, of the six operating units included from a printing company, two units each fell into Groups 1, 3 and 4. Similarly, two units of a home and personal products company were members of Group 2, and a further two were members of Group 3. This variation in green supply activity supports evidence on environmental policies (Ketola 1997), that the link between corporate policy and implementation at operating units can be weak. Operating units seem to have considerable discretion over whether and how to implement environmental policies (Bowen 2001). Thus it is more useful to examine the implementation of green supply initiatives at the operating unit level, than at the more aggregated business unit level[4].

The next section examines the typical organisational characteristics, corporate environmental context, and purchasing and supply context for typical members of

---

[3] We calculated standardised factor scores for each type of green supply (product-based, greening supplier management, and advanced) and then undertook a cluster analysis using the K-means method as outlined in Kaufman and Rousseeuw (1990) and implemented in SPSS. The four cluster solution was retained as it fitted best with the theoretical discussion which follows.

[4] Given the comparatively small sample size in this study, it was difficult to control for business unit level effects statistically. All the analyses reported here are based on the operating unit level sample. In addition, we conducted several of the analyses excluding duplicate facilities from within the same business unit. Where this yielded substantively different results, from the reported operating unit sample, we describe this in the text.

each group[5]. By examining the differences between the groups we will argue that the identified groups are coherent, and that the characteristics of the groups can go some way to explain the green supply approach followed by units in that group. We will later examine the overall, environmental and purchasing performance outcomes of typical units in the groups.

9.4.2 Firm Size and Industry

Table 9.4 illustrates the size and industry characteristics of the average unit in each strategic group. We used two measures of size. The first is an absolute measure of the number of employees at the operating unit (item 4.1). On this measure, there was a significant difference between the groups, with Group 1 units being significantly smaller than both Group 2 and Group 4 units. The least proactive group, Group 1, contained the smallest units. This finding is in line with several studies which have found a positive relationship between firm size and environmental proactivity (e.g. Aragon-Correa 1998; Sharma and Nguan 1999; Bowen 2001). Larger units may generate more organisational slack which then facilitates the search for solutions to environmental problems (Sharma et al. 1999; Bowen 2002). Alternatively, larger units may be more visible in society and therefore more prone to pressure for improvement on environmental issues (Bowen 2000). In either case, there is no evidence in our sample of units which are larger relative to other sister units within their business unit undertaking a more proactive green supply approach (item 4.2).

We deliberately selected a broad range of industrial activity in order to capture a wide variety of green supply activity. Most notable among the industry characteristics of the groups is that all the Group 4 units were in manufacturing industries which are not considered "high impact". Several of the Group 4 units were part of a large glass manufacturer which had built environmental improvements into product and process developments with suppliers as part of their cleaner technology efforts. Group 3 units were significantly more likely to belong to high impact manufacturing, but less likely to belong to other manufacturing. These findings together suggest that while there may be a relationship between industry and green supply, our study did not have a finely grained enough approach. Further work is required on the green supply activity of members of the same, or closely related industries (building on e.g. Carter and Carter 1998).

---

[5] We used one way ANOVA to assess whether there are significant differences between scores across groups.   Since we conducted several simultaneous ANOVA tests, we encountered the difficulty of alpha inflation: the more tests we conducted, the more likely we were to incorrectly identify significant differences between scores by chance (i.e. our alpha values would be inflated beyond the usually acceptable 0.05 or 0.01 levels). We therefore adjusted the significance level for each set of tests using Bonferroni's correction (Brown and Melamed 1993). Furthermore, although ANOVA can identify whether there is a significant difference between group scores, it cannot tell us where this difference lies. We therefore conducted post hoc Tukey HSD tests on the significant ANOVAs to identify which group scores were significantly different from each other (Brown and Melamed 1993; Bryman and Cramer 1997).

Table 9.4. Size and industry characteristics of green supply groups

Organisational Characteristic	G1	G2	G3	G4	sig. (p)
Size					
4.1 Log number of employees	3.62	4.83	4.42	5.31	0.007*
4.2 Size relative to other units in the business unit	3.00	3.13	3.60	4.00	0.350
Industry					
4.3 High impact manufacturing	17%	25%	80%	0	0.003**
4.4 Other manufacturing	67%	75%	0	100%	0.000**
4.5 Other non-manufacturing	17%	0	20%	0	0.162
4.6 Highly regulated	3.89	3.67	3.00	4.67	0.396

Notes: item 4.1: log number of employees at the operating unit; item 4.2: scored on 1: much smaller than other units to 7: much larger than other units; item 4.6: scored 1: strongly disagree to 7: strongly agree for the statement "Our industry is highly regulated"; see Bowen (2001) for industry group definitions; * $p<0.05$; ** $p<0.01$ after Bonferroni's correction (critical values are 0.023 and 0.0050 respectively for 6 comparisons).

The extent of regulation and legislation is often cited as a predictor of environmental management activity (Peattie and Ringler 1994; Green et al. 1996; Hampson and Johnson 1996; Min and Galle 1997). However, our results concur with other studies which have found a lack of significance between regulation and environmental activities (e.g. Cater and Carter 1998) (item 4.6). We could argue that since the environmental supply behaviours we observed are those which are undertaken above and beyond those required by regulation and legislation, the current level of regulation is irrelevant. A more likely explanation for this finding from our interviews, however, is that it is not current regulation which effects green supply activity, but the threat of future regulation (Hampson and Johnson 1996).

9.4.3 Corporate Environmental Approach

Firms differ in their corporate approach to the management of environmental issues. Some merely comply with environmental laws and regulations, while others move beyond compliance to a more proactive approach (Hunt and Auster 1990; Roome 1992). Table 9.5 shows the corporate environmental context of units in each group. We found a positive relationship between middle managers' perceptions of corporate environmental proactivity and green supply (Bowen et al. 2001). The further the middle managers' perceptions of the corporate attitude to environmental issues is in advance of legislation, regulation, and other firms in the industry, the more likely the unit is to implement green supply initiatives (see item 5.2 – the scores for Groups 2, 3 and 4 are significantly higher than for Group 1).

The general pattern seems to be for units with high environmental commitment and interest among its personnel to follow a more proactive green supply strategy (items 5.3-5.6). This pattern is especially pronounced for top and middle manager commitment, and is most likely a reflection of the corporate environmental approach. Environmental attitudes of general managers varied as widely as those of the purchasing managers outlined in Table 9.2. For example, when asked whether environmental issues always represented a net cost to the business (item 5.8), roughly the same number of respondents strongly agreed as strongly disagreed (29.1% to 23.6%).

Table 9.5. Corporate environmental context of green supply groups

Corporate Environmental Context	G1	G2	G3	G4	sig. (p)
Corporate Environmental Proactivity					
5.1 Top managers' perception	4.18	2.57	3.93	4.90	0.012*
5.2 Middle manager's perception	3.41	4.44	4.83	5.11	0.000**
Environmental commitment					
5.3 Top management commitment	3.56	4.43	5.00	5.00	0.005*
5.4 Middle management commitment	2.30	3.87	4.60	5.33	0.000**
5.5 Top purchasing management commitment	2.89	3.62	3.17	4.00	0.275
5.6 Employee interest	3.33	3.60	4.20	4.67	0.163
Environmental attitudes					
5.7 Only profitable companies can afford environmental initiatives	3.22	1.47	2.20	0.33	0.002**
5.8 Environmental initiatives always present a net cost to the business, however well intentioned	3.00	2.80	5.20	3.33	0.003**
5.9 Environmental initiatives always pay off in the long run	4.89	5.60	5.80	5.67	0.221
5.10 Improving our environmental performance can make us more profitable	5.00	5.20	5.40	5.00	0.928

Notes: all items are scored on a scale of 1: strongly disagree to 7: strongly agree with statements like "Our top management are committed to environmental issues" (item 5.3); * $p<0.05$; ** $p<0.01$ after Bonferroni's correction (critical values are 0.014 and 0.003 respectively for 10 comparisons).

Of interest here is the difference in environmental mind-set of middle managers in units belonging to different environmental supply groups. Most striking is the Group 3 unit respondents' belief that environmental initiatives always present a net cost to the business however well intentioned (item 5.8 – Group 3 score is significantly higher than both Group 1 and 2). Recall that Group 3 units tend to focus their green supply efforts on greening the supply process, that is efforts to green the supplier base through introducing environmental criteria into vendor assessment, ranking suppliers, supplier questionnaires etc. Such initiatives do not necessarily lead to cost reductions. Indeed, units in Group 3 may be willing to undertake certain initiatives with longer term risk management or competitive positioning in mind rather than immediate commercial gain. It is hardly surprising, therefore, that the middle managers in these firms have an ambivalent attitude to the potential immediate cost gains of environmental initiatives. Despite this, Group 3 managers have confidence that environmental issues will pay off in the longer term. It is also notable that Group 1 companies were significantly more likely than each of the other groups to believe that only profitable companies can afford environmental initiatives (item 5.7).

9.4.4 Purchasing and Supply Context

Thus far we have examined the organisational and environmental characteristics of the different groups of units based on their green supply activity. We will now turn our attention to variations in the purchasing and supply context across units belonging to our different green supply groups (see Table 9.6). We might expect the most strategic purchasing and supply processes, and those with the highest level of capabilities appropriate for green supply, to belong to the groups undertaking the most proactive green supply activity (Bowen et al. 2001).

Table 9.6. Purchasing and supply context of green supply groups

Purchasing and Supply Context	G1	G2	G3	G4	sig. (p)
Strategic Purchasing					
6.1 Purchasing knowledge and skills	3.26	4.81	3.83	3.78	0.000**
6.2 Purchasing resources	3.75	4.69	4.58	3.00	0.001**
6.3 Purchasing status	2.96	4.67	4.33	4.22	0.000**
6.4 Purchasing foresight	2.52	3.52	3.72	3.11	0.017
Green Supply Management Capabilities					
6.5 Liaison between purchasing and other functions	3.43	4.87	4.67	5.00	0.031
6.6 Detailed purchasing policies and procedures	3.14	5.06	4.67	5.00	0.000**
6.7 Partnership approach with suppliers	3.29	4.94	4.00	4.33	0.015*
6.8 Technical skills of purchasing professionals	2.86	5.31	3.83	5.00	0.000**
6.9 Advanced understanding of environmental issues and how they affect supply	2.57	4.06	3.33	4.33	0.001**

Notes: all items are scored on a scale of 1: strongly disagree to 7: strongly agree with statements like "We have advanced capabilities in liaison between purchasing and other functions" (item 6.5); * $p<0.05$; ** $p<0.01$ after Bonferroni's correction (critical values are 0.016 and 0.003 respectively for 9 comparisons).

Examination of Table 9.6 reveals a more complex pattern. Group 1 units, which implement very few green supply initiatives, did indeed score lowest on the various aspects of strategic purchasing, and on all of the capabilities required for a green supply approach. Group 1 units scored significantly less than Group 2 units for all items in Table 9.6. In these units, purchasing lacks the knowledge, skills, resources, status, and the foresight to sell its initiatives in other parts of the organisation. Purchasing managers in these units also felt that they possessed supply capabilities to a lesser extent than their equivalents in units belonging to other groups. This inhibits their ability to implement the green supply measures implemented by Groups 2, 3 and 4.

Group 2 exhibited the highest level of strategic purchasing for all indicators other than purchasing foresight (item 6.4). They, along with Group 4 units, also tended to possess significantly higher levels of supply management capabilities than Group 1. Yet Group 2 units tend to implement only product-based green supply initiatives such as packaging waste reduction or recycling. Group 2 are not particularly proactive green supply units since they do not aim very far in advance of current legislation, do not require much specific environmental effort in their supply operations, and do not show much organisational commitment to environmental issues.

Why, then, do Group 2 units apparently waste the opportunity to implement the most proactive green supply measures when they clearly have purchasing functions and a supply process which is strategic and capable enough to implement them? The answer may lie precisely in the strategic nature of their purchasing and supply activities. Product-based green supply is aimed at eliminating excess resources and cost from the sourcing process. Minimising waste and excess cost is the main focus of many strategic purchasing units. Units which are strategic in their purchasing approach are more likely to seek materials efficiency and cost reduction regardless of any environmental motive. They are more used to the language of efficient resource use and waste minimisation, and so it is not surprising that they are more likely to implement product-based green supply initiatives.

9.4.5 Summary of Green Supply Practice

In summary, our examination of current green supply practice found a wide range of activity aimed at making supply chain management more environmentally sound. The most prevalent were product-based initiatives such as waste reduction in the customer-supplier interface, and the least prevalent were advanced green supply initiatives such as building environmental criteria into risk and reward sharing agreements.

We found four main groups of units which approached green supply in different ways, supporting our "horses for courses" view. Group 1 were the smallest on average, undertook very little green supply, were the least proactive in their corporate environmental approach, were the least strategic in their approach to supply, and possessed the lowest level of supply capabilities. At the opposite end of the environmental proactivity spectrum, Group 4 were the largest units which implemented advanced green supply initiatives to the greatest extent. They were all manufacturing units with the most proactive corporate environmental approach and highest level of managerial commitment to environmental issues. Although they were highly strategic in their purchasing activity, they were not necessarily the best prepared with supply capabilities to implement a green supply approach.

Group 2, however, scored very highly on both strategic purchasing and possessing supply capabilities despite not being in a corporate context which gave environmental issues a high priority. This led them to implement product-based green supply to a more than average extent, since these initiatives fitted best with their purchasing strategy and capabilities, and promised efficient materials use and cost reductions.

The final group of units, Group 3, were in a proactive and committed corporate environmental context, but did not have as strategic a focus in purchasing and supply management capabilities as Group 2. This group attempted to green the supply process which did not necessarily lead to cost reductions in the short run, but would, the middle managers believed, lead to long term commercial benefit. It is likely that the impetus for Group 3 initiatives lay outside the purchasing function (unlike Group 2), with greening the supply process being undertaken for corporate risk management or competitive positioning reasons rather than purely for effective and efficient supply management (as in Group 2).

9.5 Green Supply and Firm Performance

Having outlined the potential to society and to individual firms of greening purchasing and supply activity, and identified types of green supply practice, a key question remains: which of the environmental supply practices yield the best performance outcomes? Which of the Groups are currently most profitable and are commercially and environmentally viable in the longer term? In order to explore these questions, we asked the middle management respondents to assess their current performance on several dimensions against business unit headquarters' expectations, their competitors' performance, and their performance during the same period in the previous year. Table 9.7 presents their responses in the form of the average economic, environmental and supply performance of each group of units.

Compared with the targets set by the business unit headquarters, Group 1 units were significantly more profitable than Group 3 units (item 7.1). This may be a disappointing result for the advocates of the commercial potential of green supply measures, since it suggests that the units which are doing least on green supply are better economic performers than those engaged in greening the supply process. However, the post hoc analysis also suggests that there is no significant difference between the profitability performance of Group 2 or Group 4 units compared with Group 1. Thus the implementation of a portfolio of green supply initiatives including several product-based green supply initiatives does not seem to harm profitability. This represents a contribution to the debate on how environmental initiatives can affect the bottom line in firms (e.g. Hart and Ahuja 1996). Firms attempting to implement costly environmental initiatives, or those without the capabilities to implement them effectively (i.e. Group 3) are less likely to see the bottom line benefits of environmental improvement. On the other hand, a focus on win-win environmental initiatives can permit improved environmental performance without jeopardizing financial profitability.

There was no significant difference in the environmental management or supply management performance of the units (items 7.3, 7.4 and 7.6). Thus there is no evidence to support the espoused benefits to the purchasing and supply process outlined in Table 9.1 (Russel 1998; Lamming et al. 1996; Carter and Carter 1998).

Table 9.7. Performance of green supply groups

Performance Indicator	G1	G2	G3	G4	sig. (p)
Compared with HQ expectations					
7.1 Profitability	4.44	3.67	2.60	3.00	0.009*
7.2 Sales performance	4.00	3.29	2.60	2.33	0.029
7.3 Environmental management	4.11	4.20	4.00	4.00	0.930
7.4 Supply management	3.67	4.07	4.00	3.67	0.620
Compared with competitors					
7.5 Profitability	4.33	4.73	3.60	3.33	0.035
7.6 Environmental management	3.22	4.27	5.00	4.00	0.023
Compared with this time last year					
7.7 Profitability	4.33	3.80	4.80	4.00	0.416
7.8 How busy	5.11	4.20	4.40	4.67	0.155
7.9 Closeness to capacity	4.00	3.20	3.80	2.67	0.335
7.10 Efficiency	3.89	4.67	5.20	5.00	0.016
Other performance indicators					
7.11 We effectively manage the environmental risks that affect our business	3.89	4.73	5.00	5.67	0.001**
7.12 We have the capabilities to continue to improve on environmental issues	4.89	4.93	4.60	6.00	0.041

Notes: all items are scored on a scale of 1: strongly disagree to 7: strongly agree with statements like "Compared with this time last year we are much more profitable" (item 7.7); * $p<0.05$; ** $p<0.01$ after Bonferroni's correction (critical values are 0.012 and 0.0025 respectively for 12 comparisons).

Group 4 units, the most proactive on green supply, do not consider their overall performance in environmental management to be superior to their competitors (item 7.6). Referring to our interview data reveals a potential explanation for this anomaly. The glass manufacturer, whose units dominate Group 4, are indeed considered a leader on environmental matters (Business in the Community 1997), but this reputation is built more on technological innovation than 'green' initiatives per se. The top manager we interviewed in this company explained that they undertake green supply for technological and commercial reasons rather than for environmental ones. Thus it is unsurprising that middle managers do not see green supply implementation as giving them an environmental lead position in their industry, only a technological and commercial one. We therefore suspect that Group 4's low score on item 7.6 is more an artifact of our question than a true reflection of Group 4's environmental management performance compared with units in the other groups.

As well as the short term performance indicators described above, we also attempted to gain a picture of the longer term performance implications of following different courses of green supply. We asked middle managers about their efficiency compared with this time last year (item 7.10), whether they feel that they effectively manage the environmental risks that affect their business (item 7.11) and whether they feel their unit has the capabilities to continue to improve on environmental issues (item 7.12). Item 7.11, on environmental risk management, was the only one of these indicators to show significant differences across the Groups. Post hoc tests revealed that Group 1 scored significantly lower than all of the other three groups on this measure. Although they are the most profitable in the short term, Group 1 units are the least used to the language of efficient and effective resource use and of the environmental risks faced in supply management. The implementers of green supply initiatives (Groups 2, 3 and 4) do reap the benefit of more effective risk management as predicted in Table 9.1 (Drumwright 1994; Bowen et al. 1998). Indeed, managing environmental risk is exactly what their green supply approaches are designed to do, with greening the supply process and advanced green supply measures particularly aimed at proactively managing potential environmental liabilities rather than the resource and cost efficiency gains of product-related measures.

Our examination of green supply performance outcomes suggests that the commercial potential of green supply measures is clearly not being reaped in short term profitability and sales performance. Despite this, there is evidence to suggest that a proactive green supply approach can prepare firms for superior longer term performance through improved management of environmental risks and the development of capabilities for continuous environmental improvement. An exception to this general pattern is made by Group 2 units which have achieved the highest level of profitability compared with their competitors by systematically concentrating on the win-win product-based green supply options.

9.6 Horses for Courses in Green Supply

Our analysis suggests that understanding the practice of greening supply depends on delving beneath the headline implementation rates across firms. We have identified four groups of green supply approach, and shown how each approach is a rational response to a variety of corporate environmental and purchasing and supply contexts. We have also linked each group's approach with its likely performance outcomes. This study has yielded several insights on the location of private benefits to firms of green supply, and begun to resolve the gap between theory and practice.

Firstly, some types of environmental supply initiatives hold more potential for immediate private gain than others. Product-based green supply initiatives in particular can be very effective eliminating not only waste from an environmental perspective, but also excess cost from an economic view. They are therefore more popular and widespread than the other types of initiatives. Efforts to green the supply process, on the other hand, can be costly in the short term as additional systems to collect data, process the information and even aid suppliers in meeting environmental objectives are required. Many of the private benefits of green supply are derived from product-based initiatives (see Table 9.1), while the prevalence of green supply is more often assessed based on either greening the supply process or advanced green supply initiatives (e.g. Business in the Environment, 1996, 1997, 1998, 2000).

Secondly, there is a link between green supply and corporate environmental objectives. In our sample, Group 3 and 4 units undertook the most proactive green supply measures and also had the most proactive corporate stance on environmental issues. In these groups there was a high level of managerial commitment to environmental issues despite a recognition by middle managers that environmental initiatives often present a net cost to the business in the short term. Where the corporate environmental stance was proactive, the private benefits to green supply resided more in the long term effect of the overall corporate approach than in any particular green supply initiative. This suggests a less direct route to the espoused benefits of green supply than theory suggests, and may explain why firms are apparently slow to adopt these practices.

Thirdly, the existence and location of private benefits from green supply were different for units not particularly proactive on green supply issues, but possessing a strategically advanced purchasing and supply function. In these Group 2 units, purchasing has the capabilities in waste and cost reduction, and the language of resource efficiency to enable a systematic search for opportunities to eliminate excess resource use in the customer-supplier interface. Group 2 units specialise in product-based green supply, whether undertaken for strictly environmental or general commercial reasons. Where the unit has the advantage of a highly capable purchasing and supply process, the private benefits of green supply reside in the particular resource and cost minimising initiatives rather than in the long term environmental programme.

Finally, the varying location of the private benefits to green supply was reflected in our description of the performance outcomes of each group. Green supply's potential private gains are being reaped by each of our groups of units in their own way. Group 1 are not undertaking much green supply activity, and are benefiting from above average short term performance in profitability and sales volume. It is unclear, however, whether this is a viable long term strategy, since these are the units least prepared to deal with environmental risks or future expansions in environmental expectations. Group 2 units exploit their purchasing capabilities and systematically search for ways to integrate environmental concerns into their supply activities. Specialising in product-based green supply allows them to generate short term, above average profit and sales performance outcomes. Group 3 units are willing to follow a proactive approach, attempting to engage in an environmental dialogue with their suppliers and green their supply bases. They do this at the cost of short term private benefit, but instead capture the private benefits of superior environmental performance in their industry and effective environmental risk management. The final group, Group 4, implement advanced green supply initiatives, but reap the private benefit of the highest level of effective risk management, and a strong belief that they have the capabilities to continue to improve on environmental issues.

Our analysis of current green supply practice thus helps explain the gap between the advantages of green supply espoused in theory and the apparently low level of integration of environmental considerations into purchasing and supply. It is not the absolute level of green supply that is the best guide to the potential private gains, but an analysis of appropriate green supply practices in context. Businesses will act on the supposed opportunities of green supply if it is broadly in their interest to do so - whether for short term profit or sales performance, or for longer term risk management or environmental capability development. Indeed, the only rational explanation for the behaviour of Group 3 and 4 units is that they see some longer term commercial benefit from their proactive approach.

We have contributed to an understanding of greening supply by identifying four archetypal patterns of practice (four green supply groups). We have shown how units with particular characteristics, within particular contexts, are likely to implement given patterns of green supply. However, our analysis has its limitations. We conducted a quantitative analysis which can only capture some of the story's complexity. Our sample is based only on units of large firms in the UK to the exclusion either of small firms or ones based in other countries. We also had difficulty in controlling for business unit level effects given the structure of our sample. Further, performance is affected by more than merely the green supply approach followed, so it is possible that the performance patterns we found in the data are caused by a variable outside our discussion.

Despite this, our work holds useful implications for both managers and researchers. The main message for business managers from this study is to approach environmental issues as they affect supply with a strategic mind-set. The most successful firms are the ones which mobilise their existing capabilities appropriately and use green supply to support other corporate and purchasing objectives. Our main message for researchers in environmental issues as they affect supply is to be aware of the contingent nature of their performance predictions. There may indeed be social benefits from green supply, but these benefits will only be achieved if business managers can be persuaded of the private benefits. The challenge for both researchers and managers is to seek out the private benefits of green supply, so that the social benefits can follow.

References

Aragon-Correa, J. A. (1998) 'Strategic Proactivity and Firm Approach to the Natural Environment', Academy of Management Journal 41, 5: 556-567.

Bowen, F. E., Cousins, P. D. Lamming, R. C. and Faruk, A. C. (1998) 'The Role of Risk in Environment-related Supplier Initiatives', in R. C. Lamming (ed) Proceedings of the 7th Annual International IPSERA Conference (London : IPSERA).

Bowen, F. E. (2000) 'Environmental Visibility : A Trigger of Green Organisational Response?', Business Strategy and the Environment, 9, 2 : 92-107.

Bowen, F. E. (2001) Does Size Matter? Organisational slack and visibility as alternative explanations for environmental responsiveness, unpublished PhD thesis, University of Bath.

Bowen, F. E., Cousins, P. D., Lamming, R. C. and Faruk, A. C. (2001) 'The Role of Supply Management Capabilities in Green Supply', Production and Operations Management, vol. 10, no. 2, pp. 174-189.

Bowen, F. E. (2002) 'Organizational Slack and Corporate Greening: Broadening the Debate', British Journal of Management, forthcoming

Brown, S. R. and Melamed, L. E. (1993). 'Experimental Design and Analysis' in Lewis-Beck, M. S. (ed.), Experimental Design and Methods, Sage, London

Bryman, A. and Cramer, D. (1997). Quantitative Data Analysis with SPSS for Windows, Routledge, London.

Business in the Community (1997). Buying into a Green Future : Partnerships for Change. (London : Business in the Environment and The Chartered Institute of Purchasing and Supply).

Business in the Environment (1996, 1997, 1998, 2000). The Index of Corporate Environmental Engagement. (London : Business in the Community, AEA Technology and SustainAbility).

Carter, C. R. and J. R. Carter (1998) 'Interorganisational Determinants of Environmental Purchasing : Initial evidence from the consumer products industries', Decision Sciences 29, 3: 659-684.

Carter, C. R., L. M. Ellram, and Ready, K. J. (1998) 'Environmental Purchasing : Benchmarking our German Counterparts', International Journal of Purchasing and Materials Management Fall: 28-38.

Drumwright, M. E. (1994) 'Socially Responsible Organisational Buying : Environmental Concern as a Noneconomic Buying Criterion', Journal of Marketing 58: 1-19.

Epstein, M. and M.-J. Roy (1998) 'Managing Corporate Environmental Performance : A Multinational Perspective', European Management Journal 16, 3: 284-296.

Green, K., B. Morton, and New, S. (1996) 'Purchasing and environmental management : interactions, policies and opportunities', Business Strategy and the Environment 5: 188-197.

Hampson, J. and R. Johnson (1996) 'Environmental legislation and the supply chain', in Proceedings of the 6th Annual International IPSERA Conference (Eindhoven : IPSERA).

Hart, S. L. (1995) 'A Natural-Resource-Based View of the Firm', Academy of Management Review 20, 4: 986-1014.

Hunt, C. B. and E. R. Auster (1990) 'Proactive Environmental Management : Avoiding the Toxic Trap', Sloan Management Review Winter: 7-18.

Kaufman, L. and P. J. Rousseeuw (1990) Finding Groups in Data : An Introduction to Cluster Analysis. John Wiley & Sons, New York.

Ketola, T. (1997) 'A map of neverland : the role of policy in strategic environmental management', Business Strategy and the Environment 6: 18-33.

Knight, A. (1996) 'Driving Continuous Environmental Improvement : the Role of the Retailer', in R. Lamming, A. Warhurst and J. Hampson (eds) The Environment and Purchasing : Problem or Opportunity (London: CIPS): 65-74.

Lamming, R. C., P. D. Cousins, and Notman, D. (1996) 'Beyond Vendor Assessment: The Relationship Assessment Project'. European Journal of Purchasing and Supply Management, 2, 4:

Lamming, R. C., A. C. Faruk, and Cousins, P. (1999) 'Environmental Soundness : A Pragmatic Alternative to Expectations of Sustainable Development in Business Strategy', Business Strategy and the Environment 8: 177-188.

Lamming, R. C. and J. Hampson (1996) 'The environment as a supply chain management issue', British Journal of Management 7, March Special Issue: 45-62.

Miller, J. and F. Szekely (1995) 'What is "Green"?', European Management Journal 13, 3: 322-333.

Min, H. and W. P. Galle (1997) 'Green Purchasing Strategies : Trends and Implications', International Journal of Purchasing and Materials Management Summer: 10-17.

Noci, G. (1997) 'Designing 'green' vendor rating systems for the assessment of a supplier's environmental performance', European Journal of Purchasing and Supply Management 3, 2: 103-114.

Peattie, K. and A. Ringler (1994) 'Management and the Environment in the UK and Germany : A Comparison', European Management Journal 12, 2: 216-225.

Roome, N. (1992) 'Developing Environmental Management Strategies', Business Strategy and the Environment 1, 1: 11-24.

Russel, T. (1998) Greener Purchasing, (Sheffield, UK: Greenleaf Publishing).

Sharma, S. and O. Nguan (1999) 'The Biotechnology industry and strategies of biodiversity conservation : the influence of managerial interpretations and risk propensity', Business Strategy and the Environment 8: 46-61.

Sharma, S., A. Pablo, and Vrendenburg, H. (1999) 'Corporate environmental responsiveness strategies : the importance of issue interpretation and organisational context', Journal of Applied Behavioural Science 35, 1: 87-108.

Stock, J. R. (1992) Reverse Logistics, (Oakbrook, IL: Council of Logistics Management).

White, P. (1996) 'Life Cycle Assessment : What Can it Tell a Buyer?', in R. C. Lamming, A. Warhurst and J. Hampson (Eds.) Purchasing and Environment : Problem or Opportunity, (London: CIPS): 75-86.

World Resources Institute (1999) 'The Next Bottom Line', Business Week, May 3rd 1999: 45-96.

Wycherley, I. (1999) 'Greening Supply Chains : The case of the Body Shop International', Business Strategy and the Environment 8: 120-127.

This work was conducted as part of the Environmentally Sound Supply Chain Management Project (ESSCMo), based at the School of Management, University of Bath (EPSRC Grant no. GR/L23253). The authors would like to thank London Underground Ltd. and the ESSCMo Club of companies for financial and research support.

10

Green Purchasing in Chinese Large and Medium-sized State-owned Enterprises

Qinghua Zhu and Yong Geng

School of Management, Dalian University of Technology, Dalian, Liaoning Province (116024) P.R.C., erinzhu@hotmail.com

Although large and medium-sized state-owned enterprises (LMSOEs) play an important role in China's economy, only in the last two to three years have these enterprises started to focus on environmental developments such as green purchasing in their business practices. As part of an investigation of this topic, a survey and site visits to 28 LMSOEs were conducted on the potential for green purchasing in China. This paper is both a review of the investigation's findings and an introduction to the elements of green purchasing among Chinese LMSOEs. It first introduces the importance of green purchasing for LMSOEs as it pertains to the marketing of their products. Then it discusses key elements of green purchasing for LMSOEs, which include: (1) the organizational framework; (2) the model of suppliers' selection; (3) key factors and criteria affecting supplier selection; and (4) the establishment of beneficial buyer-supplier relationships. Lastly, it analyzes the challenges of LMSOEs for green purchasing in the context of the rapidly evolving Chinese economy, and summarizes the progress that has been made. The paper concludes that both Chinese LMSOEs and foreign investment enterprises in China can benefit from green purchasing cooperation if they establish close long-term supplier-buyer relationships.

10.1 Introduction

Under pressure from governments and the general public, as well as non-governmental pressure from abroad, more and more enterprises are working to improve their environmental performance. In business, environmental issues challenge the evaluation criteria of traditional competitive advantages. In the past, aspects such as costs, quality and delivery were considered to be the critical success factors. The traditional criteria remain fundamental indicators of competitiveness while the introduction of environmental management further increases the scope to create competitive advantages (Hutchison 1998). Christopher (1993) suggested that enterprises could gain true competitive advantage through Supply Chain Management (SCM), including purchasing management. For a typical manufacturer, purchased inputs generally account for over 60% of all product costs and 50% of the quality problems that arise in operations (GEMI 2001: 30). The nature of the materials and products enterprises purchase and use are important for them to achieve their goal of reducing environmental impacts at every stage of the production system. In addition, as consumers themselves, they can exercise substantial leverage and influence over other producers (their suppliers). Therefore, green purchasing-the integration of environmental considerations into purchasing policies, programs and actions-is critical for enterprises because it leads to eco-efficiency, cost-saving and improved public perception.

In China, state-owned enterprises (SOEs) dominate the economy. There are over 6,840 large and medium-sized industrial state-owned enterprises (LMSOEs) across the country (Consulate General of the People's Republic of China in Houston 2001). LMSOEs are also responsible for much environmental degradation due to a lack of awareness and outdated practices, among others, and as a result, they are more likely to come under pressure from the Chinese government to improve their environmental performance in comparison to other Chinese enterprises. Due to increasing pressure on natural resources, a growing population and the rapid development of the economy, the Chinese government has begun to enact more and more effective environmental laws (Wang and Qin 2000). Some of the pressure to create environmental legislation has also been due to international pressure and agreements such as the WTO, which China joined in November 2001. For instance, China's entrance into the WTO will require enterprises to address and overcome "green barriers" and increase their international competitive ability (Deng and Wang 1998). Christmann and Taylor (2001) suggested that exports and sales to foreign customers drive environmental performance of firm in China. Faced with global competition, the key question for Chinese SOEs is whether they can maintain their control, influence and drive in the national economy when the market is opened to others. Thus, the main issue is whether SOEs can maintain their competitiveness while achieving sustainable improvement (Wang 2001).

The purpose of the survey and the site visits undertaken through Dalian University of Technology (DUT), one of the best Chinese universities and located in the northeast region, was to describe some of the experiences and lessons concerning supplier selection and management of LMSOEs in China in relation to green purchasing. The study focused on manufacturing and processing industries because they primarily place pressure on the environment. The questionnaire was based on information obtained from several industrial experts and the literature (Zsidisin and Hendrick 1998, Min and Galle 1997 and Carter et al. 2000). Questionnaires were sent to both purchasing or environmental protection departments of 500 Chinese enterprises. Three hundred and two manufacturing and processing enterprises, representing 60.4% of the total, sent back the questionnaires, and more than one questionnaires got from one enterprise was integrated into one questionnaire through site visits or interviews with managers. The respondent enterprises include power plants and steel plants all over the country, as well as other industries such as chemistry, pharmacy, automobile and shipyards mainly in Liaoning province and Jinlin city. Among the respondent enterprises, 216 replies were from LMSOEs, representing 71.5% of the total. The questionnaire covers a range of green purchasing, including the organizational framework, the models of supplier selection, the key factors and criteria affecting supplier selection, and the establishment of supplier-buyer relationships. In addition to the survey, data for the study were obtained during site visits and interviews with managers responsible for environmental issues at 28 typical LMSOEs. Table I provided a brief introduction for three typical LMSOEs, which were chosen according to their industries, position in supply chains and key pressures for green SCM, and will be further discussed in the following sections. Conclusions are drawn from the survey, the site visits and the interviews.

Table 10.1. Brief introduction for three typical entrprises

Name	Industry	Position in Supply Chains	Size[1]	Main Pressures for environmental management
Dalian Mero Pharmaceutical Plant	Pharmacy	Produce intermediate products	729	*Laws and regulations *Public image
Dalian Diesel Engine Plant	Automobile	Produce both final products and intermediate products	2273	*Chinese customers *Potential foreign customers
Dalian New Shipyard	Shipyard	Produce final products	4523	*Export *Laws and regulations

[1]**Note**: The sizes of three enterprises are for the year of 2001.

10.2 Marketing Products for LMSOEs

Competitive LMSOEs have an increased focus on customers' needs and the development of value-added products and services. As such, it is important to understand a LMSOE's role as part of a supply chain to understand its potential for competitive advantages. Figure 1 presents a typical supply chain. The supply chain network structure will have a different look depending on which is the focal enterprise, that is, the enterprise whose management is mapping the supply chain (Stock and Lambert 2001:57). This paper defines an enterprise that produces final products as the focal enterprise. Tier 1 suppliers are those from which the focal enterprise purchases products and service. Tier 2 suppliers are the suppliers to tier 1 suppliers and so on. Since we limit our work to manufacturing and processing LMSOEs, the enterprises studied can be classified into two types according to their customers. The first type produce intermediate products, and their customers are other manufacturing and processing enterprises. The second type of LMSOEs produce final products, and their customers are wholesalers, retailers or end customers. Both types of LMSOEs are not able to neglect supplier selection and management, which will be further discussed below.

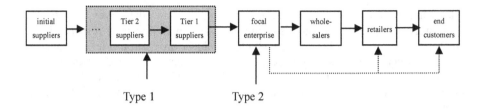

Figure 10.1. A typical supply chain

10.2.1 LMSOEs Producing Intermediate Products

For the first type of LMSOEs, those provide intermediate products to enterprises, it is advantageous to become the long-term suppliers to large downstream enterprises with large market shares and high profit margins. After joining the WTO in November 2001, China should be able to attract more foreign investment. However, the interviews conducted during the site visits highlighted that most joint ventures or Foreign Direct Investment (FDI) enterprises in China presently purchase materials and components mainly from their home countries, as well as upstream foreign enterprises already in China. In addition, it was found that the main reason for this phenomenon is that Chinese enterprises can not provide products that meet foreign enterprises' requirements on environmental standards. For example, enterprises from developed countries evaluate not only their direct suppliers but also second-tier suppliers (suppliers' suppliers). In this regard, it is noteworthy that in their investigation, Carter, Kale and Grimm (2000) put forward ten top environmental supplier evaluation criteria and that, among these, second-tier supplier environmentally friendly practice (EFP) evaluation is the second most important criterion.

In fact, supplier requirements and supplier evaluation are often mandated activities to meet with the requirements of third-party certification such as ISO 14001 (Section 4.4.6). Our survey demonstrated that Chinese enterprises think it is important to get ISO 14001 certification, which may partly result from the real or anticipated requirements of some foreign enterprises. For example, Bristol-Myers Squibb, IBM and Xerox have encouraged their Chinese suppliers to develop environmental management systems consistent with ISO 14001, while Ford, GM and Toyota have required their Chinese suppliers to obtain the ISO 14001 certification (GEMI 2000:7). From these findings, it would seem that it is very important for LMSOEs to apply environmental requirements in their supplier selection process. This is the case for LMSOEs that want to become long-term suppliers to joint ventures or FDI enterprises in China with strong environmental policies.

10.2.2 LMSOEs Producing Final Products

The second type of LMSOEs produce final products which are mostly consumed in the large Chinese market. If a LMSOE can capture a larger share of the Chinese market, higher profit should result. However, Chinese consumers have increasingly heightened environmental awareness and are starting to prefer green products (Lo and Leung 2000). In Chinese cities such as Dalian, most people (83.4%) stated that they preferred to buy green products, and 72.5% of consumers were willing to pay more for greener products (Zhu and Geng 2000). This is in line with trends in other countries such as the USA, in which an estimated 75% of consumers made their purchasing decisions with the enterprises' environmental reputation in mind and 80% of consumers were willing to pay more for environmental products (Carter *et al*. 2000). In the UK, researchers, Haji-Gazali and Simula, showed in 1994 that on average consumers were willing to pay approximately 13% more for green products (cited in Varangis *et al*. 1995:21). Although this does not reflect what consumers are actually willing to pay more for green products, it does imply that consumers are aware of and interested in green products. If a green product and non-green product are being sold at the same price, consumers tend to prefer the green option.

In China, many individuals seem to be eager for greener consumption, with numerous advertisements for green products in the media, especially TV and newspapers. This is partly because of previous environmental practices in China. In the past, most Chinese cared little about green products; but now that they have become richer, many are afraid of consuming some products that may prove harmful to their health. This is especially true for foods and medicines. On the other hand, resources such as water and electricity are relatively more expensive in China than that in some developed countries, especially compared with average incomes. Therefore, some Chinese consumers may prefer resource efficient products even if at a higher product cost. For example, people in Dalian are willing to pay more for water-saving washing machines because the water quota set for each household by the government is low during the period of a water crisis such as the summer of 2001. As the source of enterprises' internal value chains, purchasing is key to producing greener products (Wu *et al*. 2001). Therefore, LMSOEs in China should gradually highlight green purchasing and select suitable suppliers as a first step.

In summary, all manufacturing/processing LMSOEs, producing either intermediate products or producing final products, should select their suppliers carefully by considering environmental performance.

10.3 Integrating Environmental Issues into Supplier Selection and Management

Environmental supplier selection is a key competitive issue for LMSOEs, but the question is how LMSOEs should do to achieve both financial and environmental performance. From the survey and the site visits, it was found that among a number of the successful LMSOEs there were some similar experiences with supplier selection and management of green purchasing. Following, we present some common experiences in Chinese LMSOEs, such as their organizational frameworks, models for environmental supplier selection, criteria for supplier selection and supplier E management, as well as analysis of the data collected.

10.3.1 Organizational Framework for Green Purchasing

Successful implementation of green purchasing in a LMSOE depends largely on how the purchasing function is organized. If the enterprise has only a few decision-making units, it is relatively easy to carry out special training including environmental education for those purchasing managers. Though purchasers play an important role in supplier selection, in order to adopt an effective and credible position within their own enterprise, they need to understand the environmental issues deeply, especially for green purchasing.

More than 95% of Chinese enterprises surveyed and all LMSOEs interviewed have centralized most of their purchasing in a special department, namely a purchasing and supply department (PSD). Staff members in the PSD purchase most materials and resources except some special devices and a few very large-volume items. In centralized purchasing, the decision about which suppliers will be suitable is made by a decision group, an enterprise coordination commission (ECC), which is composed of senior managers and some key mid-level managers from different departments including environment, health and safety (EHS). Staff members in the PSD are chosen from different departments by rigorous evaluation undertaken over the previous several years.

The switch to environmentally preferable products is slowed somewhat by purchasing inertia. The staff of the centralized PSD may hesitate to change products or suppliers because they are open to criticism from all departments using the materials or components. However, environmental improvement is usually one of the strategic objectives and since the ECC makes the final decision, most staff in the different departments and workshops of a LMSOE do not object to trying to use new products from new suppliers, especially if they are more environmentally friendly.

Centralized purchasing also makes it possible for PSD staff to play an important role in improving environmental performance. Green purchasing requires the bridging of two separate worlds: purchasing and the environment (Consultancy and Research for Environmental Management 1997: 2). Clearly, purchasers cannot be expected to be environmental experts. Moreover, they work under time constraints and usually cannot afford to study environmental issues thoroughly. Nevertheless, purchasers occupy a strategically important position regarding environmental concerns both within their own enterprises and towards others such as suppliers (Ermen and Mielisch, 1998: 201).

To succeed, a LMSOE must have an appropriate formal organizational framework and control mechanisms to match its objective. Central purchasing makes it possible for an enterprise to integrate environmental factors into purchasing. Moreover, the ECC including managers from the EHS department ensures consideration of environmental issues during the whole purchasing process.

Enterprise A

Dalian Mero pharmaceutical plant, a large processing LMSOE, received substantial funding for relocation[1] and made obvious improvements to its production processes. Its attention to environmental issues is mainly due to strong consumer pressure for greener products. The plant began to integrate environmental criteria within supplier selection, mainly by bidding, which helped to save more than two million RMB for the supply of penicillin bottles, and at the same time decreased the number of suppliers for penicillin from ten to one. Moreover, the plant established mutually beneficial relationships with their suppliers, which also resulted in environmental improvements. The vice general manager of the plant we interviewed mentioned that they developed a symbiotic relationship between the buyer and the supplier.

The plant became one of the first of nineteen enterprises to be awarded the state certificates of Good Manufacture Praxis (GMP) designed for the pharmaceutical industry on January 12, 2001. Moreover, the plant received the certification for most of their products including medicines and intermediate materials, which helped the plant to obtain a larger share of the market in China[2]. In addition, the plant has begun to discuss potential cooperation with some foreign enterprises.

Enterprise B

An automobile engine manufacturer, Dalian Diesel Engine Plant, established in 1951, was one of the key plants for the design and manufacture of diesel engines in China. However, in the middle of 1990s,the plant met serious difficulties including a reduced market share, partly because of requirements from downstream enterprises. The automobile industry was strongly motivated to manage EHS impacts in its supply chains because it attracted more attention from stakeholders, including regulators concerned with gas mileage, emissions, and the recovery of materials from retired vehicles. The process of automobile manufacturing involves many safety and environmental risks, and manufacturers rely on multiple first-tier suppliers (direct suppliers) for a steady flow of components and a quick delivery to market.

Through a series of green purchasing measures and total environmental quality management, the plant established close and long-term relationships with more than thirty downstream enterprises, especially with First Automobile Works (FAW), one of the biggest automobile manufacturers in China. Now the plant is ready to establish a joint venture with Mercedez-Benz. The director of PSD told us that they try their best to buy all materials and components from suppliers with strong reputations for both quality and environmental credibility. One of the imaginative tests that the plant employs in this regard, is to see if suppliers can provide formal invoices[3].

Enterprise C

China is the third biggest shipbuilding nation in the world and, as a result, Chinese shipyards pay strict attention to keeping good relationships with suppliers that demonstrate strong environmental performance. On August 20, 2001, the contract signing ceremony for the Dalian New Shipyard to build five 300,000-ton super-large oil tankers for Iran took place in the Great Hall of the People in Beijing. The building of such large oil tankers is an important indication of a country's shipbuilding capabilities. In the international competitive bidding for the US$370 million contract, the Dalian New Shipyard defeated Japan's Mitsubishi and the Korea's Daewoo and some other strong competitors. The Dalian New Shipyard, founded in 1990, has now become China's largest ship export base.

Box 10.1 Three typical LMSOEs with green purchasing program

[1]Note: The difference of land value between urban areas and suburb areas is very high in some large Chinese cities. The relocating enterprises got the full land value compensation and enjoyed many preferable policies from the local government, such as lower tax and land value in the new estates.
[2]Note: In Chinese pharmaceutical industry, consumers think that a plant can produce greener products and prefer to buy its products if the plant can get the certificate of GMP.
[3]Note: All enterprises with good credit have strict financial controls, and they always provide formal invoices. However, some small enterprises do not have adequate financial controls, and sometimes can not or are not willing to provide formal invoices in order to escape income tax. Although this practice is not easily understood by many Westerners, it does occur in China

10.3.2 Models of Supplier Selection for Green Purchasing in LMSOEs

A centralized organizational framework makes it easier for LMSOEs to realize green purchasing, but the issue of how to carry out environmental supplier selection is also very important. Among the models being used, bidding and life cycle analysis (LCA) are the most popular.

Information is a key factor in the continuing development of SCM. However, access to information is a big issue for LMSOEs in China. Many LMSOEs lack information on suppliers, especially their environmental performance. In addition, many LMSOEs we interviewed estimated that their purchasing cost accounts for more than 80% of their total product cost, which is much higher than the average of 60% of all product cost (GEMI 2001: 30). How to reduce purchasing cost has become a key factor for LMSOEs to improve their competitive ability. Our survey illustrated that, in order to find more suitable suppliers and to cut costs, most LMSOEs use bidding for important purchases. This was highlighted by all LMSOEs we interviewed. For example, enterprise A in Box I holds a formal bidding meeting bi-annually. All departments participate in the bidding processes; the PSD is in charge of the bidding while the ECC makes the final decision. Bids can be simply sent by fax to long-term partners and allies, sent by mail to potential suppliers, and publicized through the Internet or other media. A typical bidding process includes three steps: searching for suppliers, selecting potential suppliers according to some requirements and limitations, and holding a formal bidding meeting.

Purchasers are key personnel for advancing environmentally preferable bidding in supplier selection. Their position means that they are the best placed and best qualified within an enterprise to argue for the adoption of greener products and more environmentally friendly purchasing practices. Purchasers have a superior vantage point as regards new supplier information, including a supplier's environmental reputation and products marketed with green credentials (Ermen and Mielisch 1998: 201). At the same time, EHS and other departments are involved in the bidding process since all staff can provide information on suppliers and all departments have members in the ECC.

Through the bidding process, all staff in the LMSOEs improve their environmental awareness. For example, to integrate environmental issues into bidding requirements, enterpise A invites environmental experts from government and consulting firms to provide further suggestions. After finishing a bidding statement which includes environmental criteria, the enterprise hands out the statements to all departments. Most departments, especially information centers, provide information on suppliers by all kind of sources including the Internet. Through such procedures, the enterprise provides education on quality, environmental performance and other aspects simultaneously, which can greatly improve staff's environmental awareness. Moreover, the enterprise also disseminates its concerns about environmental performance by adding environmental criteria into the bidding statement.

At the same time, some LMSOEs began to assess green purchasing impacts and their financial consequences through the entire product life cycle. The life cycle perspective is important because customers or downstream enterprises, regulators, and courts are increasingly concerned about the environmental impacts before and after manufacturing occurs. Normally, the PSD has primary responsibility for supplier selection and management of the product life cycle, and a support role during manufacturing and product design. Such integrated proactive purchasing focuses on whole-chain financial optimization which typically results in 10%-20% cost saving and a substantially greater increase in profits (Burt and Pinkerton 1996). For example, enterprise B extended its focus to include upstream enterprises' environmental performance issues. In order to solve the environmental problems, the enterprise added environmental considerations to strategic sourcing initiatives, and carried out strategic policy with a slogan of "green plant, environmental engine". The enterprise now has a broader focus on optimizing business performance along its supply chain considering all factors, including environmental factors contributing to cost, quality and risk. Its product life cycle analysis involves an interdepartmental team, including marketing, design and operations, strategic planning, PSD and EHS, while the ECC plays the most important role in supplier selection. PSD's participation is critical because knowledge of suppliers and well-formulated purchasing specifications can add value through the entire life cycle. The staff of PSD with the help of other departments could identify the value-adding opportunities within their control, such as improving supplier performance, and identifying input substitutions with such things as lower toxicity, easier disassembly, higher recycled content and lower regulatory burden.

Models of green purchasing are key to environmental supplier selection, but most LMSOEs are still exploring which model is the most suitable. Our survey and interviews show that bidding and LCA are two helpful models, which are often used simultaneously.

10.3.3 Key Factors and Criteria Affecting Supplier Selection and Evaluation

In addition to the organizational framework and the models for green purchasing, there are two other important aspects which need to be discussed. One is the key factors affecting environmental supplier selection by Chinese LMSOEs, and the other is the criteria by which the qualified suppliers are evaluated.

In order to understand what leads Chinese LMSOEs to select environmental suppliers, we designed a ranking and factors for the survey based on a study by Min and Galle (1997). To compare the results with those from Min and Galle (see table 10.3), we design a five-point scale. From table 10.2, we can see that the factors affecting Chinese LMSOEs' selection of suppliers with environmental consideration are similar to those in the USA, and mainly include three aspects: laws and regulations, costs for purchasing and disposal, and suppliers' environmental performance. These results are consistent with our findings from interviews. LMSOEs are always looking for new opportunities to cut costs and increase total value, and many of them recognize that green purchasing could improve their environmental performance, as well as their financial performance.

During the supplier selection, one key step is to explore evaluation criteria for supplier selection. In order to get the information on the selection criteria used, we investigated some typical LMSOEs. Ten environmental supplier evaluation criteria were identified (see table 10.3). We use a seven-point scale here so that the results can be compared with the results in the USA (Carter *et al.* 2000).

Table 10.2. Key factors that affect a Chinese firm's choice on supplies

Factors Affecting Supplier Selection[1]	China			USA[1]		
	Average Degree of Importance[2]	Raw Rank	Adjusted Rank[3]	Average Degree of Importance[2]	Raw Rank	Adjusted Rank[3]
Federal/Provincial Municipal environmental regulations	1.458 (0.644)	1	1	1.852 (1.104)	4	2
State environmental regulations	1.649 (1.029)	2	2	1.844 (1.111)	3	2
Cost of environmentally friendly goods	1.848 (1.042)	3	3	2.135 (1.046)	5	3
Cost for disposal of hazardous materials	1.890 (1.098)	4	3	1.703 (1.073)	2	2
Potential liability for disposal of hazardous materials	1.896 (0.946)	5	3	1.488 (0.966)	1	1
Cost of environmentally friendly packages	1.929 (0.944)	6	3	2.145 (1.288)	6	3
Buying firm's environmental mission	1.932 (0.898)	7	3	2.420 (1.288)	7	4
Supplier's advances in developing environmentally friendly goods	2.127 (1.190)	8	4	2.745 (1.102)	9	5
Environmental partnership with suppliers	2.127 (0.780)	9	4	2.829 (0.780)	10	5
Supplier's advances in providing environmentally friendly packages	2.411 (1.090)	10	5	2.640 (1.115)	8	5

[1]Source: The list of the factors and the data in the USA are from Min and Galle(1997).
[2]Note: Number in parentheses is standard deviations
 Scale: 5=not at all important, 1=extremely important
[3]Note: The same adjusted rank indicates no statistically significant difference in means at $p=0.05$.

Due to different enterprise results, we distinguish different enterprises in terms of industry membership and geographic location. From table 10.3, we find Chinese LMSOEs ignore some criteria that are highlighted by US enterprises. For example,

US enterprises consider the criterion of second-tier supplier environmentally friendly practice evaluation as the second important, while most Chinese LMSOEs ranked it of low importance. This also helps explain why it is difficult for Chinese enterprises to export their products or become suppliers to foreign enterprises in China. To understand this issue further, we interviewed during a site visit the vice general manager of enterprise C which exports all its products to foreign markets. We found the enterprise had adequate information on all the suppliers and some of the foreign ship purchasers designate specific suppliers for some special or core components.

Table 10.3. Top ten environmental supplier evaluation criteria

Environmental supplier evaluation criteria[1]	Rank	Average Degree of Importance[2] (rank)					
	USA	Jinlin City	Power plants[3]	Power plants	Steel Plants	Liaoning Province
Public disclosure of environmental record	1	4.55(7)	5.00(5)	4.80(9)	4.00(9)	4.75(6)
Second-tier supplier Environmentally Friendly Practice evaluation	2	3.90(10)	4.12(10)	4.47(10)	4.00(10)	4.63(7)
Hazardous waste management	3	5.64(3)	5.39(2)	5.70(3)	4.44(7)	5.75(3)
Toxic waste pollution management	4	6.40(1)	5.50(1)	5.53(4)	5.33(2)	5.88(2)
Products potentially conflict with laws	5	5.33(4)	4.78(7)	5.82(2)	5.50(1)	4.38(8)
ISO 14001 certification	6	4.00(9)	4.83(6)	5.15(6)	5.22(3)	4.38(9)
Reverse logistics program	7	4.11(8)	4.19(9)	4.82(8)	4.56(6)	4.38(10)
Environmentally Friendly Practice in product packaging	8	5.27(5)	4.56(10)	5.08(7)	4.11(8)	5.63(5)
Ozone Depleting Substance management	9	5.18(6)	5.00(3)	5.44(5)	4.89(4)	5.75(4)
Hazards air emission management	10	5.70(2)	5.00(4)	5.94(1)	4.78(5)	6.00(1)

[1]Source: The list of the criteria is from Carter, Kale, and Grimm (2000).
[2]Note: Number in parentheses is standard deviations
    Scale: 1=not at all important, 7=extremely important
[3]Note: These plants are joint ventures, which are investigated to be compared with other state-owned enterprises

Some LMSOEs, including enterprises A, B and C, developed criteria for supplier selection by considering environmental factors, and our interviews confirmed that most of them benefited from their effort to add environmental factors to their criteria on supplier selection. However, it seems that suitable supplier criteria is still a challenge for many LMSOEs in China.

10.3.4 Monitoring Suppliers and Building Mutually Beneficial Relationships

After selecting suitable suppliers, it is even more important for a LMSOE to maintain long-term relationships with these suppliers. Our survey showed that all LMSOEs established files for suppliers providing key materials and components. By developing close long-term relationships with suitable suppliers, LMSOEs gain business benefit either toward the bottom line (decreased cost) or the top line (increased sales and revenues). Among those, the main benefits include: (1) reduced raw materials waste; (2) reduced transportation costs; (3) reduced waste disposal costs; (4) reduced compliance costs; (5) reduced cost of incidents; (6) reduced risk of business interruption due to regulatory violation, boycott, supplier interruption, spill, toxic release, etc; (7) customer retention rate increase; and (8) increased market share as a result of enhanced reputation attracting new customers. On the other hand, most LMSOEs ceased purchasing from some former suppliers due to their failure to improve their environmental performance. Our survey showed that 23% of LMSOEs assessed environmental performance and stopped purchasing from former suppliers on environmental grounds. However, further discussion with those downstream LMSOEs revealed that the reasons for such actions were rarely about environmental matters alone. Rather, environmental performance was just one of several factors. In addition, we found that close buyer-supplier relationships led to a trend towards few suppliers in the supply chain. The reasons for this were: perceived benefits of single sourcing; inability of some suppliers to meet buyer certification requirements; and increased management time invested in suppliers.

As purchasers are in constant contact with suppliers' sales departments, they contribute to the perception of the suppliers' environmental credibility. They verify suppliers' environmental standards and practices, and make other enterprises aware of green purchasing-related issues(Ermen and Mielisch 1998: 201). Therefore, PSD staff are critical to maintaining close long-term relationships with suitable suppliers.

10.4 Findings and Conclusions

With the changes brought about by economic globalization in the new century, environmental performance appears to be increasingly important for Chinese LMSOEs seeking to maintain a dominant position in the national economy. This is even more pressing in view of changes such as China's accession to the WTO and the increasing environmental awareness of Chinese citizens. Some LMSOEs use interdepartmental teams to optimize whole-system performance of entire supply chains, and focus on integrating environmental issues into supplier selection and management. However, green SCM is still a new concept in China. Some Chinese enterprises have recognized its importance and tried to put it into practice, but most of these enterprises lack experiences as well as necessary tools and management skills. The site visits to selected enterprises revealed that the understanding of green purchasing is high among senior managers in LMSOEs, but they still lack ways to integrate it into green supply chains. However, mid-level managers or employees have a poor understanding of green purchasing. In addition, our survey and interviews showed that most LMSEOs care little about some aspects of green purchasing such as second tier supplier evaluation and product packaging.

Fortunately, centralized purchasing establishes the foundation to overcome these barriers. It is possible to carry out green purchasing and improve performance by educating and training staff in the purchasing and supply departments. Many LMSOEs have established coordination commissions composed of senior managers and some key mid-level managers, which can increase internal communication among different departments leading to the successful implementation of green purchasing. By developing some criteria for environmental supplier selection, and by developing strategic partnerships with suppliers, LMSOEs have radically reduced the number of suppliers and gained more control over purchasing cost. This was also demonstrated by our survey and interviews. Although it is still difficult for most Chinese LMSOEs to establish long-term relationships with upstream and downstream enterprises, bidding involving environmental criteria can be a bridge to finding more suitable suppliers and maintaining good relationship with them.

LMSOEs have made some progress on environmental supplier selection and management for green purchasing, so they will have opportunities when competing in global markets. As mentioned above, following China's acceptance into the WTO, there should be more joint ventures and FDI enterprises. Presently these enterprises purchase materials and components mainly from their home countries or upstream foreign enterprises in China. If LMSOEs can strengthen their evaluation of suppliers and ensure quality products, they should have a competitive advantage and become stable and excellent suppliers to these foreign enterprises in China. In the case of joint ventures and FDI enterprises in China, purchasing from home countries increases their costs associated with transportation and human resources, while establishing upstream joint ventures or FDI enterprises needs a great deal of investment and increases risk. If foreign enterprises can encourage and support more Chinese suppliers by providing new and essential knowledge about technology and management, and establish good relationships with these Chinese suppliers, they should derive many benefits.

Acknowledgements

This work is supported by the Canadian-Chinese Scholarship Exchange Program (CCSEP). It is also the results of the project supported by National Natural Science Foundation of China (70202006) and the project supported by Liaoning Provincial Doctoral Startup Foundation (2001102090). The authors are grateful to two professors at Dalhousie University for constructive comments on earlier versions of this paper, namely, Prof. Donald J. Patton, and Prof. Raymond P. Cote. We are also grateful to Miss Zoe Kroecker at Dalhousie University for reviewing the paper. Meanwhile, special thanks to Yiping Zhao, Wei Li and Xiuzhen Lan at Dalian University of Technology for participating in the investigations. Finally, we would like to thank the editor and the anonymous reviewers for their helpful comments.

References

Alexander, F. (1996) 'ISO 14001: What does it Mean for IE's', IIE Solution 28.1:14-18.

Baily, P., and D. Farmer (1990) Purchasing Principal and Management (London: Pitman).

Beamon, M.B. (1999) 'Designing the Green Supply Chains', Logistics Information Management 12.4:332-42.

Burt, D., and R. Pinkerton (1996) A Purchasing Manger's Tool to Strategic Proactive Procurement (American Management Association, New York: 8-10)

Carter, R.C., and J. R. Carter (1998) 'Interorganizational Determinants of Environmental Purchasing: Initial Evidence from the Consumer Products Industry', Decision Sciences 29.3: 28-38.

Carter, R.C., R. Kale, and C.M. Grimn (2000) 'Environmental Purchasing and Firm Performance: an Empirical Investigation', Transportation Research Part E, 36: 219-88.

Christmann, P. and G. Taylor (2001) 'Globalization and the Environment: Determinants of Firm Self-regulation in China', Journal of International Business Studies 32.3: 439-458.

Christopher, M. (1993) 'Logistics and Competitive Strategy', European Management Journal 11.2: 258-61.

Consulate General of the People's Republic of China in Houston (2001) 'Doing Business With China: Reform of the State Owned Enterprises' (from http: www.chinahouston.org Oct. 7, 2001)

Consultancy and Research for Environmental Management (ed.) (1997) Green Procurement: Opportunities for Stiching Milieukeur (CREM Report, 97.187; Amsterdam: Consultancy and Research for Environmental Management).

Deng, H., and H. Wang (1998) 'How do Chinese Enterprises Overcome Green Barrier', Business Studies 12.10: 45-47.

Ermen, R. V., and A. Mielisch (1998) 'The European Green Purchasing Network: Addressing Purchasers across Sectors and Boudaries', in T. Russel (ed.), Green Purchasing: Opportunities and Innovations (Sheffield, UK: Greenleaf Publishing): 199-207.

GEMI (Global Environmental Management Initiative) (2001), New Paths to Business Value March 2001.

Gray, V., and J. Guthrie (1990) 'Ethical Issues of Environmentally Friendly Packaging', International Journal of Purchasing and Materials Management 20.8: 31-36.

Huang, C., and W. He (1998) Information Economics (Beijing, China: Economy Science Publisher).

Hill,K.E. (1997) 'Supply-Chain Dynamics, Environmental Issues, and Manufacturing Firms', Environment and Planning A 29: 1257-74.

Hutchison, J. (1998) 'Integrating Environmental Criteria into Purchasing Decision: Value Added?', in T. Russel (ed.), Green Purchasing: Opportunities and Innovations (Sheffield, UK: Greenleaf Publishing): 199-207.

Klassen, R., and C. Mclaughlin (1996) 'The Impact of Environmental Management on Firm Performance',. Management Science 42.8: 1199-1214

Lo, C. W., and S.W. Leung (2000) 'Environmental Agency and Public Opinion in Guangzhou: The Limits of a Popular Approach to Environmental Governance', The China Quarterly (2000): 677-704.

Min, H., and W.P. Galle (1997) 'Green Purchasing Strategies: Trends and Implications', International Journal of Purchasing and Materials Management (Summer 1997): 10-17.

Murphy, P., R. Poist, and C. Braunschwieg (1995) 'Role and Relevance of Logistics to Corporate Environmentalism: An Empirical Assessment', International Journal of Physical Distribution and Logistics Management 25.2:5-19.

Pratt, K.M. (1997) 'Environmental Standards could Govern Trade', Transportation and Distribution 38:68-76.

Sarkis, J., M. Liffers, and S. Malette (1998) 'Purchasing Operations at Digital's Computer Asset Recovery Facility', in T. Russel (ed.), Green Purchasing: Opportunities and Innovations (Sheffield, UK: Greenleaf Publishing): 270-81.

State Economic and Trade Commission, China (2001) 'Several Problems that Require Full Attention in Deepening: Reform of State-owned Enterprises', (From: http://www.sect.gov.cn, June 4, 2001).

Stock, J.R. and D.M. Lambert (2001) Strategic Logistics Management (New York, USA: McGraw-Hill Higher Education)

The National People's Congress (2000) 'Report on the Outline of the Tenth Five-year Plan for National Economic and Social Development (VII)' (From: http://www.china.org.cn, Oct. 7, 2001)

Varangis, P.N., R. Crossley, and C.A. Primo Brago (1995) Is there a Commercial Case for Tropical Timber Certification? (World Bank Policy Research Working Paper, WPS 1479; Washington, DC: World Bank).

Vermeer, E.B. (1995) 'Management of Environmental Pollution in China: Problems and Abatement Policies', China Information 5.1: 39.

Verschoor, A.H., and L. Reijinders (1997) 'How the Purchasing Department can Contribute to Toxics Reduction', Journal of Cleaner Production 5.3:187-91.

Walton, S.V., R.B. Handfield, and S.T. Melnyk (1998) 'The Green Supply Chain: Integrating Suppliers into Environmental Management Process', International Journal of Purchasing and Materials Management Spring: 2-11.

Wang, W. (Vice Chairman of the State Economic and Trade Commission) 'Talking about China State-owned Enterprises in the 21st Century', (From http://www.chinahouston.org, Oct. 7, 2001).

Wang, X., and T. Qin (2000) 'Analysis on Practical Effect of Chinese Environmental Laws: Considering from Decision Mechanism', Environmental Protection 23.8: 8-10.

Wu, C.Y., Q.H. Zhu, and Y. Geng (2001) 'Green Supply Chain Management and Enterprises' Sustainable Development', China Soft Science, 16.3:67-70.

Zhu, D., H. Lin, and X. Zeng (2000) 'Decision for Selection of Suppliers', Management Moderlization 7.2: 25-27.

Zhu, Q.H., and Y. Geng (2000) 'Comparison of Environmental Awareness of Residents in Hongkong and Dalian' ('Academic Report'; Dalian: Dalian University of Technology Aug. 5, 2000).

Zsidisin, G.A., and T.E. Hendrick (1998) 'Purchasing's involvement in environmental issues: a multi-country perspective', Industrial Management & Data Systems, 7: 313-20.

11

Greening of Suppliers/In-bound Logistics – In the South East Asian Context

Purba Rao

Asian Institute of Management, 123 Paseo de Roxas, Makati City, 1260 Philippines, purba@aim.edu.ph

This chapter presents findings on the greening of the supply chain, which is taking place in the growing manufacturing region that is South East Asia. Empirical research using a survey questionnaire as the research instrument is undertaken for a regional area encompassing the Philippines, Indonesia, Malaysia, Thailand, and Singapore. The objective of this paper is to present the findings of this survey research which identify major issues in this greening process and business practice. The focus of this chapter, from a larger study, is primarily on the in-bound logistics practices of various South East Asian manufacturing firms.

11.1 Introduction

It is expected that a major portion of the world's manufacturing will be taking place in South East Asia creating many job and service opportunities in this part of the world. Yet, this growth in manufacturing will also bring about substantial environmental burden in terms of the generation of various environmentally harmful wastes. This phenomenon poses tremendous challenges for South East Asian governments, communities, and industry of this region. While economic growth is so urgently needed in this region, environmental sustainability and the health of millions of people in this part of the world can no longer be ignored. This challenge is even made larger by Asia's expanding population with a resultant increase in consumption and pollution. Yet, there is another trend, which is being observed in Asia today – this refers to the large manufacturing organizations subcontracting and delegating smaller modules of the manufacturing process to small and medium enterprises. These small and medium enterprises, which constitute critical elements of the supply chain, do not have the resources and capabilities, which the large organizations have and thus often end up being the largest polluters of the environment. This happening recently has been observed and many large corporations, who pride themselves on environmental and socially responsibility, have started encouraging, guiding and even funding their suppliers to be green.

The importance of supply chain management in improving the overall corporate environmental management has long been recognized in environmental standards such as BS 7750, ISO 14000 systems of standards, and the parallel European Union (EU) regulation on eco-management and audit. Also research has observed that the pollution and waste generated by the supply chain can contribute to global problems like climate change, global warming, ozone layer depletion, and acid rain if not addressed properly [Min and Galle, 1997]. One of the most effective ways of addressing environmental problems in the supply chain is to focus on waste prevention and control at the source through the greening of the purchasing function. This focus refers to the greening of suppliers, making sure that the raw materials they use are environmentally friendly, the production process they employ are clean, and green, the use of energy minimized, and the mode of delivery and transportation they adopt are relatively pollution free.

A major part of the in-bound logistics of the greening of supply chain could be achieved if the suppliers could be encouraged to turn become green. This is also a desired feature on the part of customers and other stakeholders of the industrial customer, who often do not draw a line between a company and its suppliers and often hold the end company responsible for the environmental liabilities of their suppliers. The greening of suppliers is also a major concern among many purchasing managers in leading edge companies. In a 1994 survey of U.S. companies, purchasing managers considered environmental and regulatory costs as their second most important economic concern. [Min and Galle, 1997] and this environmental concern has led to the greening of suppliers as well.

Greening the supply chain may be understood by industry as screening suppliers for their environmental performance and then doing business with only those that meet the regulatory standards. In fact, the driving forces for implementing the concept into the company operations comprise "reactive regulatory reasons to proactive strategic and competitive advantage reasons" (Sarkis, 1999). These evolving concepts also include working collaboratively with suppliers on green product designs, holding awareness seminars, helping suppliers establish their own environmental program and so on.

Formulation of a strategy to achieve a greening of suppliers and in-bound logistics is a challenge. This may be partly due to increased material cost and limited availability of qualified suppliers because of the need for non-traditional material and parts. Also, sometimes the recycled materials become more costly than the original virgin material because of the processing involved. Many times, industrial customers, with quality in mind, require the materials they buy to be made out of new or virgin material rather than out of reused material. Cox et al. (1998) observed that managers, including purchasing managers felt that recycled materials were more expensive. Over and above this barrier, many felt that they would have to continue to use new materials because their customers would like them to use it.

11.2 Brief Review of Literature

In their case based research on integrating suppliers into environmental management process, Walan, Handfield and Melnyk (1998) have observed two evolving trends in the business today. The first trend pertains to how environmental issues are becoming an intrinsic part of strategic planning agendas, perhaps due to stricter regulation and stronger requirement for environmental accountability. The second trend refers to the companies integrating their supply chain to bring down the operating costs and improve customer service. Combining these two trends it appears that companies must now involve suppliers and purchasers to contribute towards the environment performance of the system as a whole and thereby address the purchasing function's impact on the environment by using and obtaining environment friendly materials for the product design, improve and green suppliers' production process, evaluate suppliers using environment criteria and greening of all of the inbound logistics process. They have also explored the importance of management's commitment to the environment friendly performance in the supply chain and 'to move beyond environment compliance to achieve a proactive supply chain'.

Walton, Handfield and Melnyk (1998) have observed how companies traditionally have always tried not to take environment issues too seriously and in many cases adopted the practice of continuing to pollute and pay a small fine. When regulation got stricter following major environmental mishaps and public outcry, they adopted the practice of controlling pollution after it is produced—the-end- of- the- pipe approach. They gradually realized that instead of the above reactive approach, if they reduced/minimized the production of pollution at the source, before it was produced, it did help to save cost. In the next step they started to appreciate what tremendous marketing advantage and competitive edge it brought and started implementing proactive approaches to environmental performance in the form of EMS (Environmental Management Systems) and greening of the supply chain.

Thus to explore the role of suppliers in environmental management the authors of this paper considered five case studies in the furniture industry where the manufacturing process had significant environment implications. In these case studies the authors found that some of the companies were indeed working with suppliers to reduce emissions, monitoring the waste streams from suppliers, helping them to set up their environmental programs and even extend technical supports to suppliers to help them with conservation of natural resources.

In order to achieve overall environmental performance of the supply chain system the authors came up with a list of ten environmental   supplier evaluation criteria such as:

- Public disclosure of environmental record
- Toxic waste and pollution management
- Hazardous waste management
- ISO 14000 certification
- Reverse logistics program and so on.

In exploring green purchasing strategies Min and Galle (1997) considered a survey analysis for selected industry groups, which were heavy producers of scrap and waste materials. With a sample of 527 responses they found the key factors affecting buying firm's choice of suppliers as:

- Potential liability for disposal of hazardous materials
- Cost for disposal of hazardous materials
- State environmental regulations
- Federal environmental regulations
- Cost of environmentally friendly goods

Since purchasing is at the beginning of the green supply chain, green marketing efforts have to be tied up with the company's environmental goals and its purchasing activities .Min and Galle also found that green purchasing contributed significantly to source reduction of pollution in terms of recycling, reuse and low-density packaging and to waste elimination in terms of scrapping or dumping, recycling sorting for non toxic incineration and bio degradable packaging.

Regarding barriers and obstacles to green purchasing Min and Galle found high cost of environmental programs, uneconomical recycling and uneconomical reusing to be the top three most important barriers. Lack of management commitment, lack of buyer awareness, lack of supplier awareness, lack of company wide environmental standards or auditing programs come next. Finally the lack of state and federal regulation constitute the last set of barriers for effective green purchasing in the supply chain.

Sarkis (1999) provides a comprehensive view of the state of research in this evolving topic, tracing the work of researchers , who have investigated both conceptually and empirically , the issues involved , the reasons for incorporating these practices and also the way they have been practices in different organizations. According to him the supply chain system should include Purchasing and in-bound Logistics, Production, Distribution (outbound logistics and marketing) and reverse logistics .He also shows how firms focus on Total Quality Management (TQM) with its emphasis on improving product quality , zero defects ,customer satisfaction, training and employee empowerment etc. and integrate it with environmental management resulting in Total Quality Environmental Management (TQEM). This integration to TQEM enables the organizations to move towards the source reduction of pollution philosophies and improves environmental performance, marketing advantage and corporate image so that the company moves on to the world-class status.

For the Purchasing sector, Sarkis considers general green purchase practice, vendor selection and in-bound logistics. For green purchasing he mentions the work of Drumwright (1994) who conducted a field study of 10 organizations and attempted to determine why organizations go for green purchasing and what the characteristics of such organizations were. Green Purchasing comprises a number of environmentally based initiatives such as supplier environmental questionnaire, supplier environmental Audit and Assessments, environmental criteria on approved supplier list, requiring supplier to undertake independent environmental certification, jointly develop cleaner technology /processes with suppliers, engage suppliers in Design for environment, product/process innovation etc. (Lamming, et al. 1999; Lloyd, 1994)

To make these initiatives effective and successful the organization needs to integrate them with long-term strategic relationships of the organization, early involvement by the supplier and customer, building trust and early involvement of the supplier in designing for environment.

Again, there has been another new concept in greening of supplier (SCEM) in using the system of environmental mentoring for the suppliers, which has recently been introduced and researched in this field of study [Hines and Johns 2001]. While the usual method of using questionnaires, using environmental criteria in evaluating suppliers or demanding certified EMS from them were gaining widespread acknowledgement in the industry, this new method of bringing about the greening of suppliers is emerging to be very useful.   This concept of environmental mentoring refers to the development of a more fundamental relationship between the customer and the supplier and this culture goes beyond monitoring and evaluating and goes to the realm of guiding and supporting the suppliers. In order to make this happen, it really involves a sea change in attitudes by larger companies who try to build much closer working relationship with their suppliers, providing guidance, advice and assistance and sharing their knowledge and skill with them. That does this to improve the quality of the product and the operational efficiency of their technology and also with the goal of achieving world-class status and remaining competitive (Jenner, 1997:8).   Research also brings out the strengths and weaknesses of Mentoring and   Partnering method (Hines and Johns, 2001), the strengths being that the concept is proactive, non threatening, sharing benefits, building teamwork and so on, while the weakness comprises cost and resource investments, lack of physical facilities, need for mentoring skills, right people and so on.

Noting these research findings, various practices and issues let us take a look at how Asian-Pacific companies are dealing with greening of their supply chains.

11.2 South East Asia Issues and Study Questions

The numerous issues listed above plus some specific to the region are identified. A number of questions do arise and which will be addressed here. The following is a listing.

- What are the most prevalent initiatives, which are undertaken by companies in this region, in order to green their suppliers?
- To what extent do the companies urge their suppliers and contractors to incorporate environmental management into their operations?
- What are the driving forces/motivators for incorporating greening of the supplier chain in these companies?
- What are the problems and obstacles involved in the process of greening the suppliers?

Studying these concerns in this regional setting will help to determine how this concept can be successfully incorporated into companies.

11.4 Research Design

To investigate the concerns regarding the state of greening of the supply chain, an empirical research study was pursued. Using a survey questionnaire as the research instrument. The questionnaire was divided into different sections, encompassing basic information, environmental awareness, and greening the suppliers.

In the basic information section, the questionnaire asked about the industry the company is in, main manufacturing activity, the size of workforce and other organizational characteristics. In the awareness portion, the questionnaire investigates the company's and the employees' knowledge regarding environmental issues outside of the company. In the greening of suppliers portion the questionnaire investigates extensive issues regarding the topic starting from motivators, driving forces, obstacles and challenges encountered, to importance of the key factors that affect a buying firm's choice of suppliers.

The survey population included all manufacturing companies registered with the Management Association of the Philippines. The questionnaire was mailed to all of these companies, about 500 in all including Philippines, Indonesia, Thailand and Malaysia. The mailing was carried out in the second half of year 2000 and a second set was sent out by the first half of year 2001.

11.5 The Results From The Survey Research

The final sample size was 48 companies, with 28 respondents from the Philippines, 8 from Indonesia, 7 from Thailand 7 from Malaysia and 1 from Singapore. They were all manufacturing companies engaged in electronics and semi conductors, automation, metalworking, chemicals, food textiles and pharmaceuticals. This is not the great response rate, but some evaluation of these companies may still be carried out. 42% of the sample comprised small and medium enterprises (SMEs) 58% comprised large enterprises. Where the definition of small and medium enterprises are those having less than 500 employees.

11.5.1 Environmental Awareness

Regarding environmental awareness, the respondents were asked to rate the company's knowledge of environmental issues outside the company, the employee's knowledge of environmental issues inside the company, and the employee's knowledge of environmental issues outside the company, on a 4-point scale from Not knowledgeable to Very Knowledgeable.

85% of the companies said that they were knowledgeable enough or very knowledgeable regarding their knowledge of environmental issues outside the company.   Again 85 % of the companies said that their employees were knowledgeable or very knowledgeable regarding environmental issues inside the company. However, only 54 % of the companies said that their employees were knowledgeable or very knowledgeable regarding environmental issues outside the company. 14.5 % of the companies said that they did not have any environmental awareness programs and 8.5 % said that their environmental awareness program were under development. 3.5 % of the companies said that all of their departments have environmental awareness programs and 33.75 % said that some of their departments have initiated their own environmental awareness programs.

In industry it is often believed that larger companies have greater environmental awareness and undertake their own environmental initiatives. We wanted to test this concept based on the data collected in the research. In the questionnaire the size of the company was measured in terms of the number of employees. A Chi share test of independence was used to check if the size of the company had any effect on

- (B1) Company's knowledge of environmental issues outside the company
- (B2) Employee's knowledge of environmental issues inside the company
- (B3) Employee's knowledge of environmental issues outside the company

The results of the tests show that size did not have any impact on the company's, or employees', awareness regarding internal or external environmental issues.

11.5.2 Training Awareness Programs

Next the research proposed to check on whether the environmental awareness programs, adopted by the company have any effect on the company's as well as employee's awareness towards environmental issues both outside and inside the company. The level of implementation of environmental awareness programs in the company ranged from no awareness programs to all departments has their own awareness programs.   The Chi square test of independence was applied.

It was found that the environmental awareness programs have significant effect/ impact on company's awareness to environmental issues outside the company and also to employee's awareness of the environmental issues outside the company. Yet, the awareness programs didn't seem to have significant impact on employee's knowledge of environmental issues inside the company.   This result could be explained by the fact that most of the training programs talk about environmental issues which are global and general in nature and do not really deal with what is happening inside their own operations.

11.5.3 Greening of Suppliers

For the greening of suppliers issue, first an evaluation of the significant driving forces, which lead companies in South East Asia to go for greening of suppliers, was carried out. For this purpose, first all of the important motivators for incorporating greening of the suppliers were identified from the literature and various practicing companies. Table 11.1 summarizes the motivators.

Table 11.1. Motivators and driving forces for greening the supply chain

1. Customer Pressure	8. Improved relations with communities
2. Avoid potential export limitations	9. Enhanced brand image and reputation
3. Environmental improvement	10. Competitiveness
4. Reduced operating costs	11. Corporate Social and Environmental responsibility
5. Increased productivity and quality	
6. Capturing workers' knowledge	12. Access to capital
7. Improved relations with authorities	13. Financial performance improvement
	14. Increase in market share

The significance of each of the motivators and their impact upon the greening of suppliers initiative was also studied. Linear Discriminant Analysis (LDA) was used to study these relationships. LDA is briefly described in the next section, with results following. For each of the above motivator indicators the respondents were asked to give their ratings on a 4-point scale, from not important to very important. The respondents were then asked to check which of the items shown in Table 11.2, would the company have undertaken in their effort to green their suppliers.

Table 11.2. Initiatives on greening of suppliers

1. Holding environmental awareness seminars for suppliers
2. Guiding/helping suppliers to establish their own environmental programs
3. Bringing together suppliers in the same industry to share their know how and problems
4. Informing suppliers about the benefits of environment-friendly production technologies
5. Urging suppliers to take environmental actions
6. Choice of suppliers by environmental criteria
7. Requiring suppliers to adopt environment friendly practices
8. Arranging funds to help suppliers for their environment programs
9. Sending company auditors to appraise environmental performance and compliance of suppliers

While the importance and significance of the greening of suppliers is gaining wide spread acceptance and acknowledgement, it is also true that the process does face many hurdles and obstacles, especially in developing countries. To understand which obstacles become critical to the companies who want to implement the concept, the respondents were also asked to assess the role played by the obstacles listed in Table 11.3.

Table 11.3. Obstacles to green supply chain management

1. Lack of interest on the part of suppliers
2. Lack of supplier financial resources
3. Lack of supplier manpower resources
4. No competitive advantage
5. No resources to help suppliers
6. Difficult to organize
7. Lack of governmental support
8. Lack of technical knowledge to help supplier
9. Too costly

For evaluating the critical obstacles, here also we would use a LDA, which is now briefly discussed.

11.6 Linear Discriminant Analysis

This statistical technique is used for analyzing data in a format that is similar to traditional regression analysis. There is a criterion or dependent variable and a number of independent variables. The dependent variable, in a two-group discriminant analysis scenario, can take up only two values, 1 or 0. For instance, let us say that we would like to determine, out of all of the items included in the list of possible motivators, which are the ones, which significantly affect the implementation of greening of suppliers in a company. In such a scenario all of the items in the list of motivators become the independent variables and the dependent variable can be called a "greening of suppliers index", which can take up a value of '1' if the company has implemented the greening of suppliers and a value of '0' otherwise, that is if it has not.

The discriminant analysis model involves a linear combination of variables of the following form:

$$D= b_o + b_1X_1 + b_2X_2 + b_3X_3 + \ldots B_kX_k .$$

Where X_1, X_2 ...X_k are motivators such as customer pressure, reduced operating costs, increase in market share, etc., each of which has been rated by the respondent companies on the 4-point scale.

The dependent variable, D, which is also called discriminating variable or grouping variable may be measured as follows:

Consider the 9 items in greening of suppliers, Table 11.2. Each of these items are checked by the respondent company if it had implemented one of these items. If the respondent did check it , the item would have a measure of value 1 and 0 otherwise. We added up all of these 9 items, the total would have a maximum value of 9 and a minimum value of 0. We then defined our "greening of suppliers index", D, as '1' if this total came out to be more than 50%, that is more than 4.5 and '0' if the total is less than 4.5. Hence we were considering a two-group discriminant analysis scenario.

The model is run on SPSS 10.0, the software gives a statistic called Wilk's Lambda, for the overall model along with a chi-square value .The overall significance of the Wilk's Lambda must be less than 5 % for the model to be acceptable. A multistage approach is used where if any independent variable has a Wilk's Lambda with a significance of more than 5 %, it is dropped from the model and the model is run again. After all independent variables having Wilk's Lambda with unacceptable significance of more than 5 % are dropped, one looks at the Wilk's Lambda of the overall model again. If this has changed to an acceptable value, having a significance less than 5 %, the model is accepted and the surviving independent variables are accepted as critical independent variables, significantly affecting the grouping variable. These critical independent variables are able to most effectively distinguish the dependent variables into the two categories of '1' or '0'.

11.6.1 Applying Linear Discriminant Analysis on the Motivators/Driving Forces

Using all of the items in Table 11.1 as independent variables and using the "greening of suppliers index" as 'D' the grouping variable we ran the LDA. For the overall model the Wilk's Lambda had a significance of .432. This result is not statistically significant. For the individual independent variables, the Wilk's Lambda and the significances emerged as shown in Table 11.4.

Looking at the Wilk's Lambda for independent variables in Table 4, there are many independent variables Authority (Improved relations with authorities), Reduce (Reduced operating costs) etc, which have Wilk's Lambda with significance of more than 5 %. So the model was run again, this time dropping the independent variables, which had significance of more than 5 %. The final value of the Wilk's Lambda, for the overall model after dropping insignificant variables came out to have an acceptable significance of .02.

The final individual independent variables with their Wilk's Lambda's are shown in Table 11.5.

Table 11.4. Tests of equality of group means for LDA of independent variables

	Wilks' Lambda	F	df1	df2	Sig.
AUTHORITY	.979	1.001	1	46	.322
BRAND	.898	5.218	1	46	.027
CAPITAL	.983	.772	1	46	.384
COMMUNIT	.938	3.034	1	46	.088
COMPETE	.972	1.305	1	46	.259
CUSTOMER	.909	4.586	1	46	.038
ENVIRON	.926	3.661	1	46	.062
EXPORT	.964	1.698	1	46	.199
FINANCL	.975	1.182	1	46	.283
MKTSHARE	.958	2.003	1	46	.164
PRODUCE	.841	8.670	1	46	.005
REDUCE	.971	1.374	1	46	.247
SOCIAL	.921	3.968	1	46	.052
WORKER	.895	5.397	1	46	.025

Table 11.5. Final LDA model for discriminating motivator factors

	Wilks' Lambda	F	df1	df2	Sig.
Customer	.909	4.586	1	46	.038
Produce	.841	8.670	1	46	.005

Since both the independent variables have significance levels less than 5 % and the overall Wilk's Lambda is also acceptable, the finalized model emerges as:

$$D= \quad -2.291 + .087 \text{ Customer} + .764 \text{ Produce}$$

Hence the critical motivator variables that significantly affect the greening of supply chain initiative are Customer Pressure and Increased Productivity and Quality.

11.6.2 Applying Linear Discriminant Analysis for Obstacles to Greening of Suppliers

In the process of implementing greening of suppliers, even leading edge companies face many challenges and obstacles, as summarized in Table 11.3. In order to determine the critical discriminatory items LDA is used here too. Before, for the Motivators /Driving forces we had used as the grouping variable a variable called "greening of suppliers index" which took a value of '1' if companies did demonstrate a propensity towards greening of suppliers, and '0' otherwise. For the obstacles test, we wanted to develop a grouping variable that would capture the lack of propensity for a company to go for greening of suppliers. Hence this time we defined our grouping variable, D' = 1-D, so that when a company lacks propensity to greening of their suppliers D' = 1 and D' = 0,otherwise.

For this model to apply, we would look at the significance of the model given by the Wilk's Lambda for the overall model and also the individual Wilk's Lambda for all of the independent variables from Table 11.3.

Upon the first run the Wilk's Lambda for the overall model had a significance of .076. For the independent variables, the individual Wilk's Lambdas are shown in Table 11.6.

Table 11.6. Tests of equality of group means for independent variables of obstacles for greening the supply chain

	Wilks' Lambda	F	df1	df2	Sig.
Govt	.997	.118	1	46	.732
Interst	.846	8.382	1	46	.006
Manpower	.846	8.382	1	46	.006
Company	.900	5.086	1	46	.029
Costly	.974	1.233	1	46	.273
Organize	.948	2.544	1	46	.118
Resource	.935	3.193	1	46	.081
Technica	.984	.765	1	46	.386
Financia	.839	8.836	1	46	.005

Since the overall significance is greater than 5%, the model is not acceptable. Looking at the Wilk's lambda of the individual independent variables, we dropped from the model all of the variables whose significance levels were more than 5%, like Lack of government support, too costly lack of technical knowledge to help the suppliers etc. After some iterations, the final overall significant model had an acceptable significance of .037. Table 11.7 shows the final discriminatory obstacle factors.

Table 11.7. Final significant results of obstacle independent variables and their significance

	Wilks' Lambda	F	df1	df2	Sig.
Interest	.846	8.382	1	46	.006
Manpower	.846	8.382	1	46	.006
Financia	.839	8.836	1	46	.005

The critical obstacles which significantly lead to companies lack of initiatives to greening of suppliers are Lack of Supplier interest, Lack of supplier manpower resources and Lack of supplier financial resources.

Thus, most of the obstacles appear to be from the supplier who is lacking in interest, manpower and financial resources.

11.7 Corporate Green Supply Practice in South East Asia

The specific action plans taken by the companies in the initiative for greening of suppliers have already been summarized in Table 11.2. A frequency analysis will show us how prevalent these practices are. Table 11.8 provides the frequency distribution of these practices.

Thus we observe the most prevalent way in which companies are undertaking greening of suppliers is by evaluating and choosing suppliers using environmental criteria, where over 50% of survey companies are completing this practice. The next most prevalent action is urging the suppliers to take environmental actions, with 40% of the respondents practicing this activity.

Table 11.8. Frequency of various green supplier practices by survey respondents

Specific action in greening of suppliers	Percentage of companies who are carrying it out
Holding environmental awareness seminars for suppliers	29.17
Guiding/helping suppliers to establish their own environmental programs	16.67
Bringing together suppliers in the same industry to share their know how and problems	14.58
Informing suppliers about the benefits of environment -friendly production technologies	33.33
Urging suppliers to take environmental actions	39.58
Choice of suppliers by environmental criteria practices	52.08
Requiring suppliers to adopt environment friendly	35.42
Arranging funds to help suppliers for their environment programs	4.17
Sending company auditors to appraise environmental performance and compliance of suppliers	18.75

11.8 Concluding Remarks and Implications of the Study

The above research helps to understand the dynamics of greening of suppliers within the South East Asian context and provides many important insights; some of which were expected and some were not. For instance, the result that size of the company did not have a significant impact on the companies' as well as employees' awareness to environmental issues might appear somewhat unexpected. This result is because it is expected that larger companies which are usually more well established than smaller organizations, to indulge in going beyond their own business and environmental goals to reach out and involve their suppliers and business partners becoming green. However upon discussing this result with industry experts it emerged that there were many companies, not necessarily large, who were making significant efforts in greening of suppliers. To give an example, Nestle Philippines has started an extensive greening of the suppliers program, holding environment awareness seminars, arranging site visits, guiding them through Environmental Management Systems. In a recent presentation organized by Nestle Philippines their major suppliers (first tier suppliers) presented their environmental initiatives and the benefits they achieved. In these presentations many suppliers who were small or medium enterprises, talked about trying to green their suppliers (second tier suppliers). Hence the small size of the first tier suppliers did not have any impact on their implementing greening of suppliers on their part.

The result that environmental awareness programs had significant impact on companies' and employees' awareness to environmental issues outside of the companies is expected because it only implied that awareness programs were effective in bringing up the awareness level. Yet the programs were not effective in the employees' awareness of environmental issues within the company. This may happen because typical awareness programs emphasize global and national environmental issues and not environmental aspects and impacts within the company. The training for environmental aspects and impacts is usually confined to members of specialized steering committees while general employees are only given awareness of broad based issues. This finding may now help top management within companies to modify the general nature of awareness programs to include environmental topics specific to company the employees are working for.

The result pertaining to the significance of motivating factors from the linear discriminant analysis, brings out that customer pressure and increased productivity and quality concerns are two critical factors leading companies to greening of suppliers initiatives. From discussions with various companies operating in this region this result does make sense. In this region many forward-looking companies are now involved in doing business with European, Australian and American companies, who are their customers. These customers urge the South East Asian companies to have a proper environmentally and socially responsible image, along with requiring a well-established environmental management system (EMS) in their operations. If the South East Asian companies have a supply chain that is environmentally questionable these customers would immediately cease to do business with them. This explains why customer pressure is so important in leading companies to greening of the suppliers.

As for the urge to increase the productivity and quality, which also emerged as a critical factor in greening of suppliers, this is an internal reason leading companies towards this initiative. As has been widely observed, integrating and streamlining the company operations with those of the suppliers helps to improve productivity and quality by eliminating delays and inefficiencies and work stoppages on account of non-availability of raw materials, improper raw materials and so on. In many cases the environmentally hazardous components, if there are any, in the product can be eliminated by material substitution on the part of the supplier. Hence the urge to enhance productivity and quality in many instances does lead companies to green their suppliers and also integrate them into their business operations.

As for the obstacles in greening of suppliers the research identifies the three significant causes: lack of interest on the part of the suppliers, lack of supplier manpower resources and lack of supplier financial resources. This result indeed is very true in this regional setting because the suppliers in most cases are companies who are just attempting to survive, in the sense that their sole effort is geared to help their business performance. These companies have little interest in exerting effort that does not directly help their bottom line. Even when their customers do urge them to go for environment friendly production and operations, excuses are provided, typically relating to economic well being of the organization. It is also true that in many cases they neither have appropriate manpower nor availability of funds to carry on such initiative. This challenge, which creates an obstacle for suppliers to turn green, can be overcome only if they are convinced about the benefits of greening that again has to be carried out upon the initiative of the lead company.

In this research the motivating factors and obstacles to greening of suppliers were considered. However, research still needs to be carried out to examine the acceptability and feasibility of innovative systems of combining various companies and their suppliers in win-win situations so that production and its required inbound logistics work in unison. For instance many world-class companies are now contemplating integrating suppliers on a more permanent basis so that new product development, innovation in process design, material substitution to ensure cleaner production etc, could be achieved. Many such systems are already operating in different countries of this region and once their success is proven, these models can be replicated in other countries as well. However, before implementing such a model extensively, research has to evaluate the effectiveness of such a system.

References

Chayod, B .(1999), Impact of ISO 14000 certified companies on Business Performance. (Thailand Environment Institute ,Bangkok, Thailand).

Christmann, P. and Taylor, G. (1999) 'Globalization and the Environment', Paper presented in the Greening of Industry Network Conference, Chapel Hill North Carolina.

Cox, J., Sarkis, J, and Wells, W. (1999) 'Exploring organizational recycling market development:The Texas-Mexico border.' In: Charter M, Polonsky, M.J. (eds) Greener Marketing :A Global Perspective on Greening Marketing Practice, :Greenleaf Publishing: 381-394.

Eriksson, K., Johanson, J.,Majkgard A. and   D. Deo Sharma, (1997) 'Experiential Knowledge and Cost In the Internationalization Process'. Journal of International Business Studies.327-354.

Greeno, J.L. and Robinson, S.N. (1992) 'Rethinking Corporate Environmental Management'. Columbia Journal of World Business 27:222-232.

Hart, S.L. (1997) 'Beyond Greening: Strategies for a Sustainable World' Harvard Business Review. January - February,: 66-76.

Hayduk, L.A.(1987) Structural Equation Modeling with LISREL: Essentials and Advances. Johns Hopkins University Press. Baltimore.

Hayes, B. E. (1998) , Measuring Customer Satisfaction , ASQ Quality Press , Milwaeekee, Wisconsin .

Hocking, R.W.D. and Power, S. (1993) 'Environmental Performance: Quality,Measurement and Improvement' Business Strategy and Environment. 2:19-24.

ISO 14000 System of Standards, (1996).Environmental Management Systems–Specification and Guidance for Use . International Organization for Standardization, Switzerland .

Luken, R. (1997) 'Trade Implications for Environmental Management Systems' paper Presented in APO World Conference on Green Productivity, Manila ,Philippines

MCDonagh, P. (1994) 'Towards an Understanding of What Constitutes Green Avertising as a Form of Sustainable Communication'. Cardiff Business School, Working Paper in Marketing and Strategy.

Min, H. and Galle, W. (1997) 'Green Purchasing Strategies : Trends and Implications' .International Journal of Purchasing and Materials Management;August:10-17.

Ottman, J.E. (1992) 'Industry's Response to Green Consumerism'Journal of Business Strategy. 13 (4):3-7.

Paton, B., (1999) 'Voluntary Environmental Initiatives and Sustainable Industry' Paper presented in the Greening of the Industry Network Conference, Chapel Hill, North Carolina.

Peattie, K. and Ratnayaka, M.(1992) 'Greener Industrial Marketing' Industrial Marketing Management.21. 2:103-110.

Peattie, K.J. and Notley, D.S.(1989) 'The Marketing and Strategic Planning Interface' Journal of Marketing Management. 4.3:330-349.

Porter, M.E. and Van der Linde, C., (1995) 'Green and Competitive: Ending the Stalemate' Harvard Business Review. September – October:20-134.

Power, S.T. and Cox, C. (1994) 'Value-Driven Organizations: A look at the New Corporate Environmentalism' Greener Management International. 5: 29-35.

Rao, P. and Kestemont, M-P. (1998) 'The Environment Barometer: Doing Business in "Green" Asean' The Asian Manager. March-April:56-61.

Sarkis, J., (1999) "How Green Is the Supply Chain ? Practice and Research." Graduate School of Management. Clark University.MA.

Walley, N. and Whitehead, B. (1994) 'It's not Easy Being Green.' Harvard Business Review. 72.3: 46-52.

Walter, I. (1982) "Environmentally Induced Industrial Relocation to Developing Countries" in S.J. Rubin and T.R. Graham (eds), Environment and Trade, Allanheld, Osmun Publishers.

12

Environmental Initiatives in the Manufacturing Supply Chain: A Story of Light-green Supply

Lutz Preuss

School of Management, Royal Holloway College, University of London, Egham Hill, Egham, Surrey, TW20 0EX, United Kingdom, Lutz.Preuss@rhul.ac.uk

The research described in this chapter is concerned with the role supply chain managers of manufacturing companies can play in environmental initiatives in the supply chains of their companies. Given the increasing economic importance of the purchasing function, a detailed examination is made of the three elements that constitute supply chain management: the management of the transformation of materials, the management of information flows and the management of supply chain relationships.

12.1 Supply Chain Management and the Natural Environment

The supply chain may not be the jazziest of topics within business studies, yet it plays a vital role in any manufacturing organisation. As a result of the pressures arising from globalising competition on the one hand and advances in information technology and logistics management on the other, supply chain management has undergone a dramatic transformation in recent decades. Emanating from a clerical function which did little more than expedite other departments' orders, its focus has widened to encompass external value-adding benefits, such as ensuring that the company's supply base provides appropriate technology.

The function has changed from a tactical to a strategic orientation, which includes the integration of the supply strategy into the overall corporate strategy, the identification of threats and opportunities in the supply market and the involvement in the early stages of product design (Quinn, 2000; Schary and Skjott-Larsen, 2001; Burt et al., 2003; Preuss, 2005). Strategic management of the supply chain is linked to a longer-term perspective in the supplier-customer relationship, often referred to as partnership sourcing (Lamming, 1993; Krause et al., 1998; Gadde and Snehota, 2000), which contrasts with the traditional adversarial buyer-supplier relationship, characterised by a short-term, competitive sourcing approach and 'win-lose results' as the expected outcome.

Handfield and Nichols (1999: 2) define supply chain management as encompassing all the activities associated with the flow and transformation of goods from the raw materials stages to the final user, the associated information flows up and down the supply chain and the management of supply chain relationships with the aim of achieving a sustainable competitive advantage. The supply chain needs to be perceived as a system consisting of three interrelated elements: the actual organisations in the supply chain, their activities and the processes which link activities and organisations (Schary and Skjott-Larsen, 2001; similarly Lambert, 2001, and Hakansson and Snehota, 1995). Supply chain management, in turn, comprises three related task areas, upstream, internal and downstream functions.

The importance of supply chain management is illustrated by the fact that bought-in components and raw materials are the largest budget item for manufacturing companies. For North American companies Burt *et al.* (2003) estimate a purchasing share of some 65% of total expenditure, and similar figures have been reported for Europe. This level of expenditure provides supply chain management with a significant leverage effect. An even greater effect on company profitability may arise from joint design initiatives with suppliers, where cost is addressed by changes to the supplied item during the design stage. Supply chain management also has a boundary-spanning role and thus can influence both the profitability and the image of an organisation. It is clearly not the only function in this position, but it is well placed to start adaptation processes in response to external changes. One mobile phones manufacturer in the sample for this study, for instance, was informed by its suppliers of rising prices for palladium and instigated efforts to replace it with a artificial material. Companies have to react to such changes sooner or later, but often the signs for an imminent change could have been picked up earlier and valuable time and resources could have been saved.

The main question this chapter addresses is whether the growing importance of supply chain management in economic terms is matched by a growing importance in environmental protection. A number of authors, like Walton, Handfield and Melnyk (1998), have established a normative claim that proactive companies must integrate the supply chain into their environmental management. Indeed, supply chain managers are central to greening the supply chain: Being responsible for the entire materials flow throughout the supply chain, they are the primary change agents in making decisions about the procurement and disposition of materials (Handfield and Nichols, 1999; Preuss, 2005).

Applying the definition by Handfield and Nichols (1999) above, greener supply chain management should address three interrelated task areas, upstream, internal and downstream of the organisation. Upstream of the organisation, supply chain managers can address the environment in the supplier selection and evaluation criteria and in the specifications for components; they might be involved in green joint design activities with suppliers or product life-cycle analyses. Within the organisation, supply chain management might be involved in initiatives like design for environment, the establishment of an environmental management system or the handling of products. Downstream of the organisation supply chain management is often charged with the responsibility for disposal and the sale of excess stock, including opportunities for recovery and recycling of materials. Another area for potential environmental improvements is logistics, where a balance needs to be struck between the requirements of just-in-time delivery and environmental criteria (Lamming and Hampson, 1996; New, Green and Morton, 2000; Preuss, 2005).

Such environmental initiatives might require supply chain members to reduce the amount of harmful substances that are used in the manufacture of a subassembly or to change their processes. The respective supplier would have to impose a corresponding requirement on its own suppliers, who in turn impose it on theirs, all the way along the chain. Once the investment in facilities of a higher environmental standard is made, the suppliers at the various stages could use their green credentials to mentor other potential customers and the environmental effect would spread into the supply chains of other manufacturers. Supply chain management could achieve a green multiplier effect all the way along – and beyond – the supply chain and become an agent of change that is more effective than any other function in the corporation (Preuss, 2001, 2005). Indeed, such a multiplier effect need not be limited to environmental issues, it could apply to any form of socially responsible buying.

The general possibility of greener supply established in the conceptual models has been confirmed in a number of case reports. These are often success stories from large corporations, such as Ciba, IBM or Procter & Gamble, which are involving the supply chain in improving their environmental performance (Lamming et al., 1996; Russel, 1998). Xerox developed an *environmental common denominator*, which conflates the environmental impact of a product across the supply chain into a single measure, as otherwise environmental developments, such as recycling, optimising transport flows or ISO 14001 accreditations, might merely happen in isolation from each other (McIntyre et al., 1998; see also the studies by Green et al., 1996, and Carter and Dresner, 2001).

The literature on corporate greening is not without its critics. It has been called "technicist" in orientation (Newton and Harte, 1997), because the successful change towards a more environmentally friendly organisation is presented as being just a matter of adopting the appropriate environmental technology and management system. In reality, however, many of the instruments may not correlate with environmental improvement. Accreditation to ISO 14001 could in theory acknowledge little more than legal compliance (Freimann and Walther, 2001). While the benefits of adopting environmentally conscious measures are extensively spelled out, the disadvantages remain underexposed. Equally, the idea that there may be organisational conflict merges into the background. Furthermore, the availability of clean technology varies between industries and companies. Some industries, especially mature ones, are operating near the theoretical limits of resource efficiency and would not have much scope for further improvement. The success stories are also mostly of large corporations, whereas small companies often lack resources for environmental initiatives and are only at a stage of compliance with legislation (Howes et al., 1997).

The literature thus presents ambivalent results. There is some evidence for an extensive involvement of supply chain managers in environmental protection initiatives. Yet this concerns in particular large corporations and companies in some specific sectors, such as chemicals, which have been under intense environmental scrutiny (Hill, 1997; Russel, 1998). This raises the question whether environmental innovation may just pick the low-hanging fruits and whether the successful initial steps do indeed lead to a continuous improvement process.

12.2 Methodology and UK Sample

This chapter seeks to offer a more detailed examination regarding the link between supply chain management and environmental protection. One aim is to confirm or reject the hypothesis that the increasing economic importance of supply chain management has lead to a greater contribution of supply chain management to environmental protection. Given the ambivalent outcome of the literature review, the answer is unlikely to be a straightforward confirmation or rejection. Thus a qualitative study has been used to arrive at a richer account. The study partly follows preconceived topics, but several additional issues emerged during the interviews, leading to a more grounded approach which identifies central themes as they arise from the data.

In order to reduce complexity and to concentrate on the most important decision-making processes within the supply chain, the analysis moves below the level of the whole supply chain system and focuses on the purchasing function within the supply chain. This is justified since purchasing managers, apart from being in a critical position to influence the size of the environmental impact of a company and its ability to gain or maintain a competitive advantage from the natural environment, are the most important decision-makers within the overall supply chain. Purchasing is, after all, the management activity that links all the stages of manufacturing within a supply chain together (Schary and Skjott-Larsen, 2001; Zsidisin and Siferd, 2001).

Applying the three themes in the definition of supply chain management by Handfield and Nichols (1999) above, this study will offer a snapshot of the current situation regarding environmental protection in the manufacturing supply chain by focusing on the way purchasing and supply chain managers in manufacturing companies manage 1) the transformation and flow of materials, 2) the associated information and 3) the relationships in the supply chain. This study will thus ask to what degree the transformation and flow of materials is influenced by supplier assessment criteria which include the natural environment and whether supplier accreditation to environmental management standards plays any role in the selection of suppliers. In terms of managing the information flow in the supply chain, the study will investigate what types of environmental issues purchasing managers perceive to be important. Regarding the management of relationships, the question will be raised whether there is evidence of a shift from adversarial or arm's length to cooperative relationships in the environmental interaction.

In contrast to previous studies, which concentrated on individual companies or specific sectors (Green et al., 1996; Hill, 1997; Russel, 1998; Handfield et al., 1997), this study aimed to cover a broad spectrum of manufacturing organisations within the United Kingdom, excluding the food and drinks industry. As a secondary criterion, a variation in terms of size, ownership and location was aimed for too. Product classes were identified according to the *Standard Industrial Classification of Economic Activity 1992* (Central Statistical Office, 1992), and the actual companies were drawn from a *Digital directory of Scottish manufacturing capability*, published by Scottish Enterprise (1997), from Dunn & Bradstreet's (1999) *Scottish business register* as well as from Kompass (2000) *UK Kompass Register*.

In total, procurement managers from 50 companies were approached for an interview, and 34 agreed to participate (for details regarding products, ownership and size see Tables 12.1. and 12.2.). Their companies include manufacturers of:

- electronics, including manufacturers of components and semiconductors;
- cars and automotive components;
- mechanical engineering, hydraulic equipment, electric motors;
- cables and antennae;
- plastics, paper and textiles;
- chemicals: agrochemicals, pigments and dyes, petrochemicals;
- maritime vessels.

The interviews were conducted in three stages, a pilot stage (Companies 1 to 6) during Summer 1998, the study proper (Companies 7 to 29) between Summer 1999 and Spring 2000 and a follow-on stage (Companies 30 to 34) between Spring and Autumn 2004. The interviews lasted between 30 minutes and two hours. They were conducted on company premises, taped and later transcribed. Due to time constraints the purchasing managers of two companies offered a telephone interview only, during which extensive notes were taken, which were written up immediately afterwards. The interviews followed a semi-structured format, so that important issues and topics were covered, but they also offered sufficient flexibility to allow respondents to present their own perceptions (for the schedule of interview questions see Appendix). Interview data were supplemented with information from company-internal sources, such as purchasing policy documents, vendor evaluation forms, environmental reports and promotional documents, or material provided externally by, for example, government departments, regulators or the media.

The 16 companies that did not respond or declined to participate were in the majority small enterprises, often with no full-time purchasing manager, or companies in extraordinary financial difficulties, which could not afford the time to give an interview. The sample does still include a number of small and medium-sized companies, and the danger of bias from non-response is thus slight. In view of the overall numbers, the results should not be interpreted in any quantitative sense. Since the production processes are not unique to the United Kingdom and since the plants are owned by US, Japanese, continental European as well as UK companies, the results should be broadly representative of manufacturing in industrialised countries.

12.3 Managing Materials Transformation: Supplier Assessment Criteria and ISO 14001

The study of the UK manufacturers yields several examples where companies have addressed environmental concerns in bought-in components or in the supplier production processes. One manufacturer of electronic test equipment (Company 11) was facing new legislation banning a heavy metal, hexavalent chromate, which had been the standard treatment for the metal casing of its product. Test and measurement equipment is used to calibrate other electronic products and hence needs to be much better insulated than the average consumer electronics product. Following the publication of a draft European Union Directive on the reduction of hazardous substances in electronic and electric products, the test and measurement company asked its sheet metal supplier to develop an alternative to hexavalent chromate which satisfies the new legal requirements while still giving the necessary electronic insulation. The sheet metal supplier developed an alternative based on rare earth metal Zirconium, which is more inert and hence less carcinogenic than chromate.

The work for this environmental initiative was entirely undertaken by the supplier. It received no financial assistance from the test equipment manufacturer towards the cost of the new technology, apart from some intangible support from the marketing of successful initiatives. The initiative to replace chromate did not lead to a more positive relationship between supplier and customer. Roughly at the same time as the initiative was introduced the customer was restructuring its supply base. A number of companies which previously supplied directly to the customer were required to instead supply to systems suppliers, and these further assemble the components before delivering them to the test equipment company. Although there seemed to be an organisational fit between the test equipment manufacturer and the sheet metal business, as both organisations are trying to be seen as active in environmental issues, the innovation is unlikely to halt or reverse a 'relegation' of the sheet metal supplier to the second division of the supply base (for more details, see Preuss, 2005).

The example demonstrates that initiatives regarding environmental protection are taking place in the supply chains of manufacturing companies. Yet do such examples represent more than anecdotal evidence from a few individual organisations? Since the aim of this study is to consider the spread of environmental initiatives across the manufacturing sector, a more comprehensive approach is sought. Rather than looking at individual examples of environmental initiatives, this is achieved by studying supplier selection and evaluation criteria and any role environmental management standards play in selecting suppliers across an array of manufacturing organisations. If purchasing and supply chain managers are serious about the environment, then arguably, environmental credentials of suppliers should emerge as one criterion that strongly influences the selection and evaluation of suppliers. Otherwise purchasing would simply not be able to register the outstanding environmental performance of a supplier.

However, the sample did not produce a consistent reference to the environment in the supplier selection and evaluation criteria. As Table 12.1. shows, conventional factors centering around quality, price and delivery clearly dominate the assessment of suppliers in the sample. The environment is only addressed in the supplier evaluation criteria of ten companies. This subset includes all the paper makers and car manufacturers in the sample, making these two the industries with the most consistent attention to the environment in the supplier assessment criteria. The other companies are two electronics manufacturers, one petrochemical corporation and one manufacturer of linoleum and vinyl floor coverings. Against a backdrop of low attention to the environment in general, paper making, car manufacture, electronics and chemicals emerge as the industries where the environment is most likely to play a role in shaping the selection of suppliers and hence the materials flow in the supply chain.

It needs to be asked, however, whether the inclusion of the environment into the supplier selection criteria is effective. The materials manager of an electronic test equipment manufacturer (Company 11), one of the exemplary firms which does include the environment in the supplier assessment criteria, could not cite any cases where the environmental performance of a supplier had been the reason for ending the relationship. However, the environment does play a role in a more limited sense: "We probably have examples where we have not proceeded with an evaluation, because we didn't like the processes or the control of the processes ... we have probably not selected someone because of their environmental performance, rather than taken them off [the list of suppliers]". In other words, the environment does serve as a threshold suppliers have to meet, but where several suppliers meet the requirements the selection reverts back to the conventional criteria of price, quality and delivery conditions.

This leads to the question whether accreditation to an environmental management standard could be a way to influence the materials flow in the supply chain in a greener direction. Accreditation to an environmental management standard is becoming increasingly common, in particular for larger manufacturers (Steger, 2000; ISO, 2005), and there is evidence that supply chain pressure can encourage companies to seek accreditation. An antennae manufacturer (Company 7) underwent the first phase of inspection for ISO 14001 in August 1999, and one of the reasons was indeed enquiries from customers regarding environmental management standards. One of the automotive manufacturers (Company 33) asked its first tier suppliers to get accredited to ISO 14001 by the end of 2004. As the first tier are likewise large corporations, this requirement did not lead to any resistance. In all the other manufacturers, however, supplier accreditation is currently not mandatory. As the purchasing manager of a group of quality paper mills (Company 23) explains, "to try and insist that ... you only deal with people who are qualified to ISO 14001 is not realistic at the moment. It will come, because it is becoming very widely accepted, but not yet."

This study of UK manufacturers has once more illustrated that the natural environment is a complex topic. There are individual examples of companies addressing environmental problems in their supply chains, where they substitute hazardous materials, reduce the amount of packaging or require suppliers to reach a certain environmental threshold. At a first glance the formal inclusion of the environment into supplier evaluation seemed to be an instrument for supply chain managers to aid such a greener materials flow in the supply chain by selecting suppliers of an above-average environmental performance. At a second glance, however, the current involvement of supply chain managers in a more environmentally friendly materials flow must be considered to be rather more limited. Most companies in the sample select suppliers without any reference to their environmental performance, while the few exemplary companies are concentrated in sectors which are in the limelight due to their public image as heavy polluters.

Table 12.1. Supplier selection and evaluation criteria of UK manufacturers

Company	Position of interviewee	Ownership and headquarters	Product	Employees at site	Criteria for supply selection or evaluation
1	Purchasing manager	Family-owned	Precision sheet metal work and assembly	201-300	Price, quality, delivery, communication, technology, financial situation, location
2	Supply base manager	Multinational PLC, HQ in US	High-performance computer workstations	401-500	Quality (30 points of 100), technology (25), delivery (30), support (15); multiplied by price index; trend analysis
3	Senior buyer	Multinational PLC, HQ in Japan	Video cassette recorders	601-700	Price most important for selection; for review: delays affecting production, rejected deliveries (30 points of 100 each), flexibility, documentation, communication, punctuality (10 each)
4	Purchasing manager	Limited company, one site	Mechanical engineering for off-shore oil	ca. 100	Quality, price, delivery
5	Assistant manager	Multinational PLC, HQ in Japan	Manufacture of semiconductors	1001-2000	Quality, cost, delivery
6	Commercial manager	Limited company, one site	Self-adhesive labels and tags in wide range of	51-100	Location, support, innovation, quality, delivery, price, market awareness
7	Materials manager	Limited company; HQ in US	Antennae and cables	200-300	Price, quality, delivery
8	Purchasing and production control manager	UK-based Limited company	Specialist cables	400-500	Quality, price, service, range of items; also responsiveness and potential for development

Table 12.1. Continued

Company	Position of interviewee	Ownership and headquarters	Product	Employees at site	Criteria for supply selection or evaluation
9	Purchasing manager	Multinational PLC; HQ in EU	Tyres	ca. 800	(Centralised buying for 90% of raw materials; alternative suppliers only when lower price)
10	Supply chain development manager	Multinational PLC; HQ in UK	Military electronics	300–400	Cost, status of supplier (key, preferred, non-approved), quality, delivery,
11	Materials manager	Limited company; HQ in US	Electronic test and measurement equipment	ca. 1300	Best overall value: technology, quality, responsiveness, delivery, total cost, *environment*
12	Purchasing manager	Medium-sized limited company	Electric motors	200	Price, quality, financial situation
13	Purchasing manager	Medium-sized; Family-owned	Hydraulic equipment	ca. 300	Cost, delivery, quality or previous performance, ISO registration
14	Purchasing manager	Multinational PLC; HQ in EU	Tyres	1000	(Centralised purchasing of raw materials; local purchasing for engineering parts, site services, etc.)
15	Purchasing manager	Multinational PLC; HQ in US	Mobile phones	over 2000	Price, quality, capability, capacity, flexibility
16	Procurement manager	Multinational limited company; HQ in Japan	Printers and faxes	ca. 1000	Technology, quality, financial stability, *environment*, after these cost
17	Purchasing manager	Multinational PLC; HQ in US	Variety of printed circuit board assemblies	900–1000	Approval by customer and cost.

Table 12.1. (continued)

Company	Position of interviewee	Ownership and headquarters	Product	Employees at site	Criteria for supply selection or evaluation
18	Purchasing manager	Limited company; HQ in UK	Fine printing and speciality paper	over 800	Technical criteria, price, availability, logistics, check that supplier's *environmental* policy is in keeping with own
19	Manager of engineering purchasing	Multinational limited company	Variety of speciality chemicals	900	Best overall deal across: safety, specification, quality, delivery, price, reliability, service
20	Head of purchasing	Multinational PLC; HQ in Switzerland	Pigments and dyes	800	Cost reduction targets, stock turnover speed targets, price
21	Mill manager	Medium-sized limited (part of larger group)	Quality paper	550-600	Quality, *environment*, safety, 'Price comes into it as well, but price is not the only consideration.'
22	Purchasing manager	Limited company (part of Dutch group)	Linoleum and vinyl floor coverings	450	Quality, price; *environment*
23	Head of procurement	Multi-site PLC; HQ in UK	Quality paper	ca. 800	*Environment*, quality, performance, delivery, investment policy, financial situation
24	Purchasing manager	Multinational limited company	Commercial ships	ca. 1500	'You can't separate price, quality and delivery'; track record
26	Purchasing officer	Medium-sized (but part of larger PLC)	Cashmere yarn	250-300	Specification, price, payment terms, delivery performance, quality

Table 12.1. (continued)

Company	Position of interviewee	Ownership and headquarters	Product	Employees at site	Criteria for supply selection or evaluation
27	Purchasing manager and member of vendor management group	Multinational limited company; HQ in US	Specialist plastic fibres	ca. 460	Supplier quality assurance, service, profitability, past history, investment plans, attitude towards training
28	Procurement specialist	Multinational PLC; HQ in UK	Petrochemicals	ca. 1120	Safety, *environment*, operational reliability, price, quality
29	Senior information engineer	Small	Eye examination equipment	ca. 45	Skills, quality, cost, attitudinal issues, ability to cope with change, financial stability, capacity
31	Purchasing manager	UK owned	Packaging machines	ca. 175	'Buy top quality products at the lowest price'
32	Purchasing manager	Japanese-owned MNC	Motor vehicles	3,800	Cost competitiveness, technology, quality of management systems, financial robustness, *environment*
33	Director of purchasing	ultimately US-owned	Sports and utility cars	(11,000 for entire company)	Global price competitiveness, delivery performance, quality, technical ability, certification to *environmental* management standard from 2005
34	Supply chain and Purchasing managers	North American-owned	Tissue paper	500	Delivery, quality, price, technical capability, financial stability, investment plans, *environment*; training and development for staff

Source: Interviews with supply chain managers, Summer 1998 (Companies 1 to 6), Summer 1999 to Spring 2000 (Companies 7 to 30), Spring to Autumn 2004 (Companies 31 to 34). The supply chain managers of Companies 25 and 30 granted telephone interview only and time constraints did not allow this question to be asked.

Even in these cases environmental issues are only considered in the early stages of the supplier selection process. If several suppliers meet the threshold, or once the relationship with a supplier is established, the environment does not seem to influence the relationship any longer. The same is true for accreditation to an environmental management standard, as for most companies the pool of suppliers with ISO 14001 accreditation is still too small. For the purchasing manager of a computer peripherals manufacturer (Company 16) insisting on supplier accreditation to ISO 14001 would mean that "we would restrict ourselves beyond what I call a reasonable competitive advantage." The chances that the environmental credentials of a supplier of above-average standing are detected and that the materials flow is channelled in a more environmentally benign direction are thus rather slight at present.

12.4 Managing Information: Environmental Issues of Concern to Purchasing

A second aspect in Handfield and Nichols' (1999) definition of supply chain management concerns information, and correspondingly environmental supply chain management must be concerned with the management of information on environmental issues in the supply chain. As a starting point for such a debate, this section will enquire what the current view of purchasing and supply chain managers is regarding environmental issues in their supply chains.

A small number of purchasing managers in the sample claimed that their supply chains do not raise any environmental problems, yet the majority of managers were able to list a number of environmental issues in their supply chains which they see as important (see Table 12.2). Unsurprisingly, the environmental problems differ between industries and companies. Where the main issue in the supply chain of a computing hard- and software corporation (Company 2) is hazardous chemicals in printed circuit boards, problems for paper makers centre around clear cutting and biodiversity in the wood pulp supply or the toxicality of bought-in chemicals (Companies 18, 21 and 23), and an automotive company sees major issues in the recyclability of car components, in packaging, paint and in water consumption (Company 32).

Despite these differences, upon closer examination, similarities become visible. For example, the list of companies which reported packaging to be an environmental issue includes manufacturers of products as diverse as video cassettes (Company 3), tyres (Companies 9 and 14), electric motors (Company 12), paper (Companies 18 and 21), cars (Company 32) or specialist textile fibres (Company 27). Although concerns vary among purchasing managers in different companies and industries, four general issues are referred to with regularity: packaging, waste, hazardous production materials and paints and solvents (for more details, see Preuss, 2005; similar results are reported by Bowen et al., 2001).

Packaging is the most frequently mentioned issue and the study yields three different approaches to packaging: re-using, reducing or recycling. Prominent examples for re-using packaging come from large manufacturers in the electronics industry. A multinational mobile phones manufacturer (Company 15) replaced all cardboard packaging for phone casings with returnable plastic containers. The initiative was undertaken primarily for quality reasons, as the packaging introduced cardboard particles into the production process, which caused a significant amount of reworking at the end of the line.

It should be noted that the issues cited most often – packaging and waste – directly concern neither the production processes of the supply chain members nor the environmental characteristics of the finished product. They are also of lower financial importance than the expenditure on materials or components. The environmental impact of a procured material or component is, of course, not related to its price, but one important motivation for environmental initiatives is cost savings. In other words, where environmental issues are of low financial importance, managers – including purchasing and supply chain managers – are less likely to address these.

Another interesting result emerges when the environmental perceptions of supply chain managers are compared with those of other corporate functions. James *et al.*, (1997) studied the perceptions of environmental and production managers regarding the importance of several environmental issues in a number of medium to large UK companies in the three sectors of chemicals, high technology manufacturing and light manufacturing (see Table 12.3.). Some differences emerge between production and environmental managers, for example the use of products and packaging rank among the areas of highest importance for production but not for environmental managers, yet broadly speaking environmental and production managers agree in their evaluation of environmental issues.

Linking these findings to the purchasing function, it can be argued that there is some discrepancy between the perception of supply chain managers on the one hand and environmental and production managers on the other. The use of raw materials is a source of some concern for all three managerial groups, yet packaging, the issue mentioned most frequently by purchasing managers ranks highly for production (ranked 4th) but not for environmental managers (ranked 9th). There is a greater contrast in the attitudes towards waste, which emerged as second most important concern of purchasing and supply chain managers, but does not rank highly for either environmental or production managers (ranked between 10th and 14th). Conversely transport, which was given a high ranking by both environmental and production managers (ranked 2nd by both), does not figure among the important concerns of most purchasing managers.

The small number of respondents limits the interpretation of these results, but there seems to be tentative evidence that the perceptions of supply chain managers regarding environmental issues contrast somewhat with those held by environmental and production managers. This finding has repercussions for corporate responses to environmental challenges in the supply chain, as effective action necessitates information on which to act on. Here cross-functional teams would arrive at a more comprehensive evaluation of environmental issues than the individual managerial groups on their own. To conclude here, purchasing managers in the sample did show some awareness of environmental issues in their supply chains.

Table 12.2. Perceived environmental issues in the supply chains of UK manufacturers

Company	Position of interviewee	Ownership and headquarters	Product	Employees at site	Environmental issues in the supply chain
2	Supply base manager	Multinational PLC, HQ in US	High-performance computer workstations	401-500	Harmful chemicals in printed circuit boards
3	Senior buyer	Multinational PLC, HQ in Japan	Video cassette recorders	601-700	Packaging, waste
6	Commercial manager	Limited company, one site	Self-adhesive labels and tags in wide range of materials	51-100	Solvent-based adhesives
7	Materials manager	Limited company; HQ in US	Antennae and cables	200-300	Scrap metals, chemicals and paints, timber, packaging, plastics from cables, oil, waste disposal
8	Purchasing and production control manager	UK-based Limited company	Specialist cables	400-500	Packaging, waste
9	Purchasing manager	Multinational PLC; HQ in EU	Tyres	ca. 800	Packaging, chemicals, waste (esp. scrap tyres), water-based alternatives to solvents
10	Supply chain development manager	Multinational PLC; HQ in UK	Military electronics	300-400	Replaced some heavy metals and radio-active materials; replaced solvents with water-based paint; semiconductors; packaging

Table 12.2. (continued)

Company	Position of interviewee	Ownership and headquarters	Product	Employees at site	Criteria for supply selection or evaluation
11	Materials manager	Limited company; HQ in US	Electronic test and measurement equipment	ca. 1300	Harmful materials in printed circuit boards; move away from solvent-based paint; product take-back at end of life; packaging
12	Purchasing manager	Medium-sized limited company	Electric motors	200	Packaging, replace solvent-based varnish
13	Purchasing manager	Medium-sized; Family-owned	Hydraulic equipment	ca. 300	Paper, wood, paints and solvents, packaging; energy
14	Purchasing manager	Multinational PLC; HQ in EU	Tyres	1000	COSHH controlled chemicals; safety in use; waste (esp. scrap tyres)
15	Purchasing manager	Multinational PLC; HQ in US	Mobile phones	over 2000	Packaging, waste management, paint; printed circuit boards now use organic finish; replace palladium
16	Procurement manager	Multinational limited company; HQ in Japan	Printers and faxes	ca. 1000	Pallets, packaging, product take-back
17	Purchasing manager	Multinational PLC; HQ in US	Variety of printed circuit board assemblies	900–1000	Waste recycling, recycling electronic components, consumables, packaging, substances dangerous to employee health
18	Purchasing manager	Limited company; HQ in UK	Fine printing and speciality paper	over 800	Pulp: clear cutting, biodiversity, effluents; water; chemicals: employee health; recycling; packaging

Table 12.2. (continued)

Company	Position of interviewee	Ownership and headquarters	Product	Employees at site	Criteria for supply selection or evaluation
19	Manager of engineering purchasing	Multinational limited company	Variety of speciality chemicals	900	Safety and health issues, CE mark
20	Head of purchasing	Multinational PLC; HQ in Switzerland	Pigments and dyes	800	Hazardous materials, returnable containers for chemicals, reduce landfill
21	Mill manager	Medium-sized limited (part of larger group)	Quality paper	550-600	Pulp: clear cutting, biodiversity, effluents; water; chemicals: employee health; recycling; packaging
22	Purchasing manager	Limited company (part of Dutch group)	Linoleum and vinyl floor coverings	450	PVCs and plasticizers; solvent-based paint; waste; packaging
23	Head of procurement	Multi-site PLC; HQ in UK	Quality paper	ca. 800	Pulp: clear cutting, biodiversity, effluents; water; chemicals; recycling; energy use
24	Purchasing manager	Multinational limited company	Commercial ships	ca. 1500	Paint and its containers
26	Purchasing officer	Medium-sized (but part of larger PLC)	Cashmere yarn	250-300	Chemicals, oils, packaging
27	Purchasing manager and member of vendor management group	Multinational limited company; HQ in US	Specialist plastic fibres	ca. 460	Wood in packaging; recycle chemicals left after production

Table 12.2. (continued)

Company	Position of interviewee	Ownership and headquarters	Product	Employees at site	Criteria for supply selection or evaluation
28	Procurement specialist	Multinational PLC; HQ in UK	Petrochemicals	ca. 1120	Production chemicals, solvent-based paints, Aerosols, returnable packaging, use of eco-labelled goods, fuel efficiency, landfill, sustainable forestry
29	Senior information engineer	Small	Eye examination equipment	ca. 45	Electromagnetic compliance, waste, obsolete stock
31	Purchasing manager	UK owned	Packaging machines	ca. 175	Packaging, waste, paints, solvents
32	Purchasing manager	Japanese-owned MNC	Motor vehicles	3,800	End-of-life vehicle directive, recyclability of car components, packaging, paint, power and water consumption
33	Director of purchasing	ultimately US-owned	Sports and utility cars	(11,000 for entire company)	Heavy metals, recyclability of components
34	Supply chain and Purchasing managers	North American-owned	Tissue paper	500	Chemicals, woodpulp, effluents, water use, transport

Source: Interviews with supply chain managers, Summer 1998 (Companies 1 to 6), Summer 1999 to Spring 2000 (Companies 7 to 30), Spring to Autumn 2004 (Companies 31 to 34). The question schedule for the pilot study (Companies 1 to 6) did not include this question. The supply chain managers of Companies 25 and 30 granted telephone interview only and time constraints did not allow this question to be asked. Hence companies 1, 4, 5, 25 and 30 are omitted..

Table 12.3. Environmental areas of importance to UK environmental and production managers

Environmental issue	Environmental managers	Production managers
CO$_2$ emissions	3.23	3.67
Transport	3.15	3.56
Land use/planning	3.15	3.00
Noise nuisance	2.92	2.44
Use of raw materials	2.62	2.89
Contaminated land	2.54	2.44
Use of products	2.50	3.44
Disposal/recycling of products	2.46	2.33
Packaging	2.15	3.11
Risk assessment	1.92	1.78
Use of energy	1.85	1.89
Waste water	1.85	1.67
Waste disposal	1.77	1.56
Waste recycling/recycling of materials	1.69	2.00
Emissions to air	1.46	1.11

Source: James *et al.* (1997: 36 and 37). 5-point scale: 5 = very important, 1 = not important

Yet of the four issues that were referred to most often – packaging, waste, hazardous production materials, paints and solvents – two do not concern the environmental performance of suppliers directly and thus only marginally improve the environmental performance of the entire chain.

12.5 Managing Relationships: From Adversarial Approach to Cooperation

As supply chain management moves towards an increasingly strategic orientation, it was argued in Section 12.1, the relationship a company has with its key suppliers should develop from the traditional arm's length or adversarial relationship to a more cooperative approach, which offers a longer-term outlook and a 'win-win' scenario. The call for a more cooperative relationship can be transferred to the way in which environmental issues are addressed in the supply chain. Apart from the option of not addressing the environment at all, a spectrum exists between an arm's length approach, where the customer dictates the required level of environmental improvement without much consultation of the supply chain, and a cooperative approach, where environmental issues are dealt with by interaction of the supply chain members. Such an active approach might even lead to a 'win-win-win' scenario, where both companies and the environment win (Elkington and Knight, 1991).

There were some purchasing managers in the sample who claimed that their organisations do not have a significant impact on the environment. The purchasing manager of a cable maker (Company 8) stated: "It tends not to be a big problem for us, in that the processes we have don't impinge on the environment unlike some chemical industries, where there are a lot of emissions, etc." Such claims, however, are as rare as cooperative supply chain relationships. Examples of the latter are provided by initiatives in the chemical industry to tackle environmental issues around the replacement of one-way containers for hazardous chemicals with reusable ones. Since the containers are contaminated with the chemicals, they have become hazardous themselves, and the reusable alternative significantly reduces the need for disposing of them. One such initiative has been undertaken by a pigments plant, which is part of a multinational speciality chemicals manufacturer headquartered in Switzerland (Company 20). Originally the chemical was delivered by its Japanese supplier in fibreboard kegs, which were land filled. After a development period of two to three years, the company and its supplier developed a returnable container, which now discharges into sealed equipment, cutting out the need to landfill 30,000 fibreboard kegs per year. Additionally the innovation also reduced the exposure risk for operators (for more details, see Preuss, 2005).

The technical complexity means that such environmental initiatives have repercussions for both supplier and customer technology and therefore require close cooperation with the supplier rather than just being unilaterally imposed by the customer. The purchasing manager was initially concerned that the innovation would lock his company into a relationship with the Japanese supplier, but within a few months the new kegs became the accepted solution of all major suppliers and this constraint does not apply anymore. This is a rare example of a genuine green multiplier effect in operation. It needs to be pointed out, however, that the replacement of contaminated one-way containers occurred only in two chemical companies (Companies 10 and 28) and in one shipyard for chemical products (Company 24). While they are examples of an active approach, the range of environmental initiatives pursued by the chemical industry is at the same time much narrower than those offered, for example, by electronics companies.

Across manufacturing as a whole, environmental issues in the supply chain are more usually addressed in a passive mode, by stipulating minimum criteria the supplier must not fall below of, rather than by nurturing on-going improvements. For example, a Japanese-owned manufacturer of video recorders (Company 3) stresses in its *Suppliers' Guide* that it is its policy to meet the requirements of ISO 14001 and that it will "continually improve our environmental performance and prevent pollution." Hence, "Suppliers are expected to be aware of the above principles and that as an environmentally conscious organisation, do not actively conduct their business in a manner which would contradict [the corporation's] aims."

The arm's length mode is prevalent in electronics but is not limited to certain industries, as it is applied, for example, in ship building (Company 24) and paper making too (Companies 17, 21 and 23). Paper making emerged from the sample as the industry with the most comprehensive catalogue of environmental checks, ranging from environmental policies, through various forms of documentation and site visits to Forest Stewardship Council[1] accreditation, a scheme which requires a minimum environmental standard of forest plantations together with a certification of origin so that the paper maker knows which plantation the purchased wood pulp came from. Yet, all the measures remain on the passive side of the spectrum; despite its comprehensive approach the paper industry did not produce any examples of a cooperative collaboration in the supply chain, at least not as far as this study is concerned.

Analysing environmental initiatives in terms of an arm's length or cooperative approach allows once more to illustrate the complexity of environmental issues. A small mechanical engineering company (Company 4) has undertaken a number of environmental protection initiatives by improving its products, and one recent example concerns the design of cleaning equipment for coke ovens in the steel industry. By a more efficient cleaning of the oven doors, steel makers can ensure that the doors fit closely and less pollutants escape into the atmosphere. This innovation is evidence of a cooperative approach in the downstream value chain. However, there are no signs of a cooperative approach to environmental initiatives in the upstream supply chain, as the purchasing manager states:

> What we are buying in, if it is raw material, as long as it meets the proper standard or level of quality, and we check that by reason of seeing that we get the proper certification, if it meets that, normally, it [i.e. supplier selection] would be on delivery time and price.

The study concludes that a perception that there are no environmental issues in the supply chain is rare among manufacturers; equally rare is a collaboration on environmental issues involving both suppliers and customer. The predominant mode of tackling environmental issues is an arm's length one, where the buying organisation imposes criteria which supply chain members must not fall below of. A degree of environmental protection is achieved in this mode, but companies are likely to forego opportunities for a more comprehensive approach to the environment. This echoes the earlier findings that supplier assessment criteria take little account of the natural environment (see Section 12.3) and that the environmental problems referred to most often by purchasing are of lesser concern when the whole supply chain perspective is considered (see Section 12.5). Across the three factors that constitute supply chain management – materials flow, information and relationships in the supply chain – the situation observed in the sample companies is thus sub-optimal.

Suggestions for further research should address greener supply at a number of levels (Preuss, 2002, 2005). The personal values of managers need to be studied, as

---

[1] For more details regarding the scheme, see: http://www.fsc.org/en/

in terms of an interest in environmental affairs managers seem to have fallen behind the general population in the Western world (Fineman, 1997). In particular, purchasing managers have been referred to as resisters to environmental innovation (Drumwright, 1994). Help on environmental issues provided by business school education (Finlay *et al.*, 1998; Bunch and Finlay, 1999) and professional associations, such as the Chartered Institute of Purchasing and Supply in the United Kingdom (IEMA, CIPS and NHS Supply, 2002), needs to be evaluated. Within companies, further research could address the effectiveness of cross-functional environmental teams, the required level of top management leadership or, given the complexity of environmental problems, the process of strategic assessment and decision making (Sarkis, 1998).

New forms of purchasing and supply chain management need to be investigated. An interesting approach here concerns product take-back. For example, Interface, a leading US manufacturer of commercial carpets, leases office flooring to its customers on one product range instead of selling it. This forces the carpet supplier to adopt a life-cycle perspective as the repercussions of take-back and disposal need to be considered before a contract is signed. Hence a much greater supply chain management contribution to environmental initiatives is needed much earlier (Frankel, 1998).

Another model for greater supply chain management involvement in environmental initiatives is offered by the notion of an industrial ecosystem. In analogy to a natural ecosystem, an industrial ecosystem consists of a network of organisations which use each others' waste and by-products and share and minimise the use of natural resources, thus jointly aiming to reduce environmental destruction. An example is provided by a number of companies in Kalundborg, Denmark, where a power plant, an enzyme plant, a refinery, a chemical plant, a cement plant, a wallboard plant and a number of farms coordinate the use of raw materials, energy and water as well as their waste management (Tibbs, 1993; Korhonen, 2004).

Given the traditionally lower status of supply chain management in comparison with other corporate functions, one – and perhaps the farthest reaching – opportunity for strengthening the role of supply chain management in environmental protection might be to allow the function to participate in the corporate strategic planning process and to offer it a seat on the Board of Management. Otherwise the great contribution purchasing could make will remain unharnessed (Preuss, 2002). The argument for a more active role in the formulation of corporate strategy is also supported by the growing importance of supply chain management in financial terms (Burt *et al.*, 2003).

12.6 Conclusions

Supply chain management, for long a neglected management function in manufacturing organisations, is increasingly achieving a strategic importance in economic terms. However, is the strategic contribution to the bottom line matched by an increasing contribution to environmental protection? Supply chain management could address the environment in a threefold way. Upstream, it could participate in joint design projects with suppliers or, via design and specifications, it could encourage suppliers to adopt greener manufacturing techniques. Within the organisation, purchasing could bring its functional perspective to bear on product life-cycle assessments, design for environment initiatives and environmental management systems. Downstream examples of a supply chain management input are recycling and re-use of excess materials. Given the various links and overlaps between the supply chains of different companies, environmental excellence can easily be transferred into the supply chains of other manufacturers. Supply chain management could thus initiate a green multiplier, which could make it a more effective change agent than any other corporate function.

The reality is, however, much less sanguine. A sample of manufacturing organisations from the United Kingdom revealed a sub-optimal approach to environmental issues in the three key areas of supply chain management, the management of the flow and transformation of materials, of information and of the relationships within the supply chain. In terms of managing the flow and transformation of materials, individual initiatives to replace hazardous materials or to reduce packaging contrast with a non-inclusion of the environment in the supplier selection and evaluation criteria for many manufacturers. Even if supply chain management were keen to raise the environmental profile of their supply chains, the lack of knowledge regarding the performance of potential and actual suppliers would prevent the environmental credentials of suppliers from being detected and used to influence the materials flow.

Purchasing managers in the sample cited four issues – packaging, waste, hazardous production materials, paints and solvents – as environmental problems in their supply chains. While this is evidence of environmental awareness, packaging and waste do not significantly concern the environmental performance of the whole supply chain and are of low financial importance. The four issues furthermore only partially overlap with the perceptions held by environmental and production managers. Where environmental issues are addressed, an arm's length mode prevails: suppliers are given minimum threshold criteria and are expected not fall below of these. Cases where managers claim their supply chain do not contain any environmental issues and cases where an environmental issue is approached by joint action of supplier and customer are both rare. The environment is thus primarily addressed in arm's length mode, and the potential for a more comprehensive approach to the environment often remains underused.

These findings lead to the conclusion that environmental initiatives in the manufacturing supply chain currently amount to little more than 'light-green' supply. Furthermore, the most active industries – electronics, automotive, chemicals and paper making – are those that are already in the public limelight over their environmental performance. In other words, regulation and public pressure emerge as the most important motivations for addressing environmental issues, which has implications for the debate between proponents of state regulation and self-regulation by industry. The study also underlines the complexity of environmental problems, as even the exemplary companies offer an uneven environmental performance across the three areas of supply chain management. They may, for example, display a comprehensive approach to the environment in their downstream relationships with customers but not address it upstream in their dealings with suppliers.

Further research into greener supply would benefit from addressing several avenues. The values supply chain managers hold regarding the natural environment should receive further attention, and the support offered by business schools and professional bodies, such as the Institute for Supply Management in the US and the Chartered Institute of Purchasing and Supply in the UK, should be evaluated. Given the complexity of environmental issues, attention to strategic assessment and decision making and to new forms of supply chain management is particularly warranted. The greatest contribution to greener supply might be achieved by allowing supply chain management greater access to strategic decision making by, perhaps, offering the supply chain function a seat on the Board of Management.

12.7 References

Bailey P, Farmer D, Jessop D, Jones D (1998) Purchasing principles and management. Financial Times Management, London

Beamon BM (1999) Designing the green supply chain. Logistics Information Management, 12: 4, 332-342

Bowen FE, Cousins PD, Lamming RC, Faruk AC (2001) Horses for courses: Explaining the gap between theory and practice of greener supply. Greener Management International, 35, 41-59

Bunch R, Finlay J (1999) Environmental leadership in business education: Where's the innovation and how should we support it?. Corporate Environmental Strategy, 6: 1, 70-77

Burt D, Dobler DW, Starling SL (2003) World class supply management: The key to supply chain management. McGraw-Hill/Irwin, New York

Business in the Environment and Chartered Institute of Purchasing and Supply (1997) Buying into a green future: Partnerships for change. Business in the Environment, London

Carter CR, Dresner M (2001) Purchasing's role in environmental management: Cross-functional development of a grounded theory. Journal of Supply Chain Management, 37: 3, 12-23

Central Statistical Office (1992) Standard Industrial Classification of Economic Activity 1992. HMSO, London

Clayton A, Spinardi G, Williams R (1999) Cleaner technology: A new agenda for government and industry. London, Earthscan

Drumwright ME (1994) Socially responsible organizational buying: Environmental concern as a noneconomic buying criterion. Journal of Marketing, 58, 1-19

Dunn & Bradstreet (1999) Scottish business register. Dunn & Bradstreet, High Wycombe

Elkington J, Knight P (1991) The green business guide: How to take up - and profit from - the environmental challenge. Victor Gollancz, London

Fineman S (1997) Constructing the green manager, In: McDonagh P, Prothero A (eds.) Green management: A reader. The Dryden Press, London

Finlay J, Bunch R, Neubert B (1998) Grey pinstripes with green ties: MBA programs where the environment matters. World Resources Institute, Washington, DC

Frankel C (1998) In earth's company: Business, environment and the challenge of sustainability. New Society Publishers, Gabriola Island, BC

Freimann J, Walther M (2001) The impacts of corporate environmental management systems: A comparison of EMAS and ISO 14001. Greener Management International, 36, 91-103

Gadde L-E, Snehota I (2000) Making the most of supplier relationships. Industrial Marketing Management, 29, 305-316

Green K, Morton B, New S (1996) Purchasing and environmental management: Interactions, policies and opportunities. Business Strategy and the Environment, 5, 188-197

Hakansson H, Snehota I (1995) Developing relationships in business networks. Routledge, London

Hall J (2000) Environmental supply chain dynamics. Journal of Cleaner Production, 8, 455-471

Handfield RB, Nichols EL (1999) Introduction to supply chain management. Prentice Hall, Upper Saddle River, NJ

Hill KE (1997) Supply-chain dynamics, environmental issues and manufacturing firms. Environment and Planning, 29: 7, 1257-1274

Howes R, Skea J, Whelan B (1997) Clean and competitive? Motivating environmental performance in industry. Earthscan, London

IEMA, CIPS, NHS Supply (2002) Environmental purchasing in practice: guidance for organisations. Institute of Environmental Management and Assessment, Lincoln, Chartered Institute of Purchasing and Supply, Stamford, National Health Service Purchasing and Supply Agency, Reading

ISO (2005) The ISO Survey of ISO 9000 and ISO 14001 certificates – 2004. International Organization for Standardization, Geneva

James P, Prehn M, Steger U (1997) Corporate environmental management in Britain and Germany. Anglo-German Foundation for the Study of Industrial Society, London

Kompass (2000) UK Kompass Register. Reed Business Information, Sutton, Surrey

Korhonen J (2004) Industrial ecology in the strategic sustainable development model. Journal of Cleaner Production, 12, 809-823

Krause DR, Handfield RB, Scannell TV (1998) An empirical investigation of supplier development: Reactive and strategic processes. Journal of Operations Management, 17, 39-58

Lambert DM (2001) The supply chain management and logistics controversy. In: Brewer AM, Button KJ, Hensher DA (eds.) Handbook of logistics and supply-chain management. Pergamon, Oxford

Lamming R (1993) Beyond partnership: Strategies for innovation and lean supply. Prentice Hall, Hemel Hempstead

Lamming R, Hampson J (1996) The environment as a supply chain management issue. British Journal of Management, 7, S45-S62

Lamming R, Cousins P, Bowen F, Faruk A (2001) A comprehensive conceptual model for managing environmental impacts, costs and risks in supply chains. In: Erridge A, Fee R, McIlroy J (eds.) Best practice procurement: Public and private sector perspectives. Gower, Aldershot

Lamming R, Warhurst A, Hampson J (1996) The environment and purchasing – Problem or opportunity?. Chartered Institute of Purchasing and Supply, Stamford

McIntyre K, Smith H, Henham A, Pretlove J (1998) Environmental performance indicators for integrated supply chains: The case of Xerox Ltd. Supply Chain Management, 3: 3, 149-156

New S, Morton B, Green K (1999) Deconstructing green supply and demand: PVC, healthcare products and the environment. Risk Decision and Policy, 4: 3, 221-254

New S, Green K, Morton B (2000) Buying the environment: The multiple meanings of green supply. In: Fineman S (ed.) The business of greening. Routledge, London

Newton T, Harte G (1997) Green business: Technicist kitsch?. Journal of Management Studies, 34: 1, 75-98

Preuss L (2001) In dirty chains? Purchasing and greener manufacturing. Journal of Business Ethics, 34: 3-4, 345-359

Preuss L (2002) Green light for greener supply, Business Ethics: A European Review. 11: 4, 308-317

Preuss L (2005) The green multiplier: A study of environmental protection and the supply chain. Palgrave, Basingstoke

Quinn JB (2002) Core-competency-with-outsourcing strategies in innovative companies. In: Hahn D, Kaufmann L (eds.) Handbuch Industrielles Beschaffungsmanagement. Gabler, Wiesbaden, 2nd ed

Russel T (1998) Green Purchasing: Opportunities and innovations. Greenleaf Publishing, Sheffield

Sarkis J (1998) Evaluating environmentally conscious business practices. European Journal of Operational Research, 107, 159-174

Schary PB, Skjott-Larsen T (2001) Managing the global supply chain. Copenhagen Business School Press, Copenhagen, 2nd ed

Scottish Enterprise (1997) Digital directory of Scottish manufacturing capability. Scottish Enterprise, Glasgow

Steger U (2000) Environmental management systems: Empirical evidence and future perspectives. European Management Journal, 18: 1, 23-37

Tibbs H (1993) Industrial ecology: An environmental agenda for industry. Annals of Earth, XI: 1

Walton SV, Handfield RB, Melnyk SA (1998) The green supply chain: Integrating suppliers into environmental management processes. International Journal of Purchasing and Materials Management, 34:2, 2-11

Zsidisin GA, Siferd SP (2001) Environmental purchasing: A framework for theory development. European Journal of Purchasing and Supply Management, 7, 61-73

Appendix   Schedule of Questions

I would like to stress that the information gained during this interview will be treated confidentially and will only be used for the purposes of academic research.

1. Organisation of the purchasing and supply department

Could you please explain the structure of your purchasing organisation to me?
What types of materials or components do you purchase?
What does a typical purchasing procedure look like?
Do you have a formal purchasing strategy?
What share of your working time is spent on short-term and what share on longer term issues?
How is your work as purchasing or supply chain manager evaluated?

2. Supply chain interaction

What decision-making criteria do you apply for your main suppliers?
How do you measure the performance of your suppliers?
Do you see some of your suppliers as partners?
If so, what does partnership mean for you?

3. Environmental concerns in purchasing and supply

Do you see any environmental issues in the major supply chains for your product?
Does your company have an environmental policy?
Is your company accredited to ISO 14001 or EMAS? Do you require such accreditation of your major suppliers?
Would you see the environment more as opportunity or more as creating costs to you?

Are there any recent examples of initiatives to address environmental challenges in your supply chains?
If so, what were the reasons for undertaking these initiatives?
Have you ever deselected a supplier over an unsatisfactory environmental performance?
Have you had any cases where you offered additional investment, benefits or other help to suppliers to encourage them to take up environmental initiatives?
Are you involved in company-internal or downstream environmental initiatives?

Thank you very much in advance for your cooperation and time.

Part III

Case Studies

13

Environmental Supply Chain Innovation

Jeremy Hall

Haskayne School of Business, University of Calgary, 2500 University Drive N.W., Calgary, Alberta, Canada T2N 1N4, hallj@.ucalgary.ca

This chapter proposes a model describing why firms should invest in environmental supply chain innovation or 'green supply' activities. It argues that large high profile companies are under pressure from a wide range of stakeholders to improve their environmental performance. In contrast, small supplier firms are under less pressure, but are highly influenced by the demands of their customers. The model attempts to demonstrate that customer firms invest in environmental supply chain innovation because suppliers with poor environmental practices can expose the customer firm to high levels of environmental risk. However, implementation is dependent upon environmental pressure, firm capabilities and the degree to which customer firms are able to control their suppliers. The model is illustrated with a case study of UK supermarket retailer J Sainsbury Plc and five of their suppliers conducted over a four-year period in the late 1990s.

13.1 Introduction

13.1.1 Environmental Supply Chains

A number of authors have recognized the link between environmental management issues and buyer-supplier relations. For example, Lamming and Hampson (1996) draw parallels between environmental management practices (e.g. life cycle analysis, waste management and product stewardship) and supply chain management practices (e.g. vendor assessment, total quality management, lean supply and collaborative practices). Sarkis (2000) notes that there are elements such as product and operational life cycles, performance measures as well as environmentally influential organizational policy elements and interdependencies between supply chain and environmental issues. Integrating these elements is a means of reducing environmental impacts, as environment decisions by one organization may also affect the decisions of their customers and suppliers. Florida (1996) argues that "...close relationships across the production chain... facilitate the adoption of advanced manufacturing practices, creating new opportunities for joint improvements in productivity and environmental outcomes." Green et al. (1996) argue that supply chain analysis is a useful way in which environmental issues can be incorporated into industrial transformation processes. They argue that green supply, "... the way in which innovations in supply chain management and industrial purchasing may be considered in the context of the environment" (p. 188) has greater potential to address environmental concerns than such things as 'green consumerism' because it is grounded in non-altruistic market principles (Green et al, 2000). Hill (1997) recognizes the importance of supply chain dynamics and its link to environmental pressure in changing firm behavior: Environmental pressure can thus be conceptualized as moving along the supply chain through two constituent elements: through customers, those who purchase the products of the firm; and through suppliers... (p. 1259).

Many of these authors have also recognized the strategic implications of environmental supply chain innovation. For example, Hampson and Johnson (1996) argue that environmental issues can be related to overall business efficiency. They also note that interest in environmental supply chains is based upon increased awareness in environmental issues, the increasingly strategic importance of purchasing and trends towards co-operation and partnership approaches between customers and suppliers. However, Hill (1997) argues that while the potential for supply chains to exert environmental pressures exist, there were relatively few cases where it was actually occurring. Young (2000) argues that this is due in part to the lack of information sharing within the supply chain: *"Only when organizations in the supply chain exchange information backward through their channels will supply chains discover more efficient, environmentally sound, and profitable disposition solutions."* Following along similar lines, New et al. (1997) argue that a holistic approach to supply chain management can benefit environmental management practices. *"... the idea is simply that a firm operating at one point of the supply chain runs the risk of organizing its activities to achieve parochial objectives which result in sub-optimization for the chain as a whole"* (p. 2). This they claim is more important for sectors where supply chains are part of competitive advantage. They also recognize the importance of who holds the power in the supply chain.

Much of the above research is empirically based and/or normative in nature, while some suggest the mechanisms by which green supply may be implemented. However, there has yet to be a sufficient explanation as to why firms should engage in such activities. In addition to overall environmental improvement (which is often counter to tangible financial performance), this paper argues that large customer firms invest in environmental supply chain innovation as a means of reducing their exposure to risks associated with their suppliers' poor environmental performance. For the purposes of this discussion, environmental innovation is defined as a new product, process or technology developed and/or adopted by a firm to reduce environmental impacts. Environmental supply chain innovation is when a supplier, under the advice, coercion or direction of a customer firm, adopts an environmental innovation. Of relevance to this discussion is the notion that there must be some form of inter-firm innovation (i.e. an exchange of information, joint development of a technology, etc). Without inter-firm innovation, the customer firm is only acting as a regulator and leaving the onus of the innovation to the supplier.

This paper argues that large high profile firms are exposed to stakeholder pressure that goes beyond legal environmental responsibilities. Such large firms often have a highly influential 'sphere of influence' over many suppliers and a 'sphere of concern' over activities of others. A sphere of influence is where a firm clearly has the capability as well as a legitimate reason to prescribe behavior to another firm. A sphere of concern is where the customer firm is weary of potential areas of risk, but there is ambiguity in whether or not they are responsible for addressing these concerns, or have the capabilities or legitimacy to deal with them. Obviously the difficulty for the customer firm is determining the boundaries of the spheres of influence and concern. Regardless, stakeholder groups with environmental interests can challenge these boundaries and generate change throughout the supply chain by targeting the customer firm and leveraging these spheres of influence and concern.

13.1.2 Environmental Pressures

A number of authors have argued that firms often change their environmental policies in response to pressure from such sources as environmental advocacy groups, consumers, regulators, neighbors and other stakeholders (c.f. Henriques and Sadorskey, 1996, 1999, Hall, 2000). This pressure can also be triggered by perception or anticipation (Hall, 2000). For example, fears over genetically modified organisms (GMOs) have been influenced by aggressive anti-GMO lobby groups and negative web pages such as www.frankenfoods.com and www.monsantosucks.com. This was a direct challenge to the primary regulatory agency, the US Food and Drug Administration (FDA), which approved the products after finding no serious health problems (Hall and Crowther, 1998; Hall and Vredenburg, 2004).

It is important to note that not all firms are exposed to the same types of pressure or to the same extent. In general large, high profile firms are under considerable pressure to improve their environmental performance, whereas many suppliers lack incentives. Lamming and Hampson (1996), Kemp & Soete (1992), Hunt and Auster (1990) and Walley and Whitehead (1994) all argue that smaller firms usually lack the resources to deal with environmental problems, Williams *et al.* (1993) and Hall (2000) argue that smaller firms are relatively unaffected by environmental pressures and therefore do not regard environmental concerns as being important.

The above discussion has emphasized the importance of stakeholders, defined by Freeman (1984, p 46) as *"any group or individual who can affect or is affected by the achievement of the organization's objectives"*. Henriques and Sadorskey (1999) identified four critical environmental stakeholder groups:

- Regulatory stakeholders, which either set regulations or have the ability to convince governments to set standards. This includes governments, (environmental regulators), trade associations, (information collectors), informal networks, (sources of technological information) and competitors (potential environmental technology leaders that may set industry norms and/or legal mandates).
- Organizational stakeholders that are directly related to an organization and can have a direct financial impact on the company. This includes

customers (who can boycott or support a company's product), suppliers (who can stop shipments if for example, it hurts the credibility or reputation of the company), employees (of which their participation is necessary to ensure a successful environmental policy) and shareholders (who may voice concerns).

- Community groups, environmental organizations, and other potential lobbies, who can mobilize public opinion in favor of or against a firm's environmental policies.
- The media, which has the ability to influence society's perception of a firm.

As will be shown below, these stakeholder groups do not necessarily affect all companies or industries. Differences such as visibility affects the level of social pressure to which a firm will be subjected (Pfeffer and Salancik, 1978). Meznar and Nigh (1995) argue that as more stakeholders take an interest in a firm's activities (i.e. the higher the profile), the more likely they will face pressure to alter aspects of its operations. This they believe will eventually lead to compliance with external expectations. Mitchell *et al.* (1997) argue that variances in stakeholder pressure are due to differences in power, legitimacy and urgency, and are therefore not the same for all firms. Thus, in contrast to other forms of innovation, firms without pressure may be hesitant to invest in environmental innovation because it is of little concern to their situation. However, Mitchell *et al.* (1997) also recognize the dynamic nature of stakeholder pressure. As such, firms not exposed to pressure in the present may very well be exposed to it in the future.

This paper argues that non-regulatory pressures are important for environmental supply chain innovation because inter-firm environmental policies are usually not a regulatory requirement: each firm is legally responsible for their own activities and not necessarily responsible for their suppliers' activities. Customer firms may be liable for purchased products or services, but they are not legally responsible for their suppliers' other activities. If customer firms only require their suppliers to meet regulations, the onus for innovation is likely to be placed upon the supplier firm, with little interaction between buyers and suppliers (Hall, 2000). In contrast to regulations, other stakeholders such as environmental groups are often not concerned about who is legally responsible for what, but rather about bringing issues to the public's attention (Murphy and Bendell 1996). Environmental campaigning often addresses issues that would not otherwise be addressed through regulations or consumer pressure, and can focus on the broader environmental impacts of industrial systems. This is in contrast to regulatory pressure, which is dependent upon the legal obligations of firms. Environmental advocacy groups can therefore expect customer firms to coerce their suppliers into adopting environmental policies that they would otherwise not have adopted. Of course environmental advocacy groups are limited in their resources and technical capabilities. They must appeal to their donors and/or members, which means that they have to initiate campaigns that reflect what their donors consider relevant or topical. Campaign targets are therefore not necessarily the most important environmental issues. A lack of technical resources also limits environmental advocacy groups on what should be addressed (Murphy and Bendell 1996).

13.1.3 Supply Chain Pressures

Although many smaller and.or lower profile suppliers may not be under environmental pressure, they are often under considerable pressure from their customers. Trends in buyer-supplier relations indicate that there is an increasing level of integration between buyers and suppliers (c.f. Lamming, 1993, Lyons et al, 1990, Helper, 1991, Mudambi and Helper, 1998 and Sako, 1992). These changes have been driven by cost savings and increased capabilities, which have led to both technical and organizational innovations.

Closer buyer-supplier relations usually favor the large customer firm. Gules and Burgess (1996) and Imrie and Morris (1992) argue that collaborative relationships are usually dictated by buyers, controlled in a hierarchical fashion and may lead to inequitable power distribution. Barringer (1997) recognizes that in some cases small firms may have no alternative but to engage in what is deemed appropriate for the customer firm. Of course these dynamics cannot occur unless the customer firm has sufficient channel power, defined by El-Ansary and Stern (1972) as the ability of one channel member to control the decisions of another. French and Raven's (1959) taxonomy aptly describes the sources of power as being derived from rewards, coercion, expertise, reference or legitimacy. Kumar (1996) adds a dynamic element by arguing that the exploitation of power may be beneficial in the short term, but tends to be self-defeating in the long run. Victims of exploitation ultimately find ways to resist, while co-operation generally offers the greatest value. This leads to the possibility that winning at the expense of another channel member is being replaced with a 'win-win' scenario.

The dominance of a powerful customer firm has important implications for environmental supply chain innovation. Suppliers are frequently not under the same types or levels of pressure as their larger customer firms, and therefore have fewer incentives to engage in environmental innovation. Large customer firms are often under pressure (or would like to pre-empt potential pressure) to address systemic environmental issues, such as those attributed to their suppliers. This is consistent with Green *et al.* (1996) and Lamming and Hampson (1996) who found that a partnership approach towards supplier relations was necessary. Hampson and Johnson (1996) argue that shorter supply chains are probably more likely to succeed and are dependent upon their ability to exert sufficient pressure on the supply chain. New *et al.* (1997) point out that dropping suppliers can be expensive and risky, thus making supplier intervention a viable option.

In summary, the integration of supply chain pressures and environmental pressures is a potentially useful means by which environmental impacts may be reduced. This is a particularly useful mechanism for smaller, lower profile firms that often lack reasons to invest in environmental innovation. However, the dynamics of environmental supply chain innovation is dependent upon a dominant supply chain member that is under environmental pressure. The next section will discuss this phenomenon with the case of Sainsbury's, a large UK supermarket.

13.2 Research Case: J Sainsbury PLC

The explanatory case study format as outlined by Yin (1993, 1994), was used. The main subject of analysis was the mechanisms that encourage the diffusion of environmental technologies or techniques from a large, influential British retailer and five subordinate supplier firms. The research period was from 1995 to 1998, during which time there was relatively high environmental sensitivities in industry, as well as an increased focus on buyer-supplier relations. The primary means of data collection was to conduct open interviews with senior managers in the large customer firm, as well as a senior manager (usually the Managing Director) in a number of their supplier firms. Interviews with industry experts, trade association officials and environmental advocates were also conducted. In most cases interviews were over two hours long. Publications (academic and the trade press) and formal company documents were also used as supporting evidence.

It is generally accepted that UK supermarket are some of the world's most successful and profitable retailers (Wrigley, 1997, Fiddis, 1997, Harvey, 2000). This is based on technical competencies in food science (Senker, 1989, 1986), information technology, (Smith and Sparks, 1995) and channel power (Burt and Sparks 1994, Bowlby and Foord, 1995). They also have innovative own-brand[1] policies, which, unlike most other retailers, involve considerable control over technical issues such as quality control, product design and production processes. More recently, UK retailers have dominated wholesaling and distribution, where retailers were increasingly focusing on service and quality as a means of satisfying the consumer (Smith and Sparks, 1995, Harvey, 2000).

The selected case study was Sainsbury's Supermarkets and five of their suppliers. The company is one of the UK's largest and sophisticated retailers, with one of the highest sales-per-square foot ratio in the industry. They are also one of the few food retailers to have a sizeable R&D budget, spending £6 million in 1996 on R&D, ranking 143 of all UK companies on R&D expenditure (Company Reporting, 1997).

Sainsbury's has been a leader in retail innovation since 1869 (Senker, 1986). Over the years they recognized the value of their name and took an increasingly active role in the supply chain as a means of enhancing and exploiting their reputation. This strategy evolved into a sophisticated own-brand strategy that was based on high quality products and a 'hands-on' approach to product development. Through these investments they were able to offer similar or better quality products than private brands, and were often a leader in new product development (Fiddis, 1997). This was in sharp contrast to the traditional approach to own-brand policies, which were primarily concerned with offering low cost alternatives and improving distribution channel performance though cost reductions (Fiddis, 1997, Mills, 1995, Narasimhan and Wilcox, 1998). These strategies allowed Sainsbury's to accumulate capabilities in food sciences, quality control and new product development. Over the years these competencies allowed Sainsbury's to eliminate

---

[1] Own brands, often referred to as private labels, are products sold under the name of the retailer, but usually manufactured by an outside company.

intermediate wholesalers and dominate their supply chain.   Like many other industries, they also became increasingly involved in collaborative, 'hands on' relationships with their suppliers.   These strategies also gave Sainsbury's 'ownership' of the products, making them responsible for the resultant environmental impacts.

Table 13.1. Sainsbury's statistics

Number of customers per week, 1995	9 million
Average number of lines, 1997*	21,000
Average number of 'own-brands', 1997*	9,500
Average number of Depots, 1997*	18
Number of items per till per week, 1995	20,000
Number of electronic transactions per week, 1995	2 million
Average sales area per store (sq. ft), 1997*	27,479
Total sales area (sq. ft), 1997*	10,387,000
Socio-economic group, 1997*	57% in ABC
Age demographics, 1997*	56% 45+
Average spend per trip, 1997*	£23.60
Advertising expenditure,1996	£31 million

Sources: Innovation Manager, Interview, Japan, 1996; Technical Services Manager, interview, UK, 1997; *Institute of Grocery Distribution (IGD). Account Management Series, J. Sainsbury, 1997

Table 13.2 is a typology of environmental impacts for retailers drawn from interviews conducted for this research. For the purpose of this discussion, there are basically two dimensions of environmental responsibilities for supermarkets: those that are directly associated with their business operations and indirect impacts associated with the products that they sell. Between these two areas are issues shared by both the manufacturer and retailer, such as impacts from their own brand products, packaging and distribution. The direct impacts such as the development and operation of supermarkets are their only mandated (regulated) area of responsibility, while the shared impacts correspond to the Sphere of Influence and the indirect impacts correspond to their Sphere of Concern. Compared to the total environmental impacts generated by their supply chain, Sainsbury's regulated area of responsibility was small and easily managed. However, they were also involved in their suppliers' environmental activities, especially where they had legitimate reasons to influence them, such as for own-brand products. Public concern was also pressuring retailers into becoming involved in activities that were considerably distant from the retailing function, such as genetically modified soybeans, farming practices (pesticide and herbicide use, animal husbandry, policies towards BSE and Hoof and Mouth Disease, etc) and whether or not timber products came from sustainable forests (J Sainsbury Plc, 1997, *Ethical Consumer*, December, 1996). Supermarkets would need very high levels of non-retailing technical competencies to engage in these activities. To address these pressures, Sainsbury's became heavily involved in associated upstream environmental technologies and practices. This was perhaps an incremental process in that as Sainsbury's had long been involved in their suppliers' activities. From the suppliers' perspective, there was therefore no abrupt change in policy that triggered the environmental initiatives; rather it was a continued policy of meeting Sainsbury's requirements as they had done in the past for other issues such as quality, heath and safety, delivery schedules, etc.

Table 13.2. Typology of environmental impacts for food retailers

Retailer ⇐ Responsibility ⇒ *Supplier*		
Direct	*Shared*	*Indirect*
Examples	Examples	Examples[2]
- Store design, construction	- Packaging[3]	- Pesticide use
- Store operation (energy consumption, refrigerant management, water usage, etc.)	- Distribution (fuel consump-tion, traffic congestion, noise, etc.)	- Animal husbandry
	- Env. impacts generated from own brand product mfg.	- Fisheries protection.
		- Forestry management
- Access to public transport		- Genetic engineering
- City centre vs. 'greenfield' development		- General env. impacts of industrial, agricultural processes

To investigate these initiatives, five suppliers were interviewed between 1996 and 1997, as shown in Table 13.3[4]. Evidence of environmental supply chain innovation was found in four cases, with a minor example in the fifth case. There was limited environmental supply chain innovation between Sainsbury's and the sauce producer, other than the exchange of European Union regulatory information on packaging protocols. According to the Sales and Marketing Director (interview, UK, 1997), the company had a relatively low profile and was unknown outside of the industry. He stated that consumer and environmental advocacy groups had

---

[2] Given that supermarkets typically carry over 25,000 products, there are far too many issues to be covered in this paper. This list is thus meant for illustrative purposes only.

[3] While packaging has generated a considerable amount of interest for retailers and their suppliers, there was a relatively clear line of responsibility with which consumers and advocacy groups attributed to the retailer. As such, it would be expected that they were under pressure to address these issues, which was indeed the case and has been extensively researched. For this reason, along with scope limitations, empirical research was not conducted on packaging.

[4] Four of the five suppliers were producers of consumable foodstuffs, while the refrigerator manufacturer's products lasted for decades. End-of-life processing was therefore not considered relevant by the suppliers at the time of the research.

little interest in this 'unemotional' industry. However, as an own brand supplier Sainsbury's had a major influence over other aspects of the operation such as recipes, packaging design and delivery schedules.

Table 13.3. Environmental supply chain innovation

Supplier	Sauce Producer	Abattoir	Tomato Producer	Pulp Supplier	Refrigerator Supplier
Supplier type:	Own brand	Proprietary/ own-brand	Own brand	Raw material (indirect)	Capital equipment
Est. Sales	£50 million	N.A.	£50 million	N.A.	£40 Million
Employees	500	800	75 (+ 100 seasonal)	370	450
% of sales to Sainsbury's	11	+50	80	Negligible	25 (formerly 60)
Env. supply chain activities	- EU Pack-aging Proto-cols	- FABL - Animal Husbandry	- ICMP - GMO policies	- FSA - Chlorine-free pulp process	- CFC-free refrigerant technologies

Key:

FABL: Farm Assured British Beef Programme

ICMP: Integrated Crop Management Programme

FSA: Forestry Standards Association

GMO   Genetically Modified organisms

Sainsbury's had a strong influence over the tomato grower, but also had a major influence over their environmental policies. Like all growers, they had to adopt principles of Sainsbury's Integrated Crop Management System (ICMS), a pesticide and herbicide reduction program developed in conjunction with the National Farmers' Union, to remain a certified supplier. This program promotes the use of biological and natural methods to control pests and diseases, using pesticides and herbicides only as a last resort. The Managing Director also confirmed that there were discussions concerning genetic engineering, but refused to disclose any details due to the high degree of controversy.

For the abattoir industry, there was a perceived lack of government credibility due to the BSE crisis and other food scares. Because of these problems, the government at that time was trying to promote the industry rather than impose constraints that would further damage it.   However, consumer, health and environmental advocacy groups as well as the media were exerting considerable pressure due to the high profile food scares and aggressive animal rights campaigning that opposed highly intensive farming practices. To address these concerns, Sainsbury's pressured the abattoir into adopting a farm assurance and animal husbandry program based on Sainsbury's 'Partnership in Livestock' scheme. It covers animal welfare issues such as the avoidance of abnormal joint and leg weaknesses due to intensive rearing programmes, the restriction of space, unnecessary or uncontrolled use of medications, as well as freedom from distress, discomfort, disease, injury, etc. The company thus employed a specialist Farm Assurance Officer and purchased most of their stock from farms certified by Farm Assured British Beef and Lamb (FABL), a national animal husbandry program.

Retailers have considerable control over the abattoir industry, and the relationships are sometimes strained. However, the abattoir managers were sensitive of the supermarket's need to maintain high standards within their supply base. They related the following: if an abattoir sells meat to a restaurant with poor storage and handling facilities and someone contracts food poisoning, it is usually a local issue; if the media finds out that the meat was supplied by a large abattoir, it becomes a regional issue; if they find out that the abattoir also supplies Sainsbury's, it becomes a national crisis. This clearly illustrates the importance of why firms have to take their suppliers' environmental policies seriously. It also illustrates the difficulties in determining the boundaries between a customer's sphere of influence and sphere of concern.

The Canadian-based pulp supplier was under high levels of regulatory pressure, which was due to the industry having a number of visibly dirty and resource-intensive processes, such as the use of herbicides, deforestation, effluent discharges and air emissions. Other issues included bio-diversity and wildlife protection. According to the Public Affairs Manager (interview, Canada, 1997), they were also under 'tremendous pressure' from customers and lobby groups to improve their environmental performance. They thus initiated a number of environmental waste reduction programs, and were a member of the Forest Stewardship Council.

Although the pulp supplier not selling directly to Sainsbury's and being thousands of miles away, environmental supply chain innovation had occurred (albeit more from UK retailers as a whole rather than as a didactic relationship). In the 1980s it was discovered that chlorine from pulp manufacturing created dioxins, which were linked to cancer. The North American producers argued that the dioxin levels were too small to be an environmental hazard[5]. However, pressure targeted at UK supermarkets (especially Sainsbury's, according to the pulp supplier's environmental affairs officer) demanded the elimination of elemental chlorine from the manufacturing process. This was an articulation of consumer concerns and a response to aggressive protests by environmental advocacy groups at a few UK supermarket locations (interview, Environmental Affairs Officer, UK, 1997). As such, even large firms like the pulp mill had to agree with the demands of the supermarkets. However, there was at that time less concern about dioxins in North America, and the company continued to produce pulp through the less expensive chlorinated process for that market.

---

[5] While the technical implications of dioxins are beyond the scope of this paper, it should be noted that there are legitimate objections against dioxin. Exposure to high levels causes chloracne and an enhanced risk of cancer, while the accumulation of dioxins in the biosphere can end up in animal fats (The WHO Regional Office for Europe). The author would like to thank an anonymous reviewer for pointing this out. What is germane to this discussion is that there were considerably different views between the North American and European market as to what was acceptable levels.

In response to the Montreal Protocol, the capital equipment supplier developed one of the world's first CFC-free commercial refrigerator technologies in 1986, well before the phase-out period. This was primarily driven by Sainsbury's, who were facing aggressive consumer According to the General Manager of the equipment supplier (interview, UK, 1997), Sainsbury's triggered the development of this technology, as they were facing aggressive consumer and environmental advocacy pressure to address global warming issues. In addition to triggering the innovation through market demand (i.e. the manufacturer's largest purchaser), Sainsbury's had a large engineering staff of 45 that interacted with the company's R&D department. Interestingly, Sainsbury's had little interest in the supplier's other environmental impacts such as process technologies, material usage, disposal, etc. This is perhaps because at the time no stakeholder group would have made the connection between a retailer and the equipment supplier's processes.

With the exception of the pulp mill, the supplier firms were not under environmental pressure (other than normal regulatory pressures and those from Sainsbury's), yet there was still evidence of environmental supply chain innovation occurring in fours cases, and to a lesser extent in the fifth case. In all cases the focus was on high-profile issues that drew the attention of environmental advocacy groups, such as animal rights, pesticide use, sustainable forestry and the elimination of CFCs, and targeted at Sainsbury's. Also of importance is that all suppliers confirmed that they would not have adopted most of these programs without pressure from Sainsbury's or other similarly influential retailers.

13.3 The Sphere of Influence Model

Based upon the above discussion, we now introduce the Sphere of Influence Model (Figure 13.1)[6]. The 'hourglass' represents the total supply chain, with the customer firm in the centre, Supply Side Stakeholders (i.e. in this case primarily suppliers) above and Demand Side Stakeholders (i.e. primarily consumers) below. One may think of these as the organizational stakeholders as defined by Henriques and Sadorskey (1999). It is shaped in this way as the customer firm often has many suppliers and many customers. The major supermarkets in the UK for example have over 25,000 suppliers and millions of customers a week. The left side represents the environmental issues faced by the supply chain. Supply side stakeholders are generally concerned about technical issues, such as evaluating the technical and economic merits of protocols, policies or programs, the feasibility of new technology or meeting regulatory approval. Negotiations between stakeholders are therefore mostly focused on technical or economic arguments, as was the case between Sainsbury's and the suppliers.

Demand side stakeholders are sometimes less technically focused and are often influenced by social and perceptual concerns. Examples include the anti-GMO movement and the use of pesticides and herbicides. In these cases, regulatory

---

[6] This model is derived from a similar model and discussions with Sainsbury's Environmental Management Manager.

bodies such as the FDA claimed that the scientific evidence indicted that the products were safe, yet there was skepticism and resistance by some consumers. Animal rights are socially constructed phenomena that were particularly salience within the UK, yet of less concern in other countries. An 'unemotional' firm like

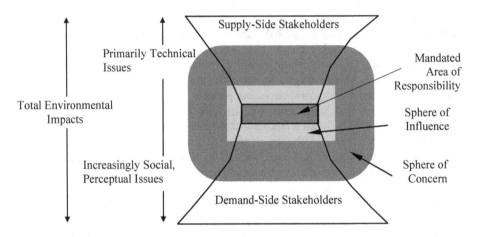

Figure 13.1. Sphere of Influence Model

the sauce producer was not at the time under any demand-side stakeholder pressure, even though they are likely to have similarly important, albeit perceptually uninteresting, environmental impacts.

The right side of the diagram lists the areas of responsibilities. The dark central part of the hourglass is the legal responsibility of the firm. This is clearly defined by regulations and easily managed, as was the case with Sainsbury's. Given that it is legally mandated, it usually does not involve supply chain issues. The next (lighter) ring is the Sphere of Influence area, where the customer firm has an influence or control over the suppliers' or customers' behavior. Own brands are one such example. Conversely, non-organizational stakeholders such as environmental groups may legitimately argue that the customer should also take responsibility for these suppliers' or customers' environmental policies. The third area is the Sphere of Concern. This is where environmental issues may be of concern to the customer firm, but they may have little or no control over these policies. For example, most UK retailers have labeling policies for GMOs, although in 1997 the US FDA issued policies counter to some of these labeling policies, putting the supermarkets under pressure from anti-GMO activists. The chlorine-free pulp issue, pesticide & herbicide use and animal husbandry programs are examples where the boundary between spheres of influence and concern become rather grey, begging the question as to whether a supermarket has a legitimate right and/or obligation to proscribe behavior in these areas. The distinction between these spheres is therefore not always clear-cut, nor is it static – what may be legitimate behavior today may not be so tomorrow. Understanding these dynamics is thus a key challenge for management.

Consistent with the literature discussed above, there is a clear relationship between the pressures to which firms were exposed and the environmental supply chain initiatives. As a high profile firm, Sainsbury's was under considerable environmental pressures. They are close to the end customer and one of the most recognized firms in the country. There was also considerable publicity generated by their suppliers' activities. For example, public pressure focused on packaging, animal rights, CFC refrigerants, pesticide use and chlorinated paper. To use Mitchell *et al.'s* (1997) argument, there was a sufficient degree of urgency and legitimacy. Sainsbury's initiatives could be traced back to these specific pressures, either in reality, the potential threat or the perception that the pressure existed. Firms allocate resources to environmental issues in response to pressures exerted upon the firms, which can be interpreted as a means of reducing risk from external agents and market conditions.

Capabilities are also needed to facilitate this process, specificlly sufficient channel power and technical capabilities. Sainsbury's 'hands on' own-brand policies allowed them to gain both channel control and technical proficiency. Consistent with Lamming (1993), Lyons *et al.* (1990), Helper (1991) and Sako (1992), Sainsbury's own-brand strategies generated inter-firm innovation dynamics, a precursor to environmental supply chain innovation, according to Young (2000), Green *et al.* (1996) and Lamming and Hampson (1996) and Hampson and Johnson (1996).

Returning to the model, large, high profile firms are often under considerable pressure from a wide range of stakeholders to address environmental concerns generated by their suppliers' activities. By targeting the supermarket, non-organizational stakeholder groups such as environmental advocates can leverage the supermarket's abilities to change their suppliers' behavior. This is obviously more effective than going after the thousands of independent suppliers, many of whom are unknown to the public and thus unlikely to draw media attention. This sort of situation is thus both a blessing and a curse for the retailer: while they are in a position to influence control over their suppliers, they are also in a position where they must take responsibility for their actions. It is thus proposed that this model will help customer firms identify and understand the following:

- The types of pressures with which the firm is exposed. This includes not only organizational and regulatory stakeholders, but also others that may be concerned about their suppliers' activities. However, not all stakeholders have the same goals, ambitions, strategies, references or ways of thinking, what Hall and Vredenburg (2003) call 'stakeholder ambiguity'. Negotiation strategies must therefore be tailored to suit each stakeholder group.
- Their exposure to their suppliers' poor environmental policies. Depending on the exposure of the customer firm, there is the risk that poor environment performance of a supplier may become the problem of the customer firm. At the same time, smaller suppliers may not have the reasons or resources to develop environmental programs. It is due to these reasons that environmental supply chain innovation becomes a sensible strategy.

13.4 Implications and Recommendations for Further Research

This paper has introduced the Sphere of Influence Model as a tool by which firms can better understand the environmental circumstances which they face. More specifically, it has been used to demonstrate why environmental supply chain dynamics occur. However, it is important to note that the model may not be appropriate for all firms. It is generally applicable to high-profile, dominant supply chain members exposed to environmental pressure, as illustrated by the apex of the 'hour glass'. Previous research by the author found that these criteria were not applicable to Japanese supermarkets, as the retailers were dominated by manufacturers and not under the same degree of environmental scrutiny as their British counterparts. Interestingly the requisite characteristics were emerging in Japanese convenience stores, and environmental supply chain dynamics were beginning to evolve (Hall, 2000), and similar dynamics are emerging in the energy sector (Hall and Vredenburg, 2003). Comparable results could be also expected in the automotive industry, both of which possess dominant supply chain members exposed to environmental pressures. It should also be noted that elements within the model are likely to vary depending on the industry context. For example, perceptual issues may be less important for certain manufacturing industries than retailers. However, firms with high profile brands such as Sony and Nike are likely to be highly exposed.

In addition to generalizability across industries (albeit cautiously), the model can also be used beyond environmental innovation. More specifically, the environmental agenda has matured to include the broader implications of sustainable development and the constraints imposed by industry, society and the environment. For example, Monsanto's development of genetic technology for agriculture involves not only environmental issues, but also considerable social implications such as the rights of individuals to choose and concerns over farmers in the developing world being dependent upon foreign multinationals. The model can also be applied to 'socially exposed' industries such as those that exploit natural resources (e.g. forestry and fisheries) or those that are dependent upon government policy (such as aerospace).

While investments in environmental innovation may be more obvious for large, high profile firms, this model also has some implications for suppliers. First, buyer-supplier relations have become increasingly important from a strategic perspective in many industries. Suppliers must demonstrate that they are both capable and reliable. Disregarding environmental issues implies that the supplier has poor management and control mechanisms, an area of increasing concern for customer firms. It also demonstrates that the supplier is out of touch with modern business issues and firm lacks the ability to adapt and innovate. Poor environmental policies or a disregard for a customer's environmental supply chain initiatives may expose the customer firm to too much risk, and may risk being dropped. Lastly, although environmental supply chain dynamics may not be applicable to all firms, it is important to note that firms operate in a dynamic environment with emerging technologies and changing social attitudes. Firms that think they are safe from environmental pressures may very well be exposing themselves to avoidable environmental risk.

References

Company Reporting (1997). The UK R&D Scoreboard 1997, DTI, London.

Ethical Consumer, December (1996). Product Report Supermarkets. pp 5-8.

Barringer, B. (1997). The Effects of Relational Channel Exchange on the Small Firm: A Conceptual Framework, Journal of Small Business Management, April, 35 (2), 65- 79.

Bowlby, S.R and Foord, J. (1995). Relational Contracting between UK Retailers and Manufacturers, The International Review of Retail, Distribution and Consumer Research, 5 (3), 334-360.

Burt, S and Sparks, L. (1994). Structural Change in Grocery Retailing in Great Britain: a Discount Reorientation? The International Review of Retail, Distribution and Consumer Research, 4, 195-217.

El-Ansary, A. and Stern, L. (1972). Power Measurement in the Distribution Channel, Journal of Marketing Research, 9, February, 47-52

Fiddis, C. (1997), Manufacturer-Retailer Relationships in the Food and Drink Industry: Strategies and Tactics in the Battle for Power, FT Management Report, Financial Times Retail & Consumer Publishing, London.

Florida, R., 1996.   Lean and Green: The Move to Environmentally Conscious Manufacturing, California Management Review, 39 (1), 81-105.

Freeman, R. (1984). Strategic Management: A Stakeholder Approach. Pitman, Boston.

French, J. and Raven, B. (1959). The Bases of Social Power, in Cartwright, D. (ed.), Studies in Social Power, University of Michigan, Ann Arbor.

Green, K., Morton, B. and New, S. (1996). Purchasing and Environmental Management: Interactions, Policies and Opportunities, Business Strategy and the Environment, 5, 188-197.

Green, K., Morton, B. and New, S. (2000). Greening Organizations, Organization & Environment, 13 (2), 206-225.

Gules, H. And Burgess, T. (1996). Manufacturing Technology and the Supply Chain. European Journal of Purchasing and Supply Management, 2 (1), 31-38.

Hall, J., (2000). Environmental Supply Chain Dynamics, Journal of Cleaner Production, 8 (6), 455–471.

Hall, J. and Crowther S., (1998).   Biotechnology: The Ultimate Cleaner Production Technology for Agriculture? Journal of Cleaner Production, 6, 313-322.

Hall, J. and Vredenburg, H. (2003).   The Challenges of Sustainable Development Innovation, MIT Sloan Management Review, 45 (1), 61-68.

Hall, J. and Vredenburg, H. (2004).   Sustainable Development Innovation and Competitive Advantage: Implications for Business, Policy and Management Education, Innovation: Management, Policy & Practice, 6 (2), 1-12.

Hampson, J. and Johnson, R. (1996). Environmental Legislation and the Supply Chain, IPSERA: International Purchasing & Supply Education & Research Association, Eindhoven University of Technology, The Netherlands.

Harvey, M. (2000). Innovation and competition in UK supermarkets Supply Chain Management; 5, (1), 15.

Helper, S. (1991). How Much has Really Changed Between US Automakers and their Suppliers? Sloan Management Review, Summer, 32 (4), 15-28.

Henriques, I. and Sadorsky, P. (1996). The Determinants of an Environmentally Responsive Firm: An Empirical Approach. Journal of Environmental Economics and Management, 30, 381-395.

Henriques, I. and Sadorsky, P. (1999). The Relationship Between Environmental Commitment and Managerial Perceptions of Stakeholder Importance, Academy of Management Journal, 42 (1), 87-99.

Hill, K. (1997). Supply Chain Dynamics, Environmental Issues and Manufacturing Firms, Environment and Planning A, 29, 1257-74.

Hunt, C. and Auster, E. (1990). Proactive Environmental Management: Avoiding the Toxic Trap, Sloan Management Review, Winter, 7-18.

Imrie, R. and Morris, J. (1992). A review of Recent Changes in Buyer-Supplier Relations, International Journal of Management Science, 20, (5/6), 641-652.

Kemp, R. and Soete, L. (1992). The Greening of Technological Progress, An Evolutionary Perspective, Futures, 24 (5), pp 437-457.

Kumar, N. (1996). The Power of Trust in Manufacturer-Retailer Relationships, Harvard Business Review, Nov.-Dec., 92-106.

Lamming, R. and Hampson, J. (1996). The Environment as a Supply Chain Management Issue, British Journal of Management, 7 (Special Issue), S45-S62.

Lamming, R. (1993). Beyond Partnership. Strategies for Innovation and Lean Supply. Prentice Hall, London.

Lyons, T., Krachenberg, A. and Henke, J. (1990). Mixed Motive Marriages: What's Next for Buyer-Supplier Relations? Sloan Management Review, Spring, 31 (3), 29-36.

Meznar, M. and Nigh, D. (1995). Buffer or bridge? Environmental and organizational determinants of public affairs activities in American firms, Academy of Management Journal, 38 (4), 975–998.

Mills, D.E. (1995). Why Retailers Sell Private Labels, Journal of Economics and Management Strategy, 3, 509-528.

Mitchell, R., Agle, B., and Wood, D. (1997). Toward a theory of stakeholder identification and salience: Defining the principle of who and what really counts. Academy of Management Review; 22 (4), 853-886.

Mudambi, R. and Helper, S. (1998). The 'close but adversarial' model of supplier relations in the U.S. auto industry, Strategic Management Journal, 19, (8), 775-792.

Murphy, D. and Bendell, J. (1997). In the Company of Partners, Policy Press, University of Bristol.

Narasimhan, C. and Wilcox, R. (1998). Private labels and the channel relationship: A cross-category analysis, Journal of Business, 71 (4), 573-600.

New, S, Green, K. and Morton, B. (1997). The Sustainable Supply Chain: Theoretical Perspectives and Practical Developments, Unpublished Working Paper. Hertfort College, Oxford and Manchester School of Management, UMIST.

Pfeffer, J., & Salancik, J. R. (1978). The External Control of Organizations. New York: Harper & Row.

Ring, P. and Van de Ven, A. (1992). Structuring Co-operative Relationships between Organizations, Strategic Management Review, 13, 483-498.

J Sainsbury Plc. (1997). Annual Environmental Report, London.

Sarkis, J., 2000. A Strategic Decision Framework for Green Supply Chain Management. Paper presented at the Academy of Management, Toronto, Canada.

Sako, M. (1992). Prices Quality and Trust: Inter-firm relations in Britain and Japan. Cambridge University Press Cambridge, UK.

Senker, J. (1986). Retail Influence on Manufacturing Innovation, D.Phil Thesis, SPRU, University of Sussex.

Senker, J. (1989). Food Retailing, Technology and its Relation to Competitive Strategy, in M Dodgson (ed.) Technology Strategy in the Firm, Longman, Oxford.

Smith, D. and Sparks, L. (1994). The Transformation of Physical Distribution in Retailing: The Example of Tesco PLC. The International Review of Retail, Distribution and Consumer Research, p 35-63.

Walley, N. and Whitehead, B. (1994). It's Not easy Being Green, Harvard Business Review, May-June, 46-53.

WHO (World Health Organization) (1999). Dioxins and their Effects on Human Health, Fact Sheet No 225, http://www.who.int/inf-fs/en/fact225.html, accessed July 28, 2002

Williams, H., Medhurst, J. and Drew, K. (1993). Corporate Strategies for a Sustainable Future, Environmental Strategies for Industry (Fischer, K and Schot, J eds.). Island Press, Washington DC.

Wrigley, N. (1997). Exporting the British Model of Food Retailing to the US: Implications for the EU-US Food Systems Convergence Debate. Agrobusiness, 13 (2),137-152.

Yin, R. (1993). Applications of Case Study Research. Sage, London.

Yin, R. (1994). Case Study Research. Design and Methods. Sage, London.

Young, R. (2000). Managing residual disposition: Achieving economy, environmental responsibility, and competitive advantage using the supply chain framework. Journal of Supply Chain Management, 36 (1), 57-66.

14

Greening Supply Chains: A Competence-based Perspective

Frank Ebinger[1], Maria Goldbach[2] and Uwe Schneidewind[3]

[1]Albert-Ludwigs-Universität Freiburg, Institut für Forstökonomie, Tennenbacher Str. 4, 79106 Freiburg, Germany. f.ebinger@ife.uni-freiburg.de
[2]Institute of Business Administration, Chair for Production and the Environment, Carl von Ossietzky University Oldenburg, PO Box 2503, 26111 Oldenburg, Germany. maria.goldbach@uni-oldenburg.de
[3]Institute of Business Administration, Chair for Production and the Environment, Carl von Ossietzky-University Oldenburg, PO Box 2503, 26111 Oldenburg, Germany uwe.schneidewind@uni-oldenburg.de

Increasing uncertainty of supply networks, globalisation of businesses, proliferation of product variety and shortening of product life-cycles have forced organisations to co-operate with their supply chain partners. Along with this, information networks and technological convergence are redefining the rules of economic and trading relationships within business. To succeed today and in the future, organisations need to create strong linkages with their business partners by using the concept of supply chain management. More and more organisations today are realising the importance of developing and implementing a comprehensive supply chain strategy and are linking this strategy equally to their overall business goals.

In addition, companies find themselves increasingly confronted with demands for green (i.e. ecologically optimised) products by an increasing number of stakeholders, such as customers, legislators or environmental groups. This greening of products requires ecological optimisation along the entire value-adding process (i.e. from cradle to grave).

This has led to the appearance of green supply chain management, a challenging task in two regards:

- In the greening of supply chains the economic objectives from conventional supply chains are extended to include ecological objectives that must be taken into consideration simultaneously. This not only increases complexity in the chain but also may equally lead to conflicting interest constellations between economic and ecological requirements.
- The greening of supply chains requires specific resources to respond to economic and ecological objectives. These resources as well as the related actors are not necessarily identical to those in conventional chains. The greening of supply chains often risks failure because of a lack of suitable resources and actors.

Therefore, the combination of material and information as well as the relationship-related dimensions of supply chain management with a strategic resource and actors perspective is of crucial importance in ensuring the greening of supply chains. In this chapter we address the question: 'what specific resources are required for the greening of supply chains and which difficulties arise in greening?'

From a theoretical perspective, we develop a framework combining the duality of eco-competitiveness (including economic and ecological objectives) with the competence-based view (Section 14.1). In analysing the competence needed for greening supply chain management, with use of Cooper and Slagmulder's (1999) two phases of supply chain 'design' and 'realisation', the resources and actors relevant in the different stages of supply chain management are discussed from a greening perspective (Section 14.2). The theoretical framework is illustrated with the example of the greening of the cotton chain at the German mail-order business OTTO (Section 14.3). Some conclusions are presented in Section 14.4.

14.1 Theoretical Framework

14.1.1 Organisational Fields in Supply Chain Management

Competitiveness can be ensured only by proactively and effectively managing the material and information flows as well as relationships along the entire value-adding process (e.g. see Handfield and Nichols 1999; Bechtel and Jayaram 1997). According to Handfield and Nichols (1999: 2), the supply chain is defined as follows:

> The supply chain encompasses all activities associated with the flow and transformation of goods from raw materials stage (extraction), through to the end-user, as well as the associated information flows. Material and information both flow up and down the supply chain. Supply chain management (SCM) is the integration of these activities through improved supply chain relationships, to achieve a sustainable competitive advantage.

This definition refers both to the material and information dimension related to the product and the relationship dimension, as concerns the actors involved.

In this context, Cooper and Slagmulder identify design and realisation as two phases in supply chain management (see Cooper and Slagmulder 1999). In greening supply chains, these two phases need to be developed, as they do not explicitly refer to the optimisation objectives of the supply chain. Just as in traditional supply chain management economic optimisation is generally implicitly assumed as the unique objective, green supply chains equally demand ecological optimisation. The concurrent optimisation of economic and ecological objectives is discussed in Section 14.1.3.

The greening of supply chains is based on the identification and activation of suitable actors and the combination of resources. With this in mind, the definition of Handfield and Nichols must be extended to a resource-based perspective, stretching the content of supply chain management from material and information flows to competence-based management along the entire chain. The following analysis is based on the latest developments of the resource-based view, also known as competence-based strategic management (see Sanchez and Heene 1997).

14.1.2 A Competence-based View

The resource-based view of a firm originates from a frustration with the structure–conduct–performance paradigm of the industrial organisation view of the firm (see Bain 1959; Porter 1980). The early resource-based theorists found the industrial organisation view (that a firm's success i wholly determined by its external environment) to be unrealistically limited, and therefore turned to the seminal work of Penrose (1959) for motivation. To counter the industrial organisation view, a number of researchers, such as Wernerfelt (1984), Dierickx and Cool (1989) and Prahalad and Hamel (1990), built a resource-based theory grounded the idea of firms' internal competences.

Since the publication of the article 'The Core Competence of the Corporation', by Prahalad and Hamel (1990), the concept of competence has attracted great attention (e.g. see Hamel and Prahalad 1994; Sanchez et al. 1996). Adopting the view that 'competence is an ability to sustain the co-ordinated deployment of assets in a way that helps a firm achieve its goals' and that competence is characterised as 'a bundle of skills and technologies rather than a single discrete skill or technology', Sanchez et al. (1996) and Sanchez and Heene (1997), in their competence-based management theory suggest an open-system view be taken of the firm, the company being connected by hierarchical and 'systemic' relations, guided by a strategic logic derived from managerial cognition and governed by management processes to co-ordinate asset stocks and flows (Van den Bosch and van Wijk 2000). As Sanchez et al. (1996) point out, firms follow their own strategic logic to achieve goals by using resources, capabilities and skills to create, produce and offer products to markets. In the model, six resources are distinguished (see Fig. 14.1):

- Strategic logic
- Management processes
- Intangible assets
- Tangible assets
- Operations
- Products

In the following discussion, we will concentrate on three resource areas, addressed as accessible external dimensions, for developing internal resources (see Table 14.1): intangible resources, tangible resources and capabilities.

14.1.2.1 Tangible Resources

Tangible resources are the fixed and current assets of the organisation that are available in the long term (Wernerfelt 1989). Their ownership properties and value are relatively easy to measure).

Tangible resources can be divided into two sub-criteria: financial and physical resources. The book value of these assets is assessed through conventional accounting mechanisms and is usually reflected in the balance sheet evaluation of companies. The other defining characteristic of tangible assets is that they are transparent (i.e. easily identifiable), and it is relatively difficult for companies to avoid competitors duplicating them (Grant 1991, Fahy and Smithee 1999).

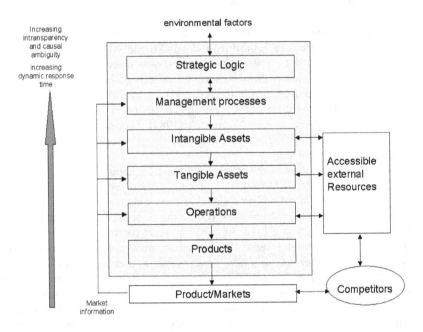

Figure 14.1. Competence-based management theory: firms as an open system. Source: Hinterhuber, et al., 2000.

14.1.2.2 Intangible Resources

Intangible resources are increasingly considered to be the ultimate roots of a company's success and are widely recognised as making a key contribution to value creation. They can be subdivided into two sub-criteria: skills and other assets. In the main, intangible assets have a relatively unlimited ability to generate competitive advantage, and firms can exploit this by using these assets in-house, renting them out (e.g. by license) or selling them (e.g. by selling a brand) (e.g. see Fahy and Smithee 1999, Wernerfelt 1989). Individual skills are not necessarily owned by the firm by contract or agreement, making them inimitable and non-substitutable. However, as noted above, they may be hired by competitors. Intangible resources are relatively difficult for competitors to duplicate.

An example of an intangible resource is intellectual property, which is afforded regulatory protection; examples of asset stocks are databases, networks and reputation (Dierickx and Cool 1989; Hall 1992). The inherent complexity and specificity of the accumulation of such assets hinders imitation and substitutability in the short run.

Table 14.1. Organizational resources

Resource	Subcriteria	Examples	Examples from a Greening Perspective
Tangible Resources	Financial Assets	debtors, financial reserves and bank deposits	Special budget for greening technology or organizational developments
	Physical Assets	plant, equipment, land, other capital goods and stocks (of raw materials)	Recycling equipment
Intangible Resources	Skills	reputation, technology, human resources, culture, the training and expertise of employees, commitment and loyalty, company reputation	Integrated environmental sound technologies
	Intangible Assets	trademarks and patents, brand, company networks and databases	Green (product) labels, Green brand
Capabilities		business forms and organization configuration, teamwork, organizational culture, trust between management and workers, production know-how, cooperativeness	Pollution prevention, Design for environment, Social development, Stakeholder dialogue

An example of an intangible resource is intellectual property, which is afforded regulatory protection; examples of asset stocks are databases, networks and reputation (Dierickx and Cool 1989; Hall 1992). The inherent complexity and specificity of the accumulation of such assets hinders imitation and substitutability in the short run.

14.1.2.3 Capabilities

Capabilities (managing operations) have proven more difficult to delineate and are often described as invisible assets or intermediate goods (Amit and Schoemaker 1993; Itami 1987). As tangible and intangible resources are not productive on their own, any analysis must also consider a firm's organisational capabilities—its abilities to assemble, integrate and manage these bundles of resources.

Essentially, capabilities encompass the skills of individuals or groups as well as the organisational routines and interactions through which all the firm's resources are co-ordinated (Grant 1991). Sanchez et al. (1996) view capabilities as 'repeatable patterns of action in the use of assets to create, produce and/or offer products to the market'. Capabilities do not have clearly defined property rights as they are seldom the subject of a transaction; this ambiguity leads to difficulties in evaluating such assets. They have limited ability to create competitive advantage in the short run owing to difficulties with learning and change but have a relatively unlimited ability to create competitive advantage in the long run (Wernerfelt 1989). Further, where capabilities are interaction-based they are even more difficult to duplicate because of the causal ambiguity.

The resource-based view literature has tended to favour capabilities as the most likely source of sustainable competitive advantage (Collis 1994).

14.1.2.4 Overview of Organizational Resources

The list of resources in any given firm is likely to be a long one, but one of the principal insights of the resource-based view is that not all resources are of equal importance or possess the potential to be a source of sustainable competitive advantage. To assess the strategic advantage of resources, a range of criteria must be considered.

Resources are distributed heterogeneously among enterprises, are at least partially immobile and must fulfil the conditions of value, rareness, transferability, non-substitutability, restricted imitability and, finally, opacity (i.e. they are difficult to identify or be reproduced as a result of their causal ambiguity, specificity, complexity and lack of 'visibility'; see Barney 1991; Collis and Montgomery 1995; Grant 1991). However, for the discussion here regarding the greening of supply chains, there is another normative criterion regarding competitive advantage: eco-competitiveness.

14.1.3 Eco-competitiveness

Even though companies are faced with demands for green products by an increasing number of stakeholders, green products currently often remain in market niches (see Meyer 2001). One major reason for this is that they are often not competitive in comparison with conventional products. This is because of the fact that in greening their products many companies take into only the account ecological aspects without respecting the economic aspects. In greening supply chains, economic and ecological objectives must be taken into consideration simultaneously. This not only increases complexity in the chain, but also equally may lead to different interest constellations between economic and ecological requirements. These constellations may be conflicting or synergetic. An ecologically optimised dyestuff for example may ensure the achievement of ecological objectives but also may be much more expensive than conventional dyes, thereby negatively affecting economic objectives.[1] The process optimisation of dyeing may, however, lead to the reduction of water and energy use (i.e. ecological improvements), while at the same time saving costs, thereby positively affecting economic objectives.

---

[1]For example the dyestuff 'indigo' is used to colour blue jeans. The German Company BASF did research into the eco-efficiency of different ways of producing this substance. In future, BASF will produce indigo in what it has sestablished to be the most eco-efficient way.

Therefore, companies must develop strategies integrating both economic and ecological aspects (see Hummel 1997). According to Hummel, these eco-strategies aim at the achievement of competitive advantages necessary to ensure survivability in the long term, reducing ecological problems along the ecological product life-cycle as far as possible.

To ensure its competitiveness, a company must respect minimum requirements in each of the three dimensions: costs, differentiation and ecology. These minimum requirements serve in assuring the survivability of the firm. The maximum requirements correspond to the benchmark of the concerned industry sector.

For the active management of the entire supply chain, Hummel's concept needs to be extended to the management of supply chains. In this chapter, the cost and differentiation dimensions will be integrated into the economic dimension. In the following, green supply chain management is defined as follows.

14.1.3.1 Definition: Green Supply Chain Management
The supply chain encompasses all activities associated with the ecological and economic flow and transformation of goods, from the raw materials stage (extraction), through to the end-user, as well as the associated information flows. Material and information flow up and down the supply chain. Green supply chain management (SCM) is the integration of these activities through improved supply chain relationships, to achieve a sustainable economic and ecological competitive advantage.

14.1.4 Conceptual Framework of Resources in Greening Supply Chain Management

The greening of supply chains is based on green innovation processes along the entire chain in two organisational phases: design and realisation. According to Handfield and Nichols (1999: 2), the main challenge in greening is activating the required resources and capabilities (i.e. material, information, know-how and other resources) and actors in the supply chain. In this chapter, the actors are also referred to as 'resource holders', as they dispose of resources (i.e. they 'hold' them).

Using Sanchez and Heene's (1997) theory, we model the greening of supply chains as a process of shared competence along the whole supply chain. In light of the perspectives discussed above in this chapter we will use the combined scheme illustrated in Figure 14.3, simplified from the view of a case study firm (in this case it is an assembly processing firm).

From this point of view, firms may share their competence either in a dyadic way with their direct partners in the chain, or by channel integration, trying to use the competence of another partner in the chain by using the (virtual) strategic resource pool (see Cooper et al. 1997, 71). This 'supply chain strategic resource pool' can be developed by one or all partners in an active or indirect way (by a normal learning process of one firm in the chain, sharing its new knowledge throughout the pool). In addition, the pool will develop by changes in the actors in the chain (i.e. through exits and entrances).

The detailed view of the chain shown in Figure 14.3 shows that the greening of supply chains is a question of using the 'right' resources from the strategic pool. (In this case, the view for an assembly firm is shown. Note that this view is valid for all other stages except the use stage. For the use stage, the eco-competitiveness scheme needs to be used; that is, choice of the right resource is also a part of the strategic logic of the firm, the strategic resource itself [value] and the absorptive capacity of the firm.)

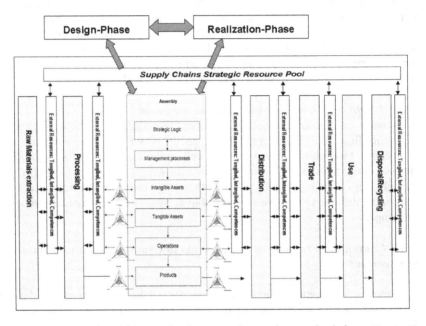

Figure 14.2. A conceptual framework of resources in greening supply chain management

In the next section, we will discuss possible resources from the view of focal firms within the two phases: design and realisation. Here, we use the assembly stage as an example and will concentrate our analysis on the three resources discussed in Section 14.1.2: tangible and intangible resources, and capabilities.

14.2 Resources and Actors in Greening Supply Chain Management

14.2.1 Resources and Actors in Greening the Design Phase

When the aim is not simply to compete successfully in product-related innovation but also to achieve the goal of greening the supply chain, a number of technical, economic and social capacities are required. As the World Business Council for Sustainable Development (WBCSD 1996: 24) has underlined:

At the design stage, the function of the product, process or service is defined, and raw materials, supplies and process chemicals are selected. These in turn determine the energy which will be consumed to create them and the waste which will be generated. In addition, for products, their durability, serviceability and energy consumption during their lifetime will also be determined.

Thus, the majority of environmental impacts arising from production, consumption and disposal of the product are determined during the product design process, which depends on actors in the whole supply chain.

Firms may include the following innovation concepts in the design phase (e.g. see Bierter 2002: S171):

- Eco-efficient process optimisation
- Eco-efficient product optimisation
- Eco-efficient design of new products
- Eco-efficient system design

Independent of the innovation concept chosen by an innovative focal firm, the firm must decide whether the resources required are to be developed internally or activated externally by co-operating with an actor external to the supply chain strategic resource pool. Our research experience suggests that every innovation strategy needs different kinds of resources and actors working on innovative solutions.

14.2.1.1 Eco-efficient Process Optimisation

Innovations in eco-efficient process optimisation have the objective of reducing the environmental burdens of firm-related production processes. Therefore, the goals are internal organisation and improved technological approaches (cleaner production). Efforts to reduce production-related waste, water, emissions and energy consumption also belong to this category. To follow this strategy, the involvement of external actors is not generally required. Sometimes, however, in special cases, external consultants or suppliers may be needed to provide some special technological or process know-how. Table 14.2 provides examples of eco-efficient process optimisation in the case of the case-study mail-order firm, Otto.

14.2.1.3 Eco-efficient Design of New Products

The eco-efficient design of a new product takes into consideration environmental burdens reduction throughout the product system and aims to prolong the use phase. To keep pace with competitive innovation, numerous technological capacities and capabilities are required, together with a critical mass of tangible and intangible technological assets. In this, and in the potentially close relationship between different technological directions (spillover), we find one of the key causes of co-operative behaviour and opportunities in the industrial innovation process. Normally, in this innovation strategy we find included all actors along the entire supply chain, including downstream actors such as traders (industrial), consumers and recyclers. In particular, traders and consumers will provide the innovative company with information and know-how relating to the 'use has and to the creation of new markets (e.g. see von Hippel 1988).

Table 14.2. Resources and capabilities in the design phase in the Otto-case

Strategy	Tangible Resource		Intangible Resource		Capability	Needed Actors
	Financial assets	Physical assets	Skills	Intangible assets		
Eco-efficient process optimisation	Training budget to support existing suppliers		Technological skills to replace heavy metal dyestuffs, chlorine bleach and easy care treatment Experience in multicultural business	Eco-management systems	Management of supply chain wide process reorganization	fiber producers, spinning companies, textile producers, clothing producers, distributors, cotton and yarn traders
Eco-efficient product optimisation	Training budget to support existing suppliers by a consultant	Flexible stocks of green cotton textiles	Know how in processing organic cotton Experience in multicultural business	Eco-Tex 100 Standard	Building and leading an innovation-team of different actors along the supply chain	fiber producers, spinning companies, textile producers, clothing producers, distributors, cotton and yarn traders

14.2.1.4 Eco-efficient System Design
From the view of eco-efficient system design, products are embedded in technological and social structures (social institutions) which should be influenced in a green way. This kind of innovation strategy considers product innovation as a means of improving individual products and the system as a whole in which they are embedded. Here, the questions that come to the fore relate to the interfacing, transfer and realisation of resources and capabilities in order to improve an entire product system in a green way. In addition to all the resources and capabilities listed as being required for the eco-efficient design of new products (see Subsection 14.2.1.3) this strategy calls mainly for capabilities in co-ordinating a system-wide innovation network, to integrate actors along the supply chain and throughout the wider product system (such as regional service providers to carry out maintenance and repair).

14.2.2 Resources and Actors in the Realisation Phase

Whereas the central resources in the design phase tend to be oriented towards resources that are both tangible and intangible, the realisation phase is concerned more with intangible resources and capabilities to support efficiency in a concretely developed supply chain. Normally, the actors involved in the design process will be selected by the focal firm to build up the supply chain.

The relevant resources follow the principles of process innovation. According to Keller (1997), the following principles are also fundamental to greening the realisation phase and apply both to material and to information flow:

- Comprehensive system optimisation in terms of market and competition
- Time efficiency
- Flow optimisation and stock reduction
- Process assurance. [2]

However, resource-oriented design principles in greening the realisation phase can be classified according to various criteria, such as:

- System modification
- Innovation for optimisation

14.2.2.1 System Modification
From the system modification view, optimisation is carried out in relation to process structure and technologies. In this case, roles are played by resources related to materials and information technology (IT; which, e.g., permit

---

[2] As a consequence of the globalisation of markets in goods, labour and information, new competitors are entering historically entrenched or even closed markets: According to Picot et al. (1997: 124-25, our translation), 'sellers' markets are transforming into buyers' markets. The buyers have become more demanding and are also no longer willing to accept problems of an organisational nature such as long delivery times or interface problems in the case of processes'.

procurement to be synchronised with production) and by manufacturing methods optimised in terms of capacity, quality, ecological considerations and cost. The skill levels of staff must be adapted to the new range of possibilities. Depending on the scope of integration, resources are chosen with reference to the extreme poles of intra-processual and inter-process resources, intra-functional and inter-functional resources, and isolated-function and combined-function resources. In Table 14.3 we give an example of system modification in relation to the case-study firm Otto, that follows after (see Section 14.3).

Table 14.3. Resources and capabilities in the realisation phase in the case of the German mail-order firm, OTTO

Resource	System modification
Tangible	
Financial assets	Investments in stocks of organic cotton
Physical assets	Established distribution channels
Intangible:	
Skills	Co-operation and communication skills Experience in multicultural business
Other	Good supplier relationships Good stakeholder relations Tools to control the supply chain (e.g. tools for supply chain costing)
Capabilities	Management and control of environmental optimisation along the supply chain
Actors required	Suppliers[a] Stakeholders

[a]Fibre producers, spinning companies, textile producers, clothing producers, distributors, and cotton and yarn traders.

14.2.2.2 Innovation for Optimisation
The second class of design principles is driven by optimisation efforts, placing potential effects at the centre of resource selection. In this case, the following goals are at the forefront when choices are made (Weber 1991: 38 ff.):

- Ensuring the relevant environmental product and process criteria are met
- Reducing costs arising from direct and indirect elements of the value chain
- Increasing flexibility in terms of quantity, product and delivery times
- Reducing throughput times, delivery times and replenishment times

- Increasing delivery reliability
- Increasing transparency and controllability
- Increasing job satisfaction and staff motivation

14.3 The Case of the German Mail-Order Business, OTTO

14.3.1 The Design Phase in Greening the Supply Chain of OTTO

The German mail-order business OTTO has managed textile chains for many years. Traditionally, its suppliers, generally fully integrated clothing manufacturers, considered ecological requirements according to the Eco-Tex 100 standard. This standard consists mainly of checking the finished apparel with regard to harmful substances. It does not, however, involve further ecological optimisation of the production process or the used fibres. In 1997, OTTO decided to go one step further by extending its ecological optimisation of cotton apparel to the entire chain, from the raw materials stage through to the finished apparel. This decision induced major changes with respect to required technological resources, know-how and management resources. In order to clarify the necessary changes in greening such a supply chain, a typical textile chain for clothes is illustrated in Figure 14.4.

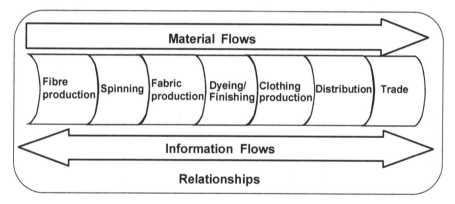

Figure 14.3. A typical textile chain for clothes

The greening of the cotton chain induces ecological optimisations on two levels. First, changes are made at the beginning of the chain (i.e. at the level of fibre production), whereby conventionally grown and produced fibres are replaced by ecologically optimised fibres (i.e.organic cotton). This is an example of eco-efficient product optimisation. The use of organic cotton requires a switch from conventional to organic cotton farming. This switch is linked to fundamental changes both at the technological level and at the know-how level (e.g. see Meyer and Hohmann 2000; Myers and Stolton 1999).

Second, changes are made at the dyeing and finishing level, so that eco-efficient process optimisations must be realised. These changes consist of substituting ecologically harmful dyestuffs and additives with ecologically optimised substances. With regard, for example, to dyeing, dyestuffs containing heavy metals must be replaced by alternatives free of heavy metals, and use of chlorine bleach must either be reduced or replaced by use of oxygen bleach. In finishing, 'easy-care' treatment must be eliminated or restricted (e.g. see Myers and Stolton 1999; Van Elzakker 1999a). These ecological optimisations induce changes and therefore require specific technological resources and know-how at the levels of fibre production and dyeing and finishing.

The question arising next is whether OTTO is able and willing to provide these resources internally (i.e. within its existing supply chain), or externally (i.e. choosing new supply chain partners). This decision relates to the required management resources for carrying out the innovation. OTTO's suppliers that make clothing from cotton grown and produced by conventional means do not usually also produce clothes from organic cotton and therefore lack the necessary technology and know-how to carry out such innovations. Therefore, OTTO was confronted with the necessity of identifying actors capable of realising this innovation. Within this context, two alternatives emerged. First, it could find different suppliers already possessing the required resources and, second, it could develop and evaluate existing suppliers in order to acquire the resources internally.

The choice to find different suppliers having the required resources was soon given up because of existing market conditions. The worldwide amount of organic cotton is still relatively small in comparison with conventional cotton. Therefore it cannot be purchased on spot markets. Moreover, the textile industry is traditionally a particularly dynamic industry, with quickly changing market cycles. Short-term, transaction-based relations occurring on a spot market generally characterise the relationships between suppliers and buyers. There is hardly any co-operation between the various actors along the supply chain. However, ecological optimisations along supply chains are oriented towards the middle or even long term, requiring close co-operation with all partners along the chain. Based on this, OTTO gave up on the option of finding new suppliers.

Therefore, the option of evaluating and developing existing suppliers to acquire the necessary resources was pursued. OTTO's partners, located in Turkey, are often fully integrated clothing manufacturers carrying out weaving and knitting, dyeing and finishing, and clothing production. The relationships between OTTO and its supply chain partners are limited to these clothing producers. In contrast to the mainstream textile industry, OTTO built up a pool of suppliers characterised by middle-term to long-term relationships. Within this pool of suppliers, the different buying departments in OTTO may choose among the suppliers and have flexibility in the placing of orders for articles.

In the context of these stable relationships, it had already been possible to realise accreditation to the Eco-Tex 100 standard. As the co-operation with these existing suppliers worked well, OTTO opted for the strategy of carrying out the greening of dyeing and finishing operations with existing suppliers. The pool of suppliers for 'green' cotton clothes was further reduced, and co-operation intensified.

As to the substitution of conventional cotton with organic cotton at the level of fibre production, suitable partners had to be found here as well. Traditionally, OTTO had not co-operated with cotton or yarn traders at all. Its suppliers (i.e. the fully integrated clothing producers) also had not traditionally co-operated with organic yarn or cotton traders. Therefore, partners had to be found possessing both the know-how and the technology to provide organic cotton yarn. Regarding the yarn, an existing supplier to the clothing producers was chosen for which the technology and know-how remained to be developed in terms of the processing of organic cotton. Regarding the cotton, an entirely new partner already possessing the required resources in trading organic cotton was chosen. In contrast to what had happened previously in the conventional textile chain at OTTO, there was co-operation not only between the organic cotton trader and the yarn trader and between the organic yarn trader and the clothing producers, but also OTTO equally co-operated directly with the cotton and yarn traders. In other words, OTTO started to co-ordinate the entire chain. This co-ordination required entirely new management resources. The greening of the cotton chain implies a strong co-operation with the organic cotton and yarn traders as well as with the clothing producer (which carries out the ecologically optimised dyeing and finishing) along the entire supply chain. The traditional co-ordination of the supply chain through market transactions was therefore replaced by hybrid co-ordination arrangements based on intense co-operation between all partners in the chain.

For the design of its 'green' cotton clothes, OTTO decided to use its conventional chain by developing and evaluating those Turkish suppliers that already possessed relatively high ecological standards and therefore potentially held the necessary resources.

The strategic concept in product development was discussed with all partners along the entire chain—a conventionally unfamiliar approach for OTTO, involving major organisational change. In conventional business, OTTO had never dealt with the cotton or yarn trader or acted in the role of a supply chain co-ordinator. Now, all partners involved in the OTTO supply chain, from cotton trader to clothing producer, had to become active and start co-operating on how to process organic cotton and realise ecological process improvements in dyeing and finishing. These changes represented an important restructuring of management resources, linked to an entirely new design of co-ordination in the supply chain.

Concerning knowledge and technology resources for processing organic cotton, the organic cotton trader had to discuss with the spinner changes likely to occur at the spinning stage. The major problem arising in this context was quality requirements for organic cotton. Owing to its limited availability, the traditional blending used in conventional cotton processing could be applied to the organic process only to a limited extent. Similar problems in the design of organic cotton fabrics occurred between the spinner and the integrated clothing producer making the fabric. The limited availability and the relatively small production quantities in organic cotton led to a limited variety of organic yarn, equally affecting the variety of organic fabrics at the level of the clothing producer. These small quantities not only have an effect on differentiation by reducing variety but also lead to higher costs that must be managed in the greening process.

The product design of 'green' cotton apparel became equally relevant at the levels of dyeing and finishing. The chosen suppliers did not have the necessary resources to realise the required process optimisations. Therefore, OTTO supported its suppliers by acquiring the technological resources and other know-how needed in dyeing and finishing. This was accomplished by providing a consultant. The challenge did not consist solely of developing and evaluating the suppliers regarding technological resources and other know-how, however. The envisaged ecological optimisations also affected the optical characteristics of the textiles (see Meyer 2001). The economic implications, especially concerning differential quality of the finished apparel, were intensively discussed between OTTO and its suppliers during the design stage. Economic implications also occurred on the cost level. The use of ecologically optimised substances were linked partly to higher costs, whereas fundamental process optimisations tend to induce cost savings (see Van Elzakker 1999b). The evaluation of these economic implications required the know-how of all actors involved in the process.

To co-ordinate the overall product design process in the chain, OTTO took over a chairing role to co-ordinate the combination of technological resources, know-how and management resources.

14.3.2 The Realisation Phase in Greening the Supply Chain of OTTO

In the formation of the production network, the final partners for the production of 'green' cotton textiles were selected. OTTO kept all the partners informed during the product design stage. The specific steps for production were finally set up, taking into consideration the existing potential of each partner. This setup included the implementation and reservation of green capacities from the total pool of production capacity at each level of the supply chain. For all partners involved, the introduction of the use of organic cotton was linked to parallel production, higher switching costs and greater administrative effort and therefore was linked to increased costs. This implies that the green production processes had to be carried out additionally and separately from conventional processes. Therefore, the movement of goods from one stage to the next had to be managed. Owing to the uncertainty of final customer demand, the chain was set up according to a pull system (i.e. production was driven by final customer demand). In this context, OTTO, being nearest to the final customer, maintained the co-ordination of the entire chain. An exception to the use of the pull system occurred at the level of fibre production. Independent of final customer demand, OTTO fixed a production quantity for organic cotton, which was stored at the spinning facilities.

The realisation of technological resources and know-how occurred mainly at the product design stage. In this phase, OTTO had to develop important management resources to fulfill its role as a supply chain co-ordinator, either managing the necessary negotiation processes entirely or supporting them in a chairing role. Here, the focus lay in the optimisation of material and information flows as well as on relationships in the existing chain set up in the preceding stages. The major resources in this stage included management capabilities, such as capacity planning, cost reduction, increased reliability of deliveries, improved response times as well as shorter delivery times. To sum up, the major challenge lay in ensuring final customer satisfaction.

This challenge is particularly difficult to handle in the 'green' cotton chain at OTTO. Demand from the final customer must go all the way back through the chain, from the fully integrated clothing producer to the spinner and cotton producer. In this particular mail-order business, capacity is particularly difficult to plan because of the high fluctuation of demand caused by the catalogue's semi-annual appearance. Only at that stage can final customer demand be identified. Extremes from little demand up to high peaks of demand for certain products are possible. These are particularly difficult to handle because of the limited availability of organic cotton and yarn and the specific resources along the entire chain. If the OTTO chain is not able to fulfill the demand, there is little possibility for OTTO to acquire alternative 'green' clothes externally. If a single partner in the chain cannot assure reliability in delivery, the whole system risks collapse as the necessary resources may be purchased only in this specific chain and not on a spot market. A solution may be a mixed 'push and pull' system along the chain that is partly realised by OTTO, but such a system is also susceptible to problems. On the one hand, in the case of capacity planning based on a push logic, there is always the risk of producing significantly over or under capacity. On the other hand, if there are no reserves at all, delivery and response times increase significantly. Any effort to shorten delivery times within a pull system is necessarily linked to cost increases, as exceptional, unforeseen material purchases are realised at a far higher price than are planned purchases.

Therefore, from a management resource point of view, the 'green' cotton chain at OTTO is particularly difficult to handle because of the limited material availability and high uncertainty in demand. This implies that process optimisation of material and information flows is possible only up to a limited extent.

14.4 Conclusions

In this chapter we have shown that the greening of supply chains is an integrated organisational process, responding at various levels in the phases of design and realisation. Owing to the increasing complexity of technological developments, access to sources of knowledge and technology external to enterprises is gaining a new status. This includes both the application and combination of knowledge and other intangible and tangible resources, thus exerting an influence not only on the intensive but also on the extensive technological possibilities. In this chapter we have discussed only actors and resources within the supply chain. However, our research results show that resources and related actors should not simply search the existing supply chain—regarding the strategic approach in the product design process, lateral co-operation with other actors (e.g. universities and consultants) should also be considered.

The theoretical reflections and the case of OTTO shows that in the design phase the most important resources are those relating to the technology, knowledge and management capabilities held by internal and external actors capable of fulfilling the ecological requirements. In the realisation phase, intangible resources become very important for green developments in the supply chain. Here, mainly co-operation and communication skills, good supplier relationships and tools for controlling the supply chain (e.g. supply chain costing) are important (see Seuring 2001).

For greening the supply chain, companies should have the following questions in mind in order to manage the re-organisation process:

- Which strategy of greening the supply chain should be followed in the design and the realisation phases?
- Which resources and capabilities are needed to implement that strategy along the supply chain?
- Which actors in the supply chain are needed to provide the required resources and capabilities?

References

Amit, R., and P.J. Schoemaker (1993) 'Strategic Assets and Organisational Rent', *Strategic Management Journal* 14 (January 1993): 33-46.

Bain, J.S. (1959) *Industrial Organisation* (New York: John Wiley).

Barney, J.B. (1991) 'Firm Resources and Sustained Competitive Advantage', *Journal of Management* 17.1: 99-120.

Bechtel, C., and J. Jayaram (1997) 'Supply Chain Management: A Strategic Perspective', *International Journal of Logistics Management* 8.1:15-34.

Bierter, W. (2002) 'System-Design: Radikale Produkt- und Prozessinnovationen', in *Jahrbuch Ökologie 2002* (Munich, Germany)

Collis, D.J. (1994) 'Research Note: How Valuable are Organisational Capabilities?' *Strategic Management Journal* 15 (Winter 1994): 143-52.

Collis, D.J., and C.A. Montgomery (1995) 'Competing on Resources: Strategy in the 1990s', *Harvard Business Review* 73.4: 118-29.

Cooper, R., and R. Slagmulder (1999) *Supply Chain Development for the Lean Enterprise: Interorganisational Cost Management* (Montvale: New Jersey, Institute of Management Accountants, Foundation for Applied Research).

Cooper, M.C., L.M. Ellram, J.T. Gardner and A.M. Hanks (1997) 'Meshing Multiple Alliances', *Journal of Business Logistics* 18.1: 67-89.

Dierickx, I., and K. Cool (1989) 'Asset Stock Accumulation and Sustainability of Competitive Advantage', *Management Science* 35: 1504-11.

Fahy, J., and A. Smithee (1999) 'Strategic Marketing and the Resource Based View of the Firm', *Academy of Marketing Science Review* 99.10.

Freiling, J. (2001) 'Entwicklungslinien und Herausforderungen des ressourcen- und kompetenzorientierten Ansatzes: Eine Einordnung in das neue Strategische Management', in H.H. Hinterhuber, S.A. Friedrich, A. Al-Ani and G. Handbauer (eds.), *Das neue Strategische Management* (Wiesbaden, Germany: Gabler, 2nd edn).

Grant, R.M. (1991) 'The Resource-based Theory of Competitive Advantage', *California Management Review* 33.3: 114-35.

Hall, R. (1992) 'The Strategic Analysis of Intangible Resources', *Strategic Management Journal* 13: 135-44.

Hamel, G., and C.K. Prahalad (1994) *Competing for the Future*, Harvard Business School Press, Cambridge Massachusetts..

Handfield, R.B., and E.L. Nichols Jr (1999) *Introduction to Supply Chain Management* (Upper Saddle River, NJ: Prentice Hall).

Hinterhuber, H.H., Friedrich, S.A., Al-Ani, A., and Handlbauer G., (2000) *Das neue Strategische Management*, 2nd edition Gabler Verlag, Wiesbaden.

Hummel, J. (1997) Strategisches Öko-Controlling (Wiesbaden, Germany: Deutscher Universitäts-Verlag).

Itami, H. (1987) *Mobilising Invisible Assets* (Cambridge, MA: Harvard University Press).

Keller, S. (1997) *Wirkungspotentiale von Prozessinnovationen* (Wiesbaden, Germany: Gabler).

Meyer, A. (2001) *Produktbezogene ökologische Wettbewerbsstrategien: Handlungsoptionen und Herausforderungen für den schweizerischen Bekleidungsdetailhandel* (Wiesbaden, Germany: Gabler and Deutscher Universitäts-Verlag).

Meyer, A., and P. Hohmann (2000) '"Other Thoughts, Other Results?": Remei's bioRe Organic Cotton on its Way to the Mass Market', *Greener Manufacturing International* 31 (Autumn 2000).

Myers, D., and S. Stolton (eds.) (1999) *Organic Cotton: From Field to Final Product* (London: Intermediate Technology Publications).

Penrose E.T. (1959) *The Theory of the Growth of the Firm* (Oxford, UK: Basil Blackwell).

Porter, M.E. (1980) *Competitive Strategy* (New York: The Free Press).

Prahalad, C.K., and G. Hamel (1990) 'The Core Competence of the Corporation', *Harvard Business Review* 3: S79-S91.

Sanchez, R., and A. Heene (1997) *Competence-based Strategic Management* (New York: John Wiley).

Sanchez, R., A. Heene and H. Thomas (1996) *Dynamics of Competence-based Competition* (AmsterdamElsevier).

Seuring, S. (2001) *Supply Chain Costing* (Munich, Germany: Vahlen).

Van den Bosch, F.A.J., and R. van Wijk (2000) *Creation of Managerial Capabilities through Managerial Knowledge Integration: A Competence-based Perspective* (draft; Rotterdam, The Netherlands: Erasmus Research Institute of Management [ERIM]).

Van Elzakker, B. (1999a) 'Organic Cotton Production', in D. Myers and S. Stolton (eds.), *Organic Cotton: From Field to Final Product* (London: Intermediate Technology Publications): 21-35.

Van Elzakker, B. (1999b) 'Comparing the Costs of Organic and Conventional Cotton', in D. Myers and S. Stolton (eds.), *Organic Cotton: From Field to Final Product* (London: Intermediate Technology Publications): 86-100.

von Hippel, E. (1988) *The Sources of Innovation* (New York: Oxford University Press).

WBCSD (World Business Council for Sustainable Development) (1996) *Sustainable Production and Consumption: A Business Perspective* (Conches–Geneva, Switzerland: WBCSD).

Weber, J. (1991) *Logistik-Controlling* (Stuttgart, Germany: Schäffer Poeschel, 2nd edn).

Wernerfelt, B. (1984) 'A Resource-based View of the Firm', *Strategic Management Journal* 5: S171-S180.

Wernerfelt, B. (1989) 'From Critical Resources to Corporate Strategy', *Journal of General Management* 14 (Spring 1989): 4-12.

Williamson, O. E. (1985) *The Economic Institutions of Capitalism: Firms, Markets, Relational Contracting* (New York, The Free Press).

'Smart' Design: Greening the Total Product System

James P. Warren and Ed Rhodes

The Faculty of Technology, Centre for Technology Strategy, The Open University, Walton Hall, Milton Keynes, MK7 6AA, UK., j.p.warren@open.ac.uk, e.a.rhodes@open.ac.uk

15.1 Introduction

When launched into the marketplace, the two-seat city car initially known as 'the smart' became iconic. In several respects, the smart car fits the criteria for an ideal city car (see Figure 15.1). Its two seat capacity matches the European average vehicle occupancy of 1.2 persons per vehicle. It is spacious inside yet still small enough to be very convenient for congested city driving. The marketing package combined within this single small vehicle; Mercedes' standards of automotive engineering and quality, a high standard of safety, distinctive styling, striking use of colour internally and externally and a high standard of interior specification. The use of highly visible sales outlets and development of fashion accessories such as 'smartware' also contributed to product strength. Capitalizing on this, its makers moved 'smart' into the realm of a brand, and the initial smart car was later redesignated the smart 'fortwo' – the first in what was intended to be a family of products. To avoid confusion between 'smart' as a brand and the initial production model, we refer to the latter throughout this chapter as the fortwo.

The production of very small automobiles has a long history in Europe, partly because of the constraints of ancient street layouts and as a means for bringing car ownership within the reach of lower income households. More recent influences include increasing congestion in European cities and growing consumer concern for the environment which, in the European Union, has been reinforced by environmental regulation of the automotive sector. Thus, early concepts for the fortwo put environmental impact at the centre of the design brief, with initial emphasis on an electric vehicle. Although electric traction was eventually abandoned, a holistic approach to the design of the product and of production processes led to comparatively low environmental impact across all stages of the cradle-to-grave product lifecycle. The innovative approach embodied in the fortwo reflects the influence of a partnership between Mercedes and Swatch during early development – see Table 15.1. Thus, the car draws strengths from Mercedes, such as in the standard of product engineering and from Swatch in such respects as the strong fashion element in interior and exterior styling. These attributes brought environmentally conscious car use closer to higher income, city-based consumers.

Figure 15.1. A fleet of fortwo cars at a dealership, awaiting collection by customers

This case study reviews how the fortwo's evolution touches on several critical areas:

- holistic concern with environmental impacts;
- use of modularity in product design;
- an intensive use of modularity in the design of the dedicated production facility developed for the fortwo;
- emphasis on participation with supply chain partners from product creation to after-sales;
- use of a highly customized build-to-order product system to 'green' the entire supply chain;
- approaches to urban transport.

We look at several contrasting aspects of the fortwo's development. One concerns the emphasis in initial marketing of the fortwo on its size, maneuverability and virtuous level of fuel consumption. In part, this was viewed in the context of new approaches to urban transport, and the potential of the fortwo in relation to the transport needs of those who do not own a vehicle, such as through vehicle sharing schemes. A second issue relates to our view of product supply as extending across the full product life cycle – in a 'total product system' (Rhodes, 2006). This concept has been given increasing meaning by the efforts of large lead companies to co-ordinate aspects of production and aftermarket support.

Table 15.1. Important events in the history of smart

1972	A radical project proposal for a 2.5 metre length car for town centres was considered by Mercedes for the year 2000. Project was shelved.
1992	The project is revived and the executive committee of Daimler Benz (Mercedes' parent company) is shown the prototype city automobile, and gives the go-ahead.
	Nicholas Hayek, Chairman of SMH (the Swiss Corporation for Microelectronics and Watchmaking Industries) coins the term Swatch-mobile and claims a working prototype exists.
Dec 1992 –Jan 1993	Hayek meets with Mercedes officials to show them what types of panel design and interior/exterior he is working on.
March 1994	The Micro Compact Car (MCC AG) joint venture is formed between Mercedes Benz AG and SMH.
June 1994	Large numbers of prototypes are built and tested.
Oct 1995	Foundation stone laid at Hambach, France, for a highly innovative manufacturing plant – smartville.
Sept 1997	The fortwo is premiered – as the smart – at the Frankfurt International Motor Show.
Oct 1997	The smartville factory officially opens.
1998	Mercedes and SMH decouple from the development, Mercedes taking full responsibility.
Oct 1998	The fortwo comes to market after some initial chassis problems and quality issues.
Nov 1998	Daimler-Benz merges with Chrysler to form DaimlerChrysler (DC).
Dec 1998	Mercedes Benz (car & truck division of DC) takes full responsibility for MCC smart.
1999	Internet sales of fortwo launched.
1999	smart centres (including sales and services network) re-launched under Mercedes stewardship.
2000	Further integration of smart network into DC.
2001- 2002	Right hand drive models launched in Japan and UK.
2003	smart roadster – a two seat sports car – scheduled for release.
2004	A new model – the 4-seat smart 'forfour' – begins production and initial launch.
2005	forfour marketed more widely.
April 2005	Plans for a fourth model – the smart 'formore' abandoned.
October 2005	Production of the roadster ceases.

It also encompasses the relationship between end-of-life reprocessing and the supply of replacement parts and recycled materials at the high end of the recycling hierarchy. The concept of a total product system raises further issues, including the relationship between the prime movers in the development of the fortwo and the key suppliers (known as system partners) and the impact of these relationships on the environmental issues inherent in production organization. This prompts comparison between process organisation at the 'smartville' industrial park and more 'traditional' approaches to vehicle manufacture. In examining these issues, we review the actual or potential reduction of environmental impact in the three main phases of vehicle life–car manufacture, car use and end-of-life vehicle processing.

15.2 Background

Overall, the development of the fortwo can be viewed as three major steps. Chronologically, these are:

- 1992-1994 – product creation;
- 1995-1998 – product realisation;
- 1999 onwards, market building and brand extension.

Some key events in the project's development, including the extension to new smart models, are indicated in Table 15.1. The global unit sales figures are shown in Table 15.2. As can be seen this indicates the slow pace of market building, an outcome that reflects initial concentration on the European market. The pattern of market development also raises questions about the longer term viability of the smart venture. But that does not detract from the significance of the smart concept or from the highly practical knowledge and ideas that it has generated.

Table 15.2. Annual unit sales of the fortwo

Year	Units sold
1998	17,000
1999	79,900
2000	102,100
2001	101,937
2002	102,302
2003	98,890
2004	72,478

15.3 Modularity in the Car – and in the Transport System?

Important elements of innovation follow from a design emphasis on modularity in contrasting modes. The physical modularity of the fortwo is immediately apparent through the visually striking use of plastic body panels in strong colours. These are a direct reflection of the influence in early design stages of SMH – the pioneers of Swatch plastic fashion watches. The panels on the fortwo are replaceable, allowing end users to re-configure the exterior colour of the vehicle for a cost that amounts to approximately 10% of the original vehicle purchase price. An example of the high use of modularity can be seen in Figure 15.2, which shows the major body component on which the other modules and sub-components are attached. This 'Tridion cell', or cage also incorporates the crumble zones required to achieve high ratings in European crash tests. In design terms, the cell gives the car a prominent line since the cell is visible on the exterior. The use of the cell within the fortwo assembly has been compared to the way aviation manufacturers make the most of aerospace frames when building aircraft.

Figure 15.2. The Tridion safety cell, after complete assembly at smartville by a main 'system partner', Magna Chassis. Here, it is being conveyed to the main fortwo assembly line.

The car was originally envisaged as a second car and as largely for city use. But it has also been associated with initiatives that seek alternative approaches to individual mobility in urban areas. These include reduced public transport fares for inter-modal journeys, reduced parking fees, preferential car rental agreements, incentives for car sharing and for using cars like the fortwo outside the prevailing owner-driver model (Mildenberger & Khare, 2000). In some parts of the EU, these initiatives have started to emerge – for example, in the case of the fortwo, in lower charges for parking and for car-wash (this reflects lower use of water, detergents etc.). The fortwo thus potentially contributes to reducing impact from personal vehicle use. So far, this potential has been limited by both the limited scope of incentives and the low sales figures shown in Table 15.2.

15.4 The fortwo and the Total Product System

Increasingly, large international companies emphasize co-ordination acoss all, or large parts of their supply chains. They attempt to co-ordinate product design as well as the highly complex flows of materials, components, support services, orders and other information that extend through the various stages in the manufacture and distribution of end products. The focus of competition is thus moving away from competition between individual companies towards competition between supply chains (Christopher, 1992) – or, more accurately, between product supply networks[1]. Combinations of methods such as quality assurance, value stream analysis, concurrent engineering, kaizen and product life planning are applied throughout a supply chain in the pursuit of continuous enhancement of competitiveness which is derived from improvement in all areas of a chain. The emphasis on supply chain co-ordination has a number of roots, including attempts to match current highly demanding, diverse market conditions, and the exploitation of information and communications technologies in business-to-business and business-to-consumer electronic commerce. The competitive impact of methods developed by some leading Japanese companies – most notably by Toyota – has been particularly influential. This influence was initially seen in manufacturing sectors, but it is now increasingly evident in service sectors.

Co-ordination across a supply chain or network emphasizes the development of long term close relationships between lead companies and key suppliers. Lead companies generally occupy strategic positions close to the point of delivery to consumers and other end customers. Their market position and purchasing power enable them to establish chain-wide systems of governance (Kaplinsky, 2000). Examples of lead companies range from vehicle brand owners (such as Mercedes) to retailers. Typically, these prime movers seek to co-ordinate activities such as product design and development, production and logistics across all the diverse stages and types of activity – from raw materials processing through to the assembly and distribution of end products. Much of this co-ordination is undertaken indirectly through collaborative relationships with core suppliers who take responsibility for the design and supply of main product systems, sub-systems or support services. Core suppliers are generally expected to manage their own parts of the supplier base – the network of companies involved in the system or sub-system for which they are responsible. Longer term, the process shifts emphasis away from performance within individual companies to performance across a network.

---

[1] There is a long standing debate about which of 'chains' and 'networks' provides the most appropriate metaphor. While preferring the more flexible connotations of networks, we use supply chain generically here.

The ideal-type model can be viewed as a development from the popularized concept of 'lean production'. Co-ordination extends across all areas and stages of activity in the total supply chain. A critical factor in this objective is the high proportion of a company's costs that are now accounted for by purchases from suppliers – 70% or more is typical. Consequently, there is emphasis on optimal use of use of human resources, production capital, space and logistics, and on paring levels of waste (such as from producing faulty parts and high levels of stocks and work-in-process) down to minimal levels. Another aspect of the model is that production flows are 'pulled' at all stages by actual demand for different product variants. Ultra-high standards of quality at each production stage, delivering products on time and the flexibility needed for low batch, high variety production are also part of the competitive equation. Consequently, the successful application across a supply chain of techniques such as 'just-in-time' (JIT) is a critical for the competitiveness of an end product. For many companies, this has involved a challenging shift from 'traditional', vertically integrated mass-production – referred to later as 'the traditional model'. In this model, the main company undertakes most types of activity internally, and may account for 70% or more of total production costs.

The environmental impact of these fundamental shifts in the organization of production appears, in general, to be positive – although systematic comparisons present considerable difficulties. For instance, the impact of JIT methods is necessarily associated with very low levels of defects in the supply of components etc. and, where suppliers are efficient, with commensurately low levels of waste of materials, energy, human effort and storage space. However, JIT supply can also be associated with relatively high environmental impact from the transport movements – needed for the collection and delivery of small part lots (Katayama & Bennett, 1996). Furthermore, JIT systems can lead to other adverse environmental impacts, such as when overall deliveries increase to a point where local congestion delays the supply of components. It is clear that various supply scenarios need to be examined carefully.

The performance of all stages of the supply chain in terms of resource use and impacts has become a focus of attention as a consequence of environmental regulation by regional, national and other bodies, and following initiatives such as ISO 9000 and ISO 14000. However, assessment of environmental impact needs to extend beyond the point of sale to take in total product life. This is partly because businesses are increasingly aiming to generate, or to increase, revenue flows in the aftermarket, extending from the point of sale through to a product's end-of-life. Companies now look beyond the supply of replacement parts in the aftermarket towards provision of a variety of services that support or enhance product use and functionality. The importance of aftermarket performance in total competitiveness is indicated by Gallagher et al's (2005) estimate that it may account for 40% of profits for a wide range of companies. Assisted by Internet based links, badge manufacturers seek to sustain long term relationships with customers, such as by providing on-line diagnosis of appliance faults and linked, rapid response from warranted support services.

Recent environmental regulation in the EU has also focused attention on the aftermarket stage of product life. Much of this relates to the performance of products in their use. In the case of non-durable products, this may aim to reduce the overall environmental impact through, say, return of packaging for reuse. For durable products from refrigerators to cars, such regulation is establishing mandatory performance targets, most obviously in terms of energy efficiency and emissions levels. Ideally, this includes the extension of product lives, such as through the refurbishment or re-manufacture of products in the later stages of their use (Guide et al, 2000). But life extension sometimes increases environmental impacts so that different products require different approaches. For instance, engines require redesign as emissions standards rise whereas a car body may provide a stable platform for a longer time, reducing the impacts associated with investments in tooling, pressing, etc.

EU standards have also focused attention on the end-of-life stage of product life. Pressure to improve environmental performance in this stage is partly indirect, for example, following from increasing restrictions on, and rising costs of, disposal in landfill sites. It also results directly from mandatory recycling targets and from requirements that manufacturers take responsibility for the collection of their products at the end of their life and for optimising their reprocessing.

Growing emphasis on the environmental characteristics of product performance, and on end-of-life reprocessing, is also linked to a reshaping of approaches within production systems – such as in product design, in materials selection and in manufacturing processes. Revenue driven approaches to product support throughout the aftermarket stage add to a need for companies to focus on the total product life cycle. This can involve long time scales since supporting a product through to the end-of-life extends, for instance, to 15 years or more after production of new models has ceased in the case of cars, and 25 to 40 years in aviation. Within these long timescales, there is scope for 're-manufacture' – of components and/or of the total product – to support life extension, for either the product owners or new users. The potential economic viability of this approach is indicated by Guide et al (2000) who provide examples of re-manufacture linked to high levels of profitability.

However, as indicated above, the environmental viability of life extension is questionable. The efficiency of individual products tends to fall off as they are used. Also, in conditions such as rising real energy costs and increasing constraints on adverse environmental impacts, the associated influences on product design tend to lead to progressive improvements in performance that surpass earlier benchmarks – as the fortwo illustrates – see Figure 5. To determine the full extent of the environmental impact linked to a product, particularly one that, for its functioning, requires additional energy and other material inputs, requires analysis that extends well beyond concern the product's supply chain. To establish the full potential for reducing those impacts requires a focus on product and performance across the total product system to encompass the full cradle to grave product cycle, and to factor in the wide range of actors and interactions involved. This focus leads to emphasis on the sustainable product systems, reshaping production organization as well as product use and recycling.

15.5 Towards a Total Product System

Production organization has a particularly important role in establishing the relatively low environmental impacts associated with the fortwo. This is associated with the Mercedes-Benz automobile division of Daimler Chrysler (DC) which took full responsibility for MCC smart in 1998 when SMH (Swatch) ceased to be involved in the project. Mercedes took the project forward by developing a new brand, a new way of production and a new method of sales distribution (Renschler, 2000). They did this in a relatively short period of time but incurred substantial costs and encountered problems in product launch. These were mitigated by strong support and risk sharing on the part of core suppliers. Their facilities are purpose built, and co-located with the main assembly factory at smartville in Hambach, France. Mercedes forged strong links with these suppliers through their involvement in product design from very early stages. These "system partners" shared in product development, taking much of the responsibility for major modules such as the cockpit and complete door assemblies, as well as sharing investment costs and financial risk. They developed their own solutions to component design and sourcing, within parameters agreed with MCC as the lead company. Some elements of organization at smartville are contrasted with a somewhat stylized 'traditional' model in Table 15.3. The 'traditional' approach represented here is a synthesis between the patterns associated with vertically integrated mass production and the approaches to relationships with workforce members and procurement from suppliers that tend to be associated with, or are inherited from mass production (see Rhodes, 2006). There are, of course, many other approaches besides smartville that contrast with 'traditional' patterns.

We emphasize the procrustean nature of Table 15.3 and that it must be viewed in context. The smart venture is not an isolated example but provides a specific, if somewhat unusual example of more general trends. First, other vehicle assemblers have developed a production organization along lines comparable with Hambach, such as by establishing suppliers parks around final assembly plants. Second, the development of the production system, which we consider further (below), needs to be viewed in conjunction with the product's characteristics, and the way that these are now 'part of the past'. The development of the smart fortwo has been overtaken by the work of other vehicle brand owners who have surpassed some of the environmental targets it set. Third, while Table 15.3 concentrates on issues of production organization, the contrast between early publicity about the fortwo's urban role and the longer term outcomes needs to be kept in mind. For instance, the introduction of fortwo has not been accompanied by the extension of mobility concepts on any substantial scale. It has not been associated with moving private users away from personal vehicle habits anymore than other mobility projects. However, this does not follow from failings on the part of MCC smart. Rather, it reflects the way that public authorities, in the main, fail to think through the issues of urban transport holistically. Within such an approach, there is scope for incentives for using smaller vehicles as well as public transport. Further thinking is needed.

Table 15.3. Process characteristics within the supply chain

Process	'Traditional'	'smartville'
1) Product design and development.	Little collaboration with suppliers. Bidding for job after product is proto-typed.	Systems partners have a large responsibility for collaborating and achieving lower costs. Work with supplier very early in design stage.
2) Ordering and purchasing.	Short term, focused on supplier price rather than suppliers' costs and capabilities.	Long term contracts emphasize cost and continuous improvement rather than price, with supply to advanced JIT standards.
3) Relationship between lead Manufacturer and suppliers.	Arm's length, low trust.	Close, classed as partners, higher levels of trust, information sharing.
4) Remuneration of production employees.	Elaborately tiered payment structures, hourly rate for most, bonus decided by upper-management.	Semi-autonomous teams, flatter organizational hierarchy, team bonus linked to team performance.
5) Main supply chain actors.	Geographically dispersed, historical layout, no room to grow.	Co-location, integrated facilities; potential for future growth.
6) Production facility.	Historically based with legacy problems.	Designed as greenfield site, best in class for environment and workers.
7) Warranty responsibility.	Largely held by final assembler.	Shared by all, traceability to individual component producers.
8) Supplier payment terms.	Delayed payment to suppliers commonplace.	When final product complete or sold to customer.
9) Supplier facilities.	Owned by supplier.	Total Hambach site managed by MCC and system partners, land owned by MCC.

Some examples amplify issues identified in Table 15.3. The extensive use of partnerships for collaborative solutions (points 1 and 2) seems to be more effective than, say, solutions developed in traditional relationships where waste of effort, energy and materials, and the financial health and innovative capabilities of a supplier are of no concern to those purchasing components etc. An example of the collaborative approach at smartville is a large cost-reduction programme (Target smart), which involved the partners in generating ideas for modifying specifications in order to reduce costs. Approximately 15% of the cost was cut from many of the components with about 60% of these savings achieved through renegotiation with suppliers (Chew, 2001a). In some cases, MCC accompanied its partners on visits to component makers to expose various 'tear-down' prices of sub-components.   MCC were able to utilise Daimler-Chrysler information on various shared sub-components in order to compare costs in a benchmarking exercise with MCC's partners.   The other 40% of savings came in the form of re-engineering suggestions from the partners and from suppliers further down the supply hierarchy (Chew, 2001a).

These savings followed from the revision of specifications to match a small car profile rather than the more expensive premium engineering solutions that characterize Mercedes. Because some MCC suppliers already manufactured components for other car manufacturers in the small car segment, it was relatively straightforward to transfer this product knowledge to the fortwo. One example included an axle assembly which was modified to reduce materials costs as well as to increase driving performance (Chew, 2001a). In another example, cost savings were made through transfer of assembly of a Bosch headlamp sub-component from smartville to the Czech Republic. The use of competitive logistic calculations also generated cost savings.

The benefits indicated by points 2), 5) and 6) in Table 15.3 include:

- the system partners assemble main modules on sites abutting MCC's final asembly lines;
- the operations of the assembler and suppliers are very tightly co-ordinated with very small buffer stocks (generally equivalent to less than one hours production);
- suppliers assure the conformance of all delivered items and there is no inspection by the assembler when parts are received by MCC;
- system partners have their own direct entry ports into MCC's assembly area with direct delivery to auto-assembly line (via extensive conveyor systems);
- other suppliers deliver via trailers parked at entry ports that feed into adjacent line locations;
- direct ordering with shared responsibility for stock control.

The gains from these aspects of supply chain organization include contributions towards greening the supply chain as well as to higher levels of production efficiency. For instance, production and delivery of the door module illustrated in Figure 15.3 minimizes energy use and avoids waste. Door panels are produced by Dynamit Nobel and are colour coated by Cubic Europe, both of which are located in linked buildings at smartville. The finished panels are supplied to a further linked building, that of Magna Doors where they are configured as part of a complete door module (which include trim, glass, mirrors, door-mounted controls, window and mirror motors). Operations in these three supplier facilities are tightly co-ordinated – with each other and with the main assembly line. Complete door modules are supplied via an enclosed link bridge, in pairs that match the build-to-order (BTO) specifications for interior and exterior colour choices and other variants. There is consistent emphasis on a very high standard of conformance with quality standards at all production stages, maximizing efficiency in the use of energy and materials. Large scale reductions in packaging waste and in transport-generated emissions become possible when components can be made locally. One estimate is that more than 95% of the transport costs for the main modules have been reduced compared to a typical automotive assembly plant (Treneman, 2001).

Figure 15.3. Detail of the left-hand-side door interior, showing fittings

With direct ordering and correct sequencing of modular units (for example engine type, wheel/tyre configuration, interior/exterior colour, door panels, etc.) a BTO production system can be achieved. In an efficient system, no vehicle is assembled without a final (named) consumer or other end-user. BTO appears likely to contribute to reduced environmental impact – for example, stocks can be minimized because manufacture is largely confined to units for which firm orders have already been placed. Using BTO incorporates consumers' actual demands as a controlling input to the production process. For instance, production of vehicles in colours or features that turn out to have limited customer appeal should, ideally, be avoided. But this carries the risk of lost sales. Where a high level of co-ordination has been developed across the various stages and branches of a supply chain, reductions in order size are achievable throughout the production network. Furthermore, experience in low inventory production systems shows clearly that the wasted energy, materials and so on that are associated with rework and mislaid or damaged components in poorly organized production systems, can be expected to reduce correspondingly. Additional gains follow from reduced transport and storage. But direct measurement of the effects of BTO is difficult - see Hines, 2002 for a discussion of the issues.

Points 7), 8) and 9) in Table 15.3 indicate risk sharing. One of the main benefits for the assembler as lead partner is that it is possible to view the entire site holistically, and to undertake a comprehensive assessment of resource use on the site. In the case of smartville, MCC and its partners were able to shape resource use from the very start through site and building design. The factory buildings were constructed from sustainable materials, and all processes within smartville are both CFC and formaldehyde free (Treneman, 2001). Efficient use of energy was emphasized in the design of buildings, and the factory recycles the heat created and water used on the site. But in the case of off-site suppliers, additional environmental burdens such as transport to smartville need to be taken into account in the overall analysis.

15.6 Panels: An Example of Modularity Within an Increased Product Range

Offering customers a product built on modular lines includes the opportunity to update or change the colour of the car by replacing door and other panels. As yet, it is not clear how frequently this will occur per car-lifetime. An indicator may be provided by a fortwo supplier who claimed that, in Italy, businesses have developed to rent replacement panels to fashion conscious consumers on a short term basis. This facility can be interpreted in several ways from an environmental perspective. For the consumer, extending a product's utility by updating its appearance is a bonus point from which, in theory, all parties gain something. Less obvious to end users is that a more extensive colour range needs to be available. Regular 'edition' changes shuffle the colour range. But previous colours need to be available for repairs, reducing some of the environmental benefits. On the other hand, the need for more comprehensive model changes may be less necessary since body colours can be readily changed and because modular construction enables easier incorporation of the latest features.

The body panels (door panels, front and rear outer-skin) play a central role in product refresh. They are comprised of Noryl GTX polyphenylene oxide resin produced by GE Plastics (Pryweller, 1998). The component manufacturers worked with MCC on the plastic body components. Part of the total product design process involved selecting materials that would behave as a rigid plastic and would limit environmental impact in manufacturing and in end-of-life reprocessing. The major panels are injection molded by Dynamit Nobel AG, including front fender, outer door, front and rear valences and wheel arch panels. Due to the high precision of the molding process, very little scrap waste is produced, and this is collected for recycling into the injection process feedstock. Panels are produced in 4 basic plastic colours and then painted by electrostatic powder paint processes that eliminate all solvents, sludge and effluent (Treneman, 2001). The panels are 100%

recyclable thermoplastic and are designed to be reversibly deformable, avoiding dents from parking bumps and impacts of up to speeds of 15 mph (Birch, 1997). This type of life-long design is important to ensure increased resistance to damage in the minimal parking spaces of many European cities.

The plastic panels have a single clear paint overcoat to enhance resistance to fading. The absence of primer and base coats saves some 50% of the costs of a typical painted body (Pryweller, 1998) reducing the resource impact of manufacture. In addition to high strength and bright colour, the use of plastics reduces overall vehicle weight. The car mass is about 725 kg – some 300 kg less than typical steel body compact vehicles (Pryweller, 1998; Wrigley, 2000). In the case of door panel manufacture, analysis of the total product system needs to consider the trade-offs between plastic and other materials. An important factor in comparison is the predicted vehicle unit volume, since the dies used for plastic panels have an expected life time of only some 200,000 vehicle platforms (Pryweller, 1998). In contrast, metal pressing dies can generally be used over a much larger volume – potentially over the entire product life time. Another consideration is consumer opinion. Steel and aluminium tend to be viewed as safer than plastic – regardless of what vehicle safety tests show. Such perceptions can be difficult to change.

15.7 Extending Model Diversity

The smart brand is being extended to other products, most notably the 4-seat, 5 door smart forfour. This has required additional production facilities. In a simplistic view, this growth is counter to green practices within the supply chain. At first glance, a doubling in component diversity could be seen as having a major adverse impact on both supply chain efficiency and the environmental burden. But the smart variants built at Hambach share a much higher level of modular components than other cars utilising platform-sharing strategy, and there is some sharing with the forfour. This use of common components for multiple products is well known and documented within the automotive industry. One example of this is the way that Volkswagen has used common vehicle platform architecture across a wide range of car brands –the VW Polo, the SEAT Ibiza and Skoda Fabia share a platform called 'PQ24' (Chew, 2001b).

Shared platforms reduce costs, but also enable cost reduction through increased levels of modularity (as in the examples of door modules and cockpit modules in the fortwo). Yet module components can be varied, supporting the production of final products that appear very different visually. The new Polo and the Fabia share over 150 major components including chassis, engines, and cooling sub-systems. This common sharing reduces labour intensive dual design pathways, achieving substantial savings in development costs (Chew, 2001b). The use of vehicle platforms is a major factor in achieving both cost savings and reduced environmental impacts.

The use of shared platforms has also been important for the development of the forfour which competes in the most competitive market segment in the European automotive industry – one in which DC has limited experience beyond the

Mercedes A-Class (Maynard, 2001). To succeed in this arena, the final product needed to be unique, cost-effective, have the highest levels of safety and to be environmentally-friendly (in terms of overall emissions) and lastly to be a high volume sales item (Maynard, 2001). This is a particularly acute point, since the fortwo was not forecast to be profitable for some time and, as was shown in Table 15.2, volumes have been disappointing. Like the fortwo, development of the forfour has involved partnership, albeit of a longer duration than that with Swatch. The forfour and the Mitsubishi Colt have been produced under a shared platform philosophy (Ostle, 2000), with DC jumping into the previously planned minicar platform that Mitsubishi was developing for the Colt. But there are important differences between the two vehicles – for instance in the forfour's use of a new version of the Tridion safety cell concept that was originally applied in the fortwo and in retention of the smart brand's distinctive plastic roof (fixed by very different methods to those used for the Colt which retains a steel roof). Nonetheless, substantial economies of scale were expected from part sharing across the fortwo, the forfour and the Colt. Production of the latter two vehicles, which share the same platform, commenced in 2004 at the NedCar factory in Born, Holland.

15.8 The 'In-Use' Product Phase

A major difference between MCC smart and its various competitors is that, while it combines fashionable design with high standards of safety and interior specification, it also contributes to reductions in environmental impacts, such as those arising from vehicle emissions. These factors are important in terms of the overall product 'package' that the end user purchases. About 80% of the environmental impact in the total product life cycle is attributable to the in-use phase (Mildenberger & Khare, 2000). Much of this impact results from the high lifetime mileage of a vehicle.

When it was launched, the fortwo set new standards for in-use emissions compared with vehicles which it competes. Most of the fortwo's competitors are larger in overall dimensions, are heavier and have higher fuel consumption. At the time of its launch, it was one of the lowest emissions cars available on the market in Europe. In the realm of fuel consumption, the diesel (which is not readily available in the UK) comes close to the 3 litre fuel target, achieving 100 km driving distance using only 3 litres of fuel. But comparisons are less favourable five years on, emphasizing how fuel consumption and emission standards tend to improve progressively, challenging earlier vehicle designs. smart's eventual response to the challenge of more recent vehicle designs may be to return to the original concept of an electric or a hybrid drive train system. MCC smart has been working with others to produce electric and diesel hybrid prototypes (Tremble, 2001). But product cost and consumer acceptance of hybrids remain a serious challenge – as they do for all car manufacturers!

Figure 15.4 summarises fuel consumption (FC) – as indicated by the fuel used to drive 100 km – and emissions. The emissions value is the sum of carbon monoxide (CO), unburnt hydrocarbons (HC), nitrogen oxides (NOx), and particulate matter (PM) that are emitted during the standard European test cycle, expressed in grams per kilometre (g/km). This represents a total emissions factor for the vehicle, where higher overall values are worse for the environment. In everyday use, the level of emissions depends on the driving style and cycle employed by the driver together with the condition of the vehicle and it's sub-components (for instance, the condition of the catalyser system).

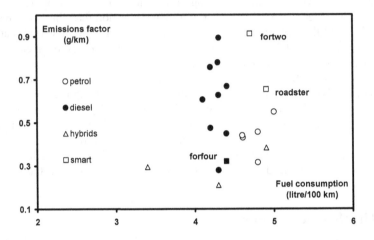

Figure 15.4. Fuel and emissions data for selected mini and small cars, Source: Vehicle Certification Agency, 2004, UK.

Note: the figure shows emissions data as a function of fuel consumption for the ten cleanest diesel cars (filled circles) and the ten cleanest petrol cars (open circles) sold in the UK. Hybrids are displayed as open triangles and all are petrol-electric. smart petrol vehicles are shown as open squares, and the diesel forfour is a filled square.

In summary, the fortwo is an efficient vehicle that, when launched, had a lower emissions impact on the environment and lower use of non-renewable resources than many cars in the mini to small car range. Two points stand out however. First, the fortwo has been overaken by subsequent developments that have improved fuel consumption and reduced emissions levels in newer cars. Second, overall sales of the car have been, so far, lower than most of its direct competitors, so that its impact is limited when looking at fleet wide effects. But the fortwo still represents a significant step towards the forward looking requirements set by various research bodies, such as the benchmarking criteria for environmentally optimised vehicles (Nieuwenhuis, 1997). Likewise, the fortwo stands up well against the stipulation of the UK's Foresight programme that mass market vehicles in 2020 need to embody the qualities shown in Table 15.5. Under the broad umbrella of requirements in Table 15.5, the fortwo scores well for its use of radical innovations that contribute environmental benefits. It compares favourably with the average vehicle parc standard.

Table 15.4. Vehicle qualities required for 2020. Source: adapted from DTI 1999, UK

Parameter	Requirement – to:
clean	have the lowest environmental impact
efficient	make best used of limited fuel resources
lightweight	use less energy to achieve mobility
telematic	communicate with other cars and network to optimize existing road structure use
intelligent	provide enhanced safety
lean	be manufactured competitively

A further critical consideration is that the fortwo constitutes a move towards more environmentally sensitive car design without, as some might fear, compromising user safety, as is indicated by its Tridion safety cell, airbags and purpose designed crumple box zones. It has also been linked to telematic programmes that reduce environmental impact as is discussed in the next section.

15.9 Extension of the Total Product System into Mobility Services

The original smart concept envisaged that a wide range of mobility services would be offered to owners through a package to be offered as 'smartmove'. In some cases, these services were to be offered with discounted rates and preferential treatment, as an incentive for customer purchase, particularly among those who are environmentally aware. Examples of these services include preferential rental rates for smart cars, and for similar vehicles, such as when renting during visits abroad. Preferential treatment can be, and has been extended to parking charges, recognizing the limited space required for a fortwo. Potentially, incentives can be extended to ferry and train tariffs, through lower prices compared to those for vehicles of more conventional lengths. The potential for links with transport providers, including vehicle hire companies, opens up the possibility that owning a car may not be the only convenient way, or the 'best' way for a person to ensure getting from one place to another.

Like other car manufacturers, MCC also looked at the use of personal digital assistants and/or mobile telephones to access internet information, including vehicle routing/navigation, weather, electronic mail services and travel bookings. This is can be accomplished more easily than in a traditional car because the roof module is plastic and is transparent to the appropriate technology. The roof is thus compatable with the potential changes and growth expected in the telematics industry. A further example is provided by the control and diagnostic systems that are installed in smart cars. But the dual benefits from this development should also be noted. Information flows to the parent company's data warehouse contribute to maintaining a vehicle's efficient use of fuel and low emissions. They also provide a flow of information on customer use of cars which can be used in planning new products, in customer retention and so on. This is also the case with telematic systems fitted in other recent cars. But the smart dealer network differs from others in the EU in that dealers are linked directly to smartville as franchisees. The overall effect is to lock all smart users into the DC/smart network. Thus, when some 3,000 UK residents acquired fortwos in advance of the launch of a righthand drive version, backed by a UK service network, their vehicles had to be taken to other European countries when they needed a service or repair. This apparent shift to a 'captive customer base' may have positive environmental benefits in so far as vehicle performance benefits from high standards of service and repair, but it also raises concerns about constrained competitiveness in the aftermarket.

Major automobile manufacturers recognize the importance of consumer demand for new types of services and the potential for increased aftermarket revenues. Accordingly, they appear to be shifting their focus further down the product system, aiming to increase revenues by offering add-on services to customers. This extension into services may signify the increasing service orientation of manufacturers, indicating an operational shift towards the total product system. The logic of this for the automotive majors is perhaps indicated by the third smart product – the forfour. MCC does not manufacture this car (this is done under contract by Nedcar in Holland). MCC's roles in this case are overall product design, operational co-ordination, distribution to the smart dealer network, marketing and supply of aftermarket services.

15.10 End-of-life Provision

Since the fortwo has only been in production since 1998, the reality of the end-of-life (e-o-l) stage of the total product system has yet to be proven. However, the design of smart as a direct response to consumers' environmental concerns and, incidentally, as an early response to the then emerging regulatory regime in the EU, is evident in three main respects. One is the choice of materials for their recycling potential – most obviously in the scale of use of plastics and the selection of plastic materials in relation to e-o-l recyclability. As a result, smart claimed at the time of launch that 80% by weight of the materials used in the construction of smart could be recycled – compared with the industry average at that time of 70%. However, this falls short of the target set subsequently by the EU Commission in its End-of-Life Vehicle Directive of 2000. This established that, by 2006, 85% of a vehicle's weight should be reused, recovered or recycled. This includes a 5% allowance for energy recovery through incineration (Seitz & Peattie, 2004), an allowance that might enable the fortwo to comply with the Directive. The fortwo's 80% figure also needs to be viewed against the 90% recyclability claimed in Toyota's 2005 advertisements for its hybrid (petrol-electric) Prius model. The latter vehicle provides another illustration of continued innovation and progress in relation to environmental targets.

Second, achievement of the fortwo's 80% target is linked to conscious design for dissembly. Its modular assembly is accomplished by simplicity in the design of fastenings and fixings. For example, the cockpit module is inserted into smart as a single unit that is secured with just two bolts. This will support easy e-o-l removal of this module and separation of its components and materials. The removal of complete modules or separation of components from individual vehicles, that irredeemably reach the e-o-l stage, supports the extension of the life of the overall pool of fortwos.

Third, partnership with suppliers, dealers and vehicle reprocessors is important in planning for the e-o-l phase and in its subsequent management.   Production partners (i.e., suppliers) are necessarily involved in the selection of recyclable materials and in design for dismantling. In relation to the other end of the life cycle, Integration of the dealer/aftermarket network through the use of information and communications technologies in the cars, at dealer/maintainer locations and at DC potentially enable an overview of the condition, use and age of the fleet. Among other things, this supports operational planning for both the e-o-l phase and its relationship to aftermarket supply through the remanufacture of components and subsystems, and their reuse in the aftermarket. Mercedes Benz (MCC's parent) was one of the early investigators of these issues when, like some other manufacturers, it explored the issues of reprocessing in partnership with specialists from outside the automotive industry (den Hond, 1998). However, the organization required to link e-o-l vehicle reprocessing with the aftermarket and to increase the use of such reprocessing is complex (Seitz & Peattie, 2004).

15.11 Conclusion

This chapter illustrates how the concept of a sustainable total product system can be advanced by utilizing the extensive interactions between a lead company – in this case MCC smart GmbH – together with its suppliers, sales/aftermarket network and e-o-l reprocessors. In order to qualify and quantify environmental impact across the total product system, it is necessary to look beyond the product life cycle and to deconstruct the entire product system. The potential benefits of deep integration within the total product system are indicated by the fortwo. This partly reflects the high level of modularity in the car's design in comparison with compared with the wider mini-car segment and the automotive industry as a whole. MCC smart has consistently emphasized the importance of building long term business relationships, in a similar way to that observed in successful Japanese automakers operating in both Japan and the United States (Liker &Yu, 2000, Dyer & Chu, 2003).

There are clear benefits from the fortwo in the in-use phase, particularly when it is compared with other vehicles in similar segments of the automotive market. What remains unclear to date however, partly because the fortwo is a relatively new vehicle and in its first generation, is how much impact it will have on extending environmentally friendly lifestyles. This includes issues of mobility that extend beyond personal car ownership to the question of comprehensive access to mobility for those without cars. The 'public/private transport' division potentially includes a range of imaginative approaches, such as the integration of the Dutch rail network and local taxis[2]. But the limitations on the exploitation of such approaches reflect rather muddled and limited thinking on the part of many of those who are in positions where they can shape the issues of urban transportation. Instead of simplistic juxtaposition of car versus public transport, more discriminating approaches to 'the car' appear to be necessary to exploit the potential of developments such as the fortwo and the more recent availability of hybrid vehicles.

Where end-life re-processing is concerned, the ability of the fortwo to deliver remains to be tested in operational conditions. Nonetheless, it is a vehicle in which these issues have been comprehensively addressed. It is to be expected that Daimler Chrysler will have learned from, and will apply the lessons gained from this experience, as from other elements of the smart venture.

Acknowledgements

The authors wish to thank their research collaborator and colleague, Ruth Carter for her comments and support in the production of this chapter.

References

Anonymous, Automotive News Europe; 2003, Vol. 8 Issue 24, p. 26

Anonymous, Automotive News Europe; 2004, Vol. 9 Issue 24, p 22 – 23, 20

ANE (Automotive News Europe) (2001) 'European Model Sales (Year to Date)', Automotive News Europe 6.24 (December 2001): 39

ANE (2001) European Model Sales (Year to Date)' Automotive News Europe 7.5 (11 March 2002):39

ANE (2000), Market Data Book, Crain Communications, London

ANE (2001), European Model Sales (Year to Date), Automotive News Europe Vol. 6, Issue 24, December.

ANE (2002), European Model Sales (Year to Date), Automotive News Europe Vol 7, Issue 5, pg. 11.

ANE (2003), European Model Sales (Year to Date), Automotive News Europe Vol 8, Issue 5, pg. 24.

ANE (2004), European Model Sales (Year to Date), Automotive News Europe Vol 9, Issue 24, pg. 22-23.

Birch, S., "Real smart, micro compact car", Automotive Engineering No. 11, Vol. 105, p. 21, Nov. 1997.

Chew, E., "How D/C cut the smart car's cost base", Automotive News Europe Vol. 6, Issue 16, page 8, 30 July 2001 (2001a).

Chew, E. " Modular approach brings cost savings for new Polo", Automotive News Europe, Vol. 6, No.24, p. 19, 3 Dec. 2001 (2001b).

Christopher, M: Logistics and Supply Chain Management: Strategies for Reducing Costs and Improving Services, Financial Times, Pitman Publishing, London, 1992

---

[2] This supports pre-booking of local taxis, available at low cost in the hinterland around railway stations, with the taxis co-ordinated with train arrivals.

den Hond, F., The "Similarity" and "Heterogeneity" Theses in Studying Innovation: Evidence from the End-of-Life Vehicle Case, Technology Analysis and Strategic Management, Vol. 10, No.4, 1998

DTI, UK Dept. of Trade and Industry, Foresight Vehicle Strategic Plan, January 1999, London. Publication no. 4402/1K/10/99/RP, UK

Dyer, J. H. & Chu, W., The Role of Trustworthiness in Reducing Transaction Costs and Improving Performance: Empirical Evidence from the United States, Japan, and Korea, Organization Science, Vol. 14, No. 1, January–February, 2003

Gallagher, T., Mitchke, M. D. & Rogers, M. C.: Profiting from spare parts, The McKinsey Quarterly, Feb 2005

Guide, V D R, Jayaraman, V, Rajesh, S & Benton, W C: Supply-Chain Management for Recoverable Manufacturing Systems, Interfaces 30, May-June 2000, pp. 125 - 142

Hines, P. and Holweg M., "Managing variability in the automotive supply chain – a foresight approach", Society of Automotive Engineers, technical paper 2002-01-0460, March 2002.

Katayama, H. & Bennett, D., Lean Production in a Changing Competitive World, International Journal of Operations & Production Management, Vol. 16, No. 2, 1996

Kaplinsky, R., Globalisation and Unequalisation: What Can Be Learned From Value Chain Analysis? Journal of Development Studies, Special Issue, 2000

Liker, J.K. and Yu Y.-C., "Japanese automakers, U.S. suppliers and supply-chain superiority", Sloan Management review, Vol. 42, Issue 1, p.81-93, 2000.

Maynard, M. "Get smart", Fortune International, p. 58, 30 April 2001.

Mildenberger, U. and Khare, A., "Planning for an environmentally-friendly car", Technovation Vol. 20, p. 205-214, 2000.

Nieuwenhuis, P. and Wells, P., " Environmentally optimised vehicle – a model for the future"., Automotive Environment Analyst", 1 Feb. 1997.

Ostle, D. and Treece, J., "DCX ready to build four-seat smart on Mitsubishi platform", Automotive News Europe Vol. 5, Issue 8, p. 4, 4 Oct. 2000.

Pryweller, J., "smart's plastic body was a thorny problem", Automotive News Vol. 73, Issue 5782, p. 22, 31 Aug. 1998.

Renscheler, A., remarks from "Opening presentation, smart", Automotive News Europe e-Business Conference, smartville, Hambach, France, 29 Nov. 2000.

Rhodes, E., From Supply Chains to Total Product Systems, in Rhodes, E., Warren, J. P. & Carter, R., Supply Chains and Total Product Systems, Blackwell Publishing, Oxford, 2006

Seitz, M. A. & Peattie, K: Meeting the closed loop challenge: the case of remanufacturing, California Management Review, Vol. 16, No. 2, 2004.

Tremble, S., "Revving at full throttle into the fast lane", The Engineer, 21 Sept. 2001.

Treneman, A., "smart new world", The Times (London), 29 Sept. 2001.

Vehicle Certification Agency (VCA) – http://www.vca.gov.uk/ accessed December 2005. (The data is also available on request from the Vehicle Certification Agency, 1 The Eastgate Office Centre, Eastgate Road, Bristol, BS5 6XX, UK).

Wrigley, A., "Rising fuel prices create tempting mart for smart cars", American Metal Market Vol. 108, Issue 187, p. 1-3, 27 Sept. 2000.

16

Environmental Management in Automotive Supply Chains: An Empirical Analysis

Anette von Ahsen

School of Business Administration, Chair of Environmental Management and Controlling, University of Essen, P.O. Box, 45117 Essen, Germany, anette.von-ahsen@uni-essen.de

This chapter addresses environmental management in supply chains of the automotive industry. In particular, a case study of BMW is presented, and the environmental management of German automotive suppliers is analysed.

Owing to the increasing number of environmental laws as well as the environmental demands of customers, car manufacturers have been improving their environmental management. At the same time, the eco-friendliness of cars depends more and more on the suppliers, who, altogether, generally represent a value-added share of more than 60% of the overall vehicle. As a consequence, environmental requirements of car manufacturers can be met only in co-operation with the suppliers.

In Section 16.1 I outline the reasons for green supply chain management from the viewpoint of a car manufacturer. A case study of BMW is given in Section 16.2. Special emphasis is put on the introduction of environmental management systems (EMSs) as well as on requirements concerning the product development and the environmental information exchange between BMW and its suppliers. In Section 16.3 some of the main results of a survey conducted with German automotive suppliers are presented. The chapter ends with concluding remarks (Section 16.4).

16.1 Reasons for Green Supply Chain Management in the Automotive Industry

Supply chain optimisation and integration have become the goal of many organisations worldwide (see Simchi-Levi *et al.* 2000). Strengthening the management of the supply chain is perceived as improving customer satisfaction as well as reducing costs. As an important part, supply chain management involves quality management: car manufacturers have a lasting effect on their suppliers' quality management and quality performance. Additionally, environmental management has been gaining increasing interest recently within supply chains (*e.g.* see Beamon 1999; Green *et al.* 1998; Handfield and Nichols 1999; McIntyre *et al.* 1998; Noci 1997; Sarkis 1998; van Hoek 1999).

The growing worldwide concern regarding the state of the environment, including pollution and resource conservation, has led to extensive environmental customer requirements such as cutting fuel consumption and reducing harmful emissions to the air. Environmental legislation and environmental voluntary agreements are also leading to immense pressure on the environmental management function of car manufacturers (for a survey of environmental voluntary agreements see Delmas and Terlaak 2001). In the USA, among other countries, the Clean Air Act (CAA, 1990), the Clean Water Act (CWA, 1977), the Resource Conservation and Recovery Act (RCRA, 1976), and the Toxic Substances Act (TSCA, 1976) are playing an important role. The European End-of-Life Vehicles Directive (CEC 2000) lays down minimum standards for environmentally acceptable recycling and disposal (e.g. the European automotive industry has to meet the 85% quota for the recoverability of end-of-life vehicles by 2006, and a 95% quota by 2015). Several European Directives set differentiated emissions limits (Northeast-Midwest Institute 2003), and the European Hazardous Substances Directive requires that information be provided about hazardous substances, preparations and products by those who manufacture and market them (CEC 2003). Germany has agreed to reduce its carbon dioxide (CO_2) emissions by 25% by the year 2005 over 1990 levels (see Trittin 2000). The law regulating waste disposal of old cars (see Gesetz über die Entsorgung von Altfahrzeugen) forces German companies to further develop their environmental management.

Car manufacturers can meet all these requirements only in co-operation with their suppliers. The degree of eco-friendliness of cars depends especially on those suppliers that are actively involved in designing products. For example the cutting of fuel consumption among other things requires the use of 'lightweight design', which in turn leads to requirements on the usage of lightweight materials by the component suppliers. As a consequence, suppliers are called on to accept greater responsibility not only for quality but also for the environment.

Most car manufacturers have implemented EMSs according to ISO 14001 (ISO 2004) or the Eco-Management and Audit Scheme (EMAS; Regulation (EC) No 761/2001) of the European Union (EU) to improve their environmental protection. The main elements of these systems include (e.g. see Cascio et al. 1996):

- A definition of environmental policy
- An environmental impact analysis of new and current articles, products and processes
- An analysis of legal and other requirements concerning environmental protection
- A definition of:
    - Environmental objectives and numerical targets at every relevant function or process as well as at the organisational level
    - Monitoring procedures for each identified objective
    - Procedures to be followed in the event of non-compliance with established environmental policies
- Procedures to ensure suppliers understand the requirements; in the case where suppliers work within or in association with customer organisational

facilities they must apply environmental standards equivalent to the customer organisation's standards
- Auditing of the EMS

Thus, the implementation of EMSs within an organisation also draws attention to the environmental management of suppliers. Thereby, for the purpose of ISO 14001, it is not necessary to impose an EMS on all suppliers or to ask for detailed information on chemicals used, potential releases to the atmosphere and so on. In contrast, EMAS emphasises that companies are to consider direct and indirect environmental aspects of their activities, products and services, where indirect environmental aspects include the environmental aspects of suppliers. EMAS requires the organisation to 'consider how much influence it can have over these aspects, and what measures can be taken to reduce the impact' (Regulation (EC) No 761/2001: Annex VI 6.4.).

Altogether, environmental co-operation with suppliers is highly rated. As a consequence, car manufacturers involve direct suppliers in extensive environmental control and follow-up systems. In the next section, I will discuss this, taking BMW as an example.

16.2 Case Study: Aspects of Green Supply Chain Management at BMW[1]

The Bayerische Motoren Werke AG (BMW) with its head office in Munich, delivered 888,181 automobiles in the year 2000; the BMW Automobiles Segment achieved a profit of € 2.380 billion. The company employs approximately 93,600 people.

Environmental management plays an important role within the organisation of BMW as well as regarding the suppliers, which represent a value-added share of 60%–70% of the overall vehicle. BMW has intensified its close co-operation with the suppliers involved in designing products with respect to quality management and, increasingly, environmental management. In the following case study, I intend to address the following questions:

- How does BMW include environmental criteria in the evaluation and selection of suppliers?
- How does BMW influence the EMSs of its suppliers?
- How does BMW involve its suppliers in meeting environmental requirements concerning product design, and what kind of environmental information is required from suppliers?

[1]This case study is based on an interview with Suzanne Dickerson, environmental manager at BMW, on 14 September 2001. The author wishes to thank her for her kind participation.

16.2.1 Evaluation and Selection of Suppliers

BMW has systematically included in its vendor rating system an environmental dimension. Thereby, the evaluation of suppliers up to now has been performed in two steps: on the one hand, the quality management and purchasing department evaluate suppliers in terms of performance measures such as quality and cost. Important criteria are the implementation of quality management systems, *e.g.* according to SN ISO/TS 16949 (ISO 2005), ratios such as the number of non-conforming parts per million (ppm), faithfulness to deadlines of deliveries and behaviour when problems occur. On the other hand, environmental management has formulated additional environmental criteria, for example the existence of EMSs, or ratios such as percentage of recyclable or re-usable materials (by volume or weight), amount of toxic or hazardous materials used and amount of toxic or hazardous waste generated. Additionally, in Spartanburg, South Carolina, the readiness of suppliers to subscribe to "supplier environmental guidelines" is an important evaluation and selection criterion (see Ahsen 2006).

Currently, a task force consisting of quality and environmental managers is working on approaches and methods for an integrated evaluation and selection of suppliers. This is part of a greater project, with the objective to integrate quality and environmental management of BMW as a whole. The idea behind this project is that environmental objectives may have no effect or may have complementary or conflicting effects on the other objectives of the company. If, for example, products or processes are changed for ecological reasons and materials are used that are less harmful to the environment, then other quality features of products might be affected (see von Ahsen and Funck 2001). Thus, BMW strives to manage environmental challenges jointly with other competitive factors such as quality, time and cost. This also leads to the integration of the quality and environmentally oriented analysis of suppliers to achieve a more comprehensive evaluation.

16.2.2 Environmental Management Systems of Suppliers

Between 1996 and 1999 BMW introduced at its own sites EMSs according to EMAS and ISO 14001. Additionally, a BMW-standard for EMSs was established, putting the special requirements of BMW into concrete terms. In 1997/98, BMW started to focus on its suppliers, beginning with those in the USA. At that time suppliers were questioned about whether they had introduced an EMS.

BMW wants each supplier to introduce an EMS (see BMW 2000: Principle 10). There is a focus on suppliers of great importance to BMW. In the beginning, some of the suppliers were rather cautious concerning the additional requests of their customer, BMW. Besides, they were not experienced in this field. Thus, 50 of the most important suppliers (those with the biggest sales volume from the viewpoint of BMW) were invited to a workshop where BMW tried to convince them to implement EMSs. First, BMW explained its own system; then, benefits as well as problems were discussed. Thereby BMW stressed that it was not only asking for its suppliers to undertake environmental activities but also supporting them. For example, for each supplier, BMW named a reference person for any problems met concerning environmental management.

Given the importance of BMW as a customer, workshops such as this can be a powerful instrument in enforcing appropriate EMSs on suppliers. Of the 25 suppliers participating in that workshop, 20 implemented an EMS afterwards. An important supplier refused to do so because of the high cost and because the main customer, a US car manufacturer accounting for nearly 95% of the supplier's sales volume, did not require such a system. After long discussions it was decided that a consultant of BMW would go to the supplier for three months to help implement the EMS. Now, the supplier maintains the system independently. From this example it becomes obvious that BMW is determined for its requirements concerning environmental management to be fulfilled.

In 1999, the German suppliers of BMW were interviewed, using the questionnaire of the Association of the German Automotive Industry (Verband der Automobilindustrie) VDA on environmental management at suppliers (see Figure 16.1). The objective of the project was to get a first overview of suppliers' activities concerning environmental protection. It was found that about 35% had already implemented an EMS (question 1 of the questionnaire); that was much more than BMW had expected.

Currently, this survey is being supplemented with additional questions and will be sent out again to monitor the progress of suppliers and to decide on necessary internal measures. Of course, this puts immense pressure on the suppliers to improve their environmental management.

This project of BMW is not an exception. Porsche AG has interviewed its suppliers about their environmental management, using the same questionnaire of the VDA (see Porsche 1999: 27). In 2001, the Ford Motor Company announced that certification to ISO 14001 by July 2003 would become a mandatory requirement for attaining 'Q1 supplier quality status', a key factor in all sourcing decisions (see Ford 2001). Similarly, the General Motors Corporation announced that by the end of 2002 all its suppliers would have to implement EMSs according to ISO 14001 (see General Motors 2001). In contrast, the VDA has decided not to enforce the implementation of such systems (see Ahsen 2006), though suppliers are encouraged to do so by German car manufacturers. Altogether, it can be presumed that as a result of these requirements from car manufacturers, most suppliers—at least in the USA—will have implemented EMSs in the near future.

16.2.3 Requests Concerning Product Development and Exchange of Information

As the eco-friendliness of cars heavily depends on supplied components, an important objective of BMW is to improve suppliers' efforts to ecodesign their products. For example, regarding the reduction of fuel consumption and emissions in the phase of product use, BMW requires its suppliers to design components according to weight-saving principles (see BMW 2001a:11). Additionally, BMW has defined a list of dangerous and therefore prohibited materials that must be avoided by suppliers.

Special emphasis is put on the suitability of the components for recycling. Suppliers are required to consider the potential of the material for economical recycling and to give preference to the use of secondary raw materials that have been recycled ('approved recycled'). Additionally, suppliers should use joining technology and modular construction to ensure that recyclable components and materials can be separated (see BMW 2001a:11). Table 16.1 shows recycling criteria within the process of product development at BMW.

VDA
Supplier Environmental Management

Supplier: _____   Site: _____
Responsible person: _____   Telephone:_____
Function: _____   Telex: _____

	Yes	No	Remarks

1. Does your company have an Environmental Management System in place in accordance with

   ☐ EMAS (EWG) 1836/93   ☐ ISO 14001   ☐ internal EMS   ☐ ☐   _____

   If yes, please provide a copy of your certificate and disregard the following questions.

   Not yet certified, but planned for ....   200__   _____
   ☐ EMAS (EWG) 1836/93   ☐ ISO 14001   ☐ ....

2. Does your company have another management system which includes environmental management?   ☐ ☐   _____
   If yes, what kind?...

3. Are environmental protection measures audited in your company?   If yes, by whom?   Internal Auditor/s ☐ ☐   _____
                                                                      External Auditor/s ☐ ☐   _____

4. In your company, are the folowing
   Production processes ....   ☐ ☐   _____
   Disposal and recycling processes ....   ☐ ☐   _____
   Produkts..   ☐ ☐   _____
   regularly inspected and assessed?

5. Are environmental aspects an integrated factor in your product planning?   ☐ ☐   _____

6. Does your company strictly adhere to relevant environmental regulations?   ☐ ☐   _____

7. Are environmental measures and the results dodumented in your company?   ☐ ☐   _____

8. Has your company established goals for continual improvement of performance and is the achievement of these goals documented?   ☐ ☐   _____

9. Are your company's employees informed and trained on environmental issues?   ☐ ☐   _____

10. Are you working with your suppliers and contractors to improve their environmental performance?   ☐ ☐   _____

11. Please briefly describe the environmental impact associated with your manufacturing processes
    _____
    _____

Date: _____   Signature: _____

Figure 16.1. Questionnaire of the association of the German automotive industry (VDA) on environmental management of suppliers

Table 16.1. Environmental criteria within the process of product development at BMW.
Source: Ahsen, 2006.

	Phase of Development			
	Goal Definition	Concept Phase	Series Production Developing Phase	Further Development
Working Packages within the Phases	Planning Task	Conceptional development	Project Approval Series Development	Release for Series Production
	Cataloque of Objetives	Selection of Alternatives	Release for Production	
	Main Criteria of Assessing Environmental Impact (Product)	Plan of Project	Set / Actual Comparison of the Recycling Rate by using Check Lists	
	Specification of a Recycling Rate	Life-Cycle Assessment of Assembly Parts		
		Dismantling Analysis		
		BMW Recycling Standard		
		Recycling Manual		

Within the goal definition phase, among others, the criteria for assessing the environmental impact of the car are defined and targets (*e.g.* recycling targets) are set. Within the concept phase the performance specifications, including detailed environmental requirements, are appointed. External suppliers as well as internal design engineers have to ensure the conformance of their components with these requirements. Additionally, the recyclable material has to be marked into each working drawing, using the BMW recycling standard: R1 meaning good recyclability; R2, sufficient recyclability; R3, insufficient recyclability. This procedure enhances the transparency of the process. Within the series production development phase the actual set points (*e.g.* concerning the recyclability) are compared with the required set points and, in the case of conformance to requirements, series production is released (see BMW 2001b: 23).

Within the product development phase, BMW performs life-cycle assessments (LCAs) on its own and supplied products (components, systems and complete vehicles) to evaluate the associated environmental loads. The assessment includes the entire life-cycle of the product (see Frankl and Rubik 2000). Thus, information about the energy and material consumption and emissions to the environment of the suppliers are necessary. Where appropriate, BMW performs LCAs in co-operation with suppliers, but mostly suppliers are urged to perform LCAs independently. To ensure a standardised and methodologically reliable use of LCAs, BMW offers to acquaint its suppliers with the procedure. The aim is to enable them to perform their own standardised assessments as elements in the complex LCAs of BMW. This also requires a shared conviction of subcontractors to co-operate in these projects to achieve optimised environmentally friendly products.

LCAs are time-consuming and thus expensive. Even more important, from the viewpoint of the suppliers, is that giving away detailed information on material and energy streams can be dangerous insofar as they can be used by the customer to carry out benchmarking processes that may lead to more difficult negotiations in the future. That is why BMW often has to convince its suppliers to co-operate at the beginning of LCA projects. The same problem arises in the context of other reporting requirements: for example, BMW expects suppliers to participate in the Automotive Industry Material Data System (IMDS).[2] The IMDS is a joint development of the Audi AG, BMW, DaimlerChrysler AG, Ford, the Adam Opel AG, Porsche, VW and the Volvo Group. Suppliers have to disclose information concerning all materials used in the system, so the car manufacturers get a detailed database that will facilitate recycling. Additionally, suppliers are to provide information on environmental reviews and audits that have been performed. The readiness to provide this information still differs from supplier to supplier. Not surprisingly, in this context the market power relationship between BMW and the individual supplier plays a key role.

The exchange of environmental information becomes even more sensitive in cases where problems occur. Several times BMW have become aware of

---

[2]See the International Material Data System (2001), available at www.mdsystem.de/html/home_en.htm.

emergency conditions at its suppliers, which in the worst case could have led to their shutdown, from the newspapers. As a consequence, BMW now requires each supplier to inform BMW within 24 hours after the appearance of any important problems (see Ahsen 2006).

Like BMW, most car manufacturers raise their requirements concerning suppliers' environmental management. Now, what does the environmental management of car manufacturing suppliers look like today? To what extent are EMSs implemented, and what kind of environmental planning and control activities are carried out? In Section 16.3 I try to answer these questions with the help of an empirical analysis.

16.3 Empirical Analysis of Environmental Management of German Automotive Suppliers

My research team and I have conducted an empirical analysis of the quality management and environmental management of automotive suppliers. Within this project, we interviewed German suppliers who are members of the VDA.[3] Altogether, between May and September 2001, 252 interviews were performed by telephone and e-mail; most importantly, quality managers and environmental managers were questioned. Nearly half of the sites where representatives were interviewed[4] have introduced an EMS; additionally, about 30% plan to do so. As Figure 16.2 shows, EMSs are implemented predominantly according to ISO 14001.[5] These results show that EMSs have become more and more common. This may be a result of corresponding requests to do so from car manufacturers, as discussed in Section 16.2.

---

[3] As there was no complete list of VDA members available we contacted those suppliers who maintained a link to their homepage on the homepage of the VDA (see http://www.vda.de). Out of the approximately 430 suppliers that are members of the VDA, we contacted 150 suppliers, with a combined total of 372 sites. From these, 104 companies, with 252 sites in all, answered our questions. We found that all the interviewed suppliers are first-tier suppliers, and sometimes, at the same time, they are also second-tier or third-tier suppliers. As sites often cannot quantify their sales volume, only the number of employees was analysed to differentiate between different sizes of site. We found the following distribution: 27.6% of the sites employ less than 200 employees; 19.6%, between 200 and 499; 23.2%, between 500 and 1,499; and 10.0%, more than 1,499 employees; 19.6% did not answer this question.

[4] For reasons of simplification, in the following 'x% of the sites' always means 'x% of the sites where representatives were interviewed'.

[5] Multiple answers were possible, because many companies implement EMSs according to EMAS and ISO 14001.

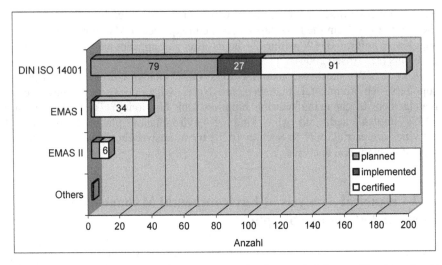

Figure 16.2. Environmental management systems implemented at the suppliers interviewed

Concerning the application of environmental planning and control instruments, we found that more than 65% of the sites use environmental ratios. This result may partly be explained by the corresponding requests for EMAS and ISO 14001 accreditation. Nearly 60% of the sites employ material and energy inventory analyses. For this purpose, all energy and materials used and waste released to the environment are identified, quantified and documented. Thus, the inventory analysis, for example, submits the comparison of planned and actual material and energy flows and items for improvement. Additionally, the results can be used as a decisive criterion according to the selection of a product or a process design concept.

From these inventory analyses a surprisingly high share of the sites carry out LCAs: more than half of them at least perform pilot projects. Again, one reason for this may be the corresponding projects of car manufacturers, as described in Section 16.2. For example, similar to BMW, DaimlerChrysler, Ford, VW and Volvo require that ecological analyses of their own as well as of supplied products are established according to ISO 14040 (ISO 1997; *e.g.* see DaimlerChrysler 2001:2). Suppliers are urged to perform the LCA independently, and part of the LCA is carried out by the car manufacturers that request such information from their suppliers, but we also found that common projects with the suppliers are carried out. Our analyses show that about 10% of the sites are engaged in such projects with their customers; there are also as many suppliers that are involved in similar projects with their subcontractors. Nearly half the sites include environmental criteria in the evaluation of their subcontractors.

Environmental protection is increasingly being carried out with use of integrated management systems. In our study we asked the suppliers if they have (or have planned) integrated management systems. The results can be seen in Figure 16.3.

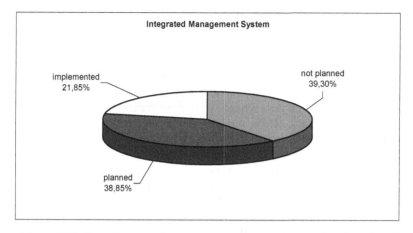

Figure 16.3. Use of integrated management systems at the suppliers interviewed

Further detailed questions showed that the majority of suppliers integrate their quality and environmental manuals as well as work briefings. Additionally, approximately 80% of the suppliers stress the importance of merging the quality and environmental departments. Not as many suppliers focus on integrated audits.

The most important reasons for integrating quality and environmental management are the achievement of better co-operation, clearly defined responsibilities and reduced costs as well as greater transparency. However, some drawbacks are also emphasised: above all, some of the staff are rather wary of integration projects. Environmental managers particularly worry that environmental management could be weakened within an integrated management system.

16.4 Conclusions

Altogether, car manufacturers are having a lasting effect on the environmental management of their suppliers. This influence is likely to intensify rather than slacken in the future, for several reasons: the automotive industry is influenced by increasing environmental legal requests as well as requirements from their customers (*e.g.* to cut fuel consumption, regarding recycling, to cut emissions and so on). Additionally, owing to the ongoing process of outsourcing not only of production but also of product design processes, the eco-friendliness of cars depends heavily on suppliers, especially on their research and development activities. As a result, car manufacturers involve their direct suppliers in extensive environmental control and follow-up systems. In this chpater I have taken a close look at the way in which the green supply chain management of BMW has been conducted over recent years. Additionally, I have described a phone and e-mail survey analysis carried out on the environmental management of automotive suppliers. The main two conclusions are as follows.

First, the case study shows that the supply chain management of BMW increasingly includes environmental activities. Thus, environmental criteria have become an important part in the process of supplier evaluation and selection: the capability and readiness of suppliers to fulfill environmental requirements (*e.g.* concerning the recyclability of components) are playing a key role in the sourcing decisions of BMW. Additionally, suppliers are urged to implement EMSs. Also, special emphasis is put on the exchange of environmental information. To facilitate the recycling of the cars, BMW requires detailed information from its suppliers concerning all materials used. Additionally, information is required, for example, for the purposes of LCAs and in case problems occur.

Last, the phone and e-mail survey highlights that automotive suppliers increasingly are implementing EMSs, especially according to ISO 14001. This implementation of EMSs is encouraged by car manufacturers. In fact, Ford and General Motors go so far as requiring that all their suppliers do so (see Ahsen 2006). Among the suppliers, environmental planning and control instruments are surprisingly widespread. Next to environmental ratios, inventory analyses and even LCAs are carried out, more often than we had expected. These projects are also heavily influenced by the car manufacturers. Considering all this together, co-operation concerning environmental management in automotive supply chains is becoming close and intense.

References

Ahsen, (2006), "Integriertes Qualitäts- und Umweltmanagement. Mehrdimensionale Modellierung und Anwendung in der deutschen Automobilindustrie", Deutscher Universitätsverlag, Wiesbaden.

Beamon, B.M. (1999) 'Designing the Green Supply Chain', Logistics Information Management 12.4: 332-42.

BMW (Bayerische Motoren Werke AG) (1999) Umweltbericht der BMW Group 1999/2000 (Munich, Germany: BMW).

BMW (Bayerische Motoren Werke AG) (2000) BMW Group Environmental Guidelines (Munich, Germany: BMW).

BMW (Bayerische Motoren Werke AG) (2001a) BMW Group International Terms and Conditions for the Purchase of Production Materials and Automotive Components (24 September 2001; Munich, Germany: BMW).

BMW (Bayerische Motoren Werke AG) (2001b) 'Sustainable Value Report 2001/2002', www.bmwgoup.com/sustainability.

Cascio, J., G. Woodside and P. Mitchell (1996) ISO 14000 Guide: The New International Environmental Management Standard (New York: McGraw-Hill).

CEC (Commission of the European Communities) (2000) End of Life Vehicles (ELV) Directive (2000/53/EC). (Luxembourg, Belgium: Office for Official Publications of the European Communities).

CEC (Commission of the European Communities) (2003) Hazardous Waste Council. Directive 91/689/EEC). (Luxembourg, Belgium: Office for Official Publications of the European Communities).

Clean Air Act (1990), 42 U.S.C 7401-7671q, P.L. 101-549, 104 Stat. 2399 (Cornell Legal Information Institute).

Clean Water Act (1977) (Federal Water Pollution Control Act), 33 U.S.C. 1251 - 1376; Chapter 758; P.L. 845, June 30, 1948; 62 Stat. 1155. (Washington, DC: US Government Printing Office).

DaimlerChrysler (DaimlerChrysler AG) (2001) Mercedes-Benz Special Terms 2001: Environmental Compatibility (Extended Enterprise Special Terms No. 36/3; September 2000; Stuttgart, Germany; Auburn Hills, Michigan, USA: DaimlerChrysler).

Delmas, M.A. and A.K. Terlaak (2001) 'A Framework for Analysing Environmental Voluntary Agreements', California Management Review 43.3 (Spring 2001): 44-63.

Ford (Ford Motor Company) (2001) 'Partnering with Suppliers to Achieve Our Environmental Goals', www.ford.com/serv...ormance&LEVEL5=partneringWithSuppliers, accessed 12 October 2001.

Frankl, P., and F. Rubik (2000) Life-cycle Assessment in Industry and Business: Adoption Patterns, Applications and Implications (Berlin and Heidelberg, Germany: Springer.

General Motors (General Motors Corporation) (2001) 'GM Teams with Suppliers to Further Resource Conservation, Prevent Pollution', www.gm.com/company/gmability/environment/partnerships/initiatives.html, accessed 11 October 2001.

Gesetz über die Entsorgung von Altfahrzeugen (Altfahrzeug-Gesetz – AltfahrzeugG), in: Bundesgesetzblatt Vol. 2002 Part I No. 41, 28.06.2002, p. 2199-2211.

Green, K., B. Morton and S. New (1998) 'Green Purchasing and Supply Policies: Do they Improve Companies' Environmental Performance?', Supply Chain Management: An International Journal 3.2: 89-95.

Handfield, R.B., and E.L. Nichols Jr (1999) Introduction to Supply Chain Management (Englewood Cliffs, NJ: Prentice-Hall).

ISO (International Organisation for Standardisation) (1997) ISO 14040: Environmental Management; Life-cycle Assessment; Principles and Framework (Berlin: ISO).

ISO (International Organisation for Standardisation) (2004) ISO 14001: Environmental Management Systems; Requirements with Guidance for Use (Berlin: ISO).

ISO (International Organisation for Standardisation) (2005): SN ISO/TS 16949: Quality management systems - Particular requirements for the application of ISO 9001:2000 for automotive production and relevant service part organizations (Berlin: ISO).

McIntyre, K., H.A. Smith, A. Henham and J. Pretlove (1998) 'Environmental Performance Indicators for Integrated Supply Chains: The Case of Xerox Ltd', Supply Chain Management: An International Journal 3.3: 149-56.

Noci, G. (1997) 'Designing Green Vendor Rating Systems for the Assessment of a Supplier's Environmental Performance', European Journal for Purchasing and Supply Management 3.2: 103-14.

Northeast-Midwest Institute (2003): Output-Based Emission Standards Advancing Innovative Energy Technologies, Washington, http://www.nemw.org/output_emissions.pdf.

Porsche (Porsche AG) (1999) Umwelterklärung 1999 (Zuffenhausen, Germany: Porsche).

Regulation (EC) No 761/2001 of the European Parliament and of the Council, 19 March 2001, Allowing Voluntary Participation by Organisations in a Community Eco-Management and Audit Scheme (EMAS), Official Journal of the European Communities L114 (24 April 2001): 1-29.

Resource Conservation and Recovery Act (RCRA), 42 U.S.C. s/s 321et seq. (1976) (Washington, DC: US Government Printing Office).

Sarkis, J. (1998) 'Evaluating Environmentally Conscious Business Practices', European Journal of Operational Research 107.1: 159-74.

Simchi-Levi, D., P. Kaminsky and E. Simchi-Levi (2000) Designing and Managing the Supply Chain: Concepts, Strategies, and Case Studies (Boston, MA: Irwin McGraw-Hill).

Toxic Substances Act (TSA), 15 U.S.C. s/s 2601et seq. (1976) (Washington, DC: US Government Printing Office).

Trittin, Jürgen (2000): Germany's New Climate Change Program, Speech at Den Haag, 22.11.2000, http://www.bmu.de/english/climate_change/doc/pdf/3314.pdf.

van Hoek, R.I. (1999) 'From Reversed Logistics to Green Supply Chains', Supply Chain Management: An International Journal 4.3: 129-34.

von Ahsen, A. (2006): Integriertes Qualitäts- und Umweltmanagement – Mehrdimensionale Modellierung und Anwendung in der deutschen Automobilindustrie –, professorial dissertation, forthcoming.

von Ahsen, A., and D. Funck (2001) 'Integrated Management Systems: Opportunities and Risks for Corporate Environmental Protection', Corporate Environmental Strategy 8.2: 165-76.

17

A Case Study of Green Supply Chain Management at Advanced Micro Devices

Philip Trowbridge, PE

Advanced Micro Devices, Inc., 5204 E. Ben White, M/S 529, Austin, Texas 78741 USA,
Philip.trowbridge@amd.com

This paper evaluates the operations of environmental supply-chain management at Advanced Micro Devices (AMD) as they existed in January 2002 The paper provides background on AMD's environmental practices and how they relate to greening the supply chain. Various practices and policies, as well as barriers and opportunities, are identified in this paper. The paper provides insights into how one organisation has been able to successfully and strategically align its supplier relationships with its environmental programmes.

17.1 Introduction

Competition is stiff, and global pressures require a quick response to competitors, customers and suppliers. This situation characterises the semiconductor industry where technology is ever-changing and demand, markets and requirements change on a regular basis. Why would any organisation in such a highly volatile and competitive situation be strategically interested in environmental issues, much less interested in other organisations' environmental performance? The answer is: because it makes business sense. To remain in business, companies need to follow regional, national and international laws and regulations while satisfying customer and stockholder requirements. They cannot remain internally focused as if they had blinkers on their corporate vision. That is why environmental issues no longer stop at the regulatory boundary but are necessarily boundary-crossing concerns. Environmental issues cross boundaries from company to customers, suppliers, competitors, the community and the environment itself. Maintaining a working 'industrial ecosystem' for electronics-based products needs to occur throughout the supply chain, particularly for corporations with operations in countries that have varying levels of environmental regulations (US AEP 2002). Advanced Micro Devices Inc. (AMD) is a critical partner in the electronics and semiconductor supply chain, especially from an environmental perspective. This case study provides a brief background on AMD and some of its corporate practices that are relevant to supply-chain management and environmental programmes. The major focus of this case study will be AMD's green supply-chain management practices: why they were developed, what they entail and how these practices have been managed thus far. Some lessons learned and future directions and issues facing AMD will conclude this case.

This case study provides a brief background on AMD and some of its corporate practices that are relevant to supply-chain management and environmental programmes. The major focus of this case study will be AMD's green supply-chain management practices: why they were developed, what they entail and how these practices have been managed thus far. Some lessons learned and future directions and issues facing AMD will conclude this case.

17.2 Background

AMD was founded in 1969, in Sunnyvale, California. The company is a global supplier of integrated circuits for the personal and networked computer and communications markets. AMD produces microprocessors, Flash memory devices and support circuitry for communications and networking applications. The company's products allow users to access, process and communicate information at ever-greater speeds and lower costs (www.AMD.com).

AMD employs approximately 14,400 people worldwide, with facilities in the USA (in Austin, Texas, and in Sunnyvale, California); Dresden, Germany; Suzhou, China; Bangkok, Thailand; Penang, Malaysia; and Singapore. The company had revenues of approximately US$3.9 billion in 2001. A truly global company, AMD derives more than half of its revenues from international markets. AMD has been publicly held since 1972 and has been listed on the New York Stock Exchange since 1979.

Semiconductors are found in a myriad of products from computers to cars. For example, Flash memory devices are components of cell phones, printers, appliances and automobiles (there may be dozens in one automobile). In order to remain strategically focused on core businesses, AMD has spun off non-core businesses in the past four years and has added a business that complements existing core products. The company also has a history of forming strategic alliances and long-term partnerships that enhance creativity and advance technology. The company manages a complex supply-chain landscape with thousands of suppliers and customers.

17.3 Environmental, Health and Safety Structure and History

From a strategic perspective, environment, health and safety is an integral part of AMD's operations. The company's commitment is evidenced by the corporate-wide environmental, health and safety (EH&S) policy:

> AMD's culture is rooted in respect for individuals. Our values commit us, as individuals and as employees, to actions that enhance the quality of life and protect the environment of the communities in which we do business. AMD's

environmental, health and safety (EH&S) programme reflects our commitment to these values. We have one set of universal standards that govern our practices worldwide. Our EHS programme is designed to provide a safe workplace for employees, protect the environment, prevent damage to property, enhance employee morale and assure compliance with applicable laws and regulations worldwide. Achievement of these objectives requires employment of best practices and a commitment to continuous improvement. We hold ourselves—management and individual employees—responsible for complying with AMD's EH&S standards and fulfilling the intent of this policy.

As stated in the policy, AMD has a set of worldwide standards that govern EH&S practices and programmes at all sites. The standards are a set of best practices, developed from internal and external experience, which promote continuous improvement. Regular assessments to determine conformance with worldwide standards are performed and often identify exemplary practices that are then shared with all sites. The worldwide standards are at the root of ISO 14001-compatible environmental management systems at each AMD site. All AMD manufacturing sites either have achieved or have been recommended for ISO 14001 certification.

AMD's environmental programme is the responsibility of the EH&S Department. The EH&S Department is organised with a corporate-wide Extended Producer Responsibility (EPR) section and site-specific EH&S functions at each manufacturing and research and development (R&D) location. Each AMD manufacturing and R&D location has dedicated EH&S personnel. The department has evolved in a similar manner to most corporate EH&S programmes. Prior to becoming a separate department, the EH&S function was part of the site's facilities or human resource organisation. In fact, at AMD's smaller manufacturing locations, the EH&S departments still reside under either the human resource or facilities departments, with dual reporting responsibility to the corporate EH&S director. The corporate EH&S director reports up through the chief administrative officer to the chief executive officer of the company.

The EPR section is strategically focused on product risk management, product design and supply-chain issues. EH&S-related international and regulatory issues with the potential to affect AMD's products and business are tracked and communicated internally.

Migration and expansion of EH&S at the corporate level occurred in conjunction with the expansion of manufacturing capacity and management's desire for a focused global directorship. Expansion of AMD's manufacturing capacity required the ability to quickly establish EH&S operations at new facilities and to share expertise between sites.

17.4 Environmental Strategy and Organisational Linkages

The EH&S strategy for AMD includes a mission statement and a multi-year corporate EH&S strategic plan that is reviewed annually. Integration with AMD's corporate strategy is critical. EH&S incorporates the plans for future growth of the corporation and the semiconductor market into the EH&S strategic plan. All worldwide EH&S groups contribute to the development of the corporate EH&S strategic plan. Using the plan as a basis for strategic direction, each manufacturing location then develops site-specific goals and strategies that support the corporate plan.

A key strategy for EH&S is maintaining the agility and flexibility to adapt to changing business conditions. Strategic plans and goals reflect the dynamic nature of the semiconductor manufacturing industry and are communicated internally to major business groups such as product-line and manufacturing groups, as well as to global supply management, finance and marketing departments so that they can incorporate relevant strategies and goals into their own strategic plans.

The linkage between marketing strategies and EH&S considerations is an important one. Emerging environmental regulatory issues that could potentially affect product design or limit AMD's access to certain markets need to be identified early so that appropriate actions can be taken. In new product design, the use of certain chemicals or elements can potentially prevent the marketing of products in various geographic regions. A specific example is the case of lead. A proposed European Union initiative calls for the elimination of lead in electronic products by 2006. Lead is used in the semiconductor industry to form critical interconnects or bonds for electrical leads. Since Europe is a large market for the industry and especially for AMD, AMD's design, manufacturing, marketing and sales functions must be aware of this issue in order to respond to and satisfy customer needs. AMD must be prepared to provide customers with lead-free products to meet customer demands and to stay competitive in the marketplace. Industry challenges, such as the elimination of lead, are important to communicate throughout the supply chain so that suppliers and customers can assist in developing solutions.

17.5 Environmental Practices at Advanced Micro Devices

17.5.1 Environmental Reporting

One EH&S practice that brings together a broad organisational overview of the company is AMD's EH&S report. AMD has been publishing an annual EH&S performance report, separate from its annual corporate financial report, since 1995, a relatively early and innovative practice for a company of AMD's size. One purpose of the environmental report is to highlight and publicise AMD's EH&S programmes and the progress that the company has made in these areas. In addition, the report is generated to inform investor groups who wish to keep apprised of the EH&S performance of the company. Thus, market and investor pressures motivated AMD to communicate organisational EH&S performance.

AMD's EH&S report is comprehensive, providing corporate and site-specific data. AMD began incorporating elements of the Global Reporting Initiative (GRI; see www.globalreporting.org) for sustainability reporting into the company's 2000 report. Although the company's 2000 and 2001 Sustainability Progress Reports contain elements of the GRI, they do not follow the GRI outline exactly.

17.5.2 Environmental Management Systems

All AMD manufacturing sites have either achieved certification, are seeking certification or have been recommended for certification to the ISO 14001 environmental management system standard. The drivers behind ISO 14001 certification can be linked to two central factors: market conditions and quality initiatives set forth by the organisation to meet anticipated customer needs. Increasingly, customers are requiring that products be manufactured under the framework of ISO (International Organisation for Standardisation) standards and other quality initiatives. The decision to certify to standards such as the ISO 9000 series, ISO 14001 and QS 9000 is based on strategic and customer market demands. Most AMD manufacturing locations are included in corporate-wide certification to ISO 9002 and QS 9000 (ISO/TS16949). Instead of seeking a single, corporate-wide ISO 14001 certification, AMD believed that site-specific certification would be a more effective strategy because of the diversity of the company's international locations and varying local pressures and requirements. With worldwide corporate environmental standards as a foundation, AMD let the sites determine the most suitable environmental management system for each locale.

An essential element in ISO 14001 is measuring continuous improvement with a solid environmental performance metrics programme. Table 1 provides an abbreviated list of environmental performance metrics collected for all AMD sites worldwide and reported in their annual sustainability reports (www.AMD.com/).

Table 17.1. Advanced Micro Devices: Example categories and performance metrics for environmental programs

Category	Metric
Resource utilization	▶ Total water used ▶ Total water conserved (reduced, re-used or recycled) ▶ Electricity consumed ▶ Fuel used
Releases and transfers	▶ Corrosive air emissions ▶ VOC air emissions ▶ Regulated hazardous waste generation ▶ Other solid waste sent for off-site disposal
Compliance	▶ Citations or notices of violations ▶ EH&S agency inspections ▶ Reportable spills or releases

VOC = volatile organic compound; EH&S = environmental, health and safety

17.5.3 Design for Environment, Health and Safety

Design for environment is practised at AMD, but in the broader sense of design for environment, health and safety (DfEH&S). Market demands have driven the creation of products with lower power consumption and higher utility. Therefore, the concentration of AMD's DfEH&S programme is on manufacturing materials and tool selection and on product packaging. AMD's EH&S and manufacturing groups work together to evaluate the tools, materials, chemicals and associated EH&S

systems required to build products safely. Upstream and downstream requirements are investigated co-operatively from the R&D phase through the implementation of new technologies.

17.5.4 Product Stewardship and Product Take-back

AMD is an electronics component manufacturer. As such, the company falls into the middle of a traditional supply chain when compared with raw materials extractors or end-user original equipment manufacturers (OEMs). Thus, from a product stewardship and product take-back perspective, OEMs (such as printed circuit board or 'box' manufacturers) currently bear the greatest burden for take-back. In most supply chains, AMD would be considered a second-tier or third-tier supplier, depending on the final product, yet AMD still maintains some responsibility regarding product take-back. OEMs often request product-specific chemical content information to assess the life-cycle impacts of the electronic goods that use AMD's products. OEMs are interested in the chemical content of both the product and the product packaging to determine if they contain banned or restricted materials. AMD responds to numerous product material content inquiries each year. Recent inquiries from customers have requested detailed product information for life-cycle inventory purposes. Requests for life-cycle inventory information are partly driven by the emerging EU Impact on the Environment of Electrical and Electronic Equipment Directive (Working Paper, Version 1, February 2001). AMD is working with various industry groups to determine what type of information may be required under these initiatives and to examine impacts on the supply chain.

17.5.5 Pollution Prevention and Resource Conservation

Pollution prevention and resource conservation are an integral part of AMD's environmental management systems at each site. AMD worldwide EH&S standards require that each site develop a pollution-prevention and resource-conservation plan that follows the basic hierarchy of reduce, re-use and recycle. The plans identify projects and incorporate a technical and economic feasibility analysis for each project. The success of AMD's programme is documented in past EH&S reports available on the Internet (www.AMD.com/). AMD does not have an official green purchasing policy at present. However, EH&S management and staff worked co-operatively with AMD's global supply management group to develop and communicate green procurement guidelines for office products and equipment. The guidelines provide the purchaser with EH&S considerations and recommendations when making the purchasing decision.

17.5.6 External Programmes

In addition to participating in industry groups, associations and R&D consortia on EH&S issues, AMD has also been active in numerous voluntary governmental environmental programmes. For example, AMD participates in environmental programmes sponsored by the US Environmental Protection Agency (EPA) such as ClimateWise,   WasteWise,   Green   Power   Partnership   and   the   PFC (perfluorocarbon)/Climate Protection Partnership (see www.epa.gov). Many AMD sites also participate in local voluntary government environmental programmes. Green supply-chain management (GSCM) at AMD covers a broad spectrum of suppliers. The components, co-ordination and related issues of the GSCM will now be discussed.

17.6 Green Supply-chain Management History and Relationships at Advanced Micro Devices

17.6.1 Drivers and Support for Green Supply-chain Management at Advanced Micro Devices

There are many drivers and champions for GSCM at AMD. AMD wants to be recognised as a sustainable organisation with strong extended producer responsibility programmes. This ambition, rising from the concepts of corporate and extended producer responsibility initiatives, has increased emphasis on EH&S issues associated with upstream and downstream stakeholders. Internally, GSCM is driven by:

- A desire to better manage the risk of a potential supply chain or business interruption arising from an EH&S issue
- A desire to work together with suppliers to identify alternative materials and equipment that minimise environmental impacts
- The recognised advantage of a strong corporate EH&S programme

External drivers for GSCM include customers' requests for more specific chemical content data and the increased interest of investors and non-governmental organisation (NGO) groups in environmental issues. AMD provides information to investor research groups such as KLD Research and Analytics (Boston, MA, USA; www.kld.com) and the Investor Responsibility Research Center (Washington, DC, USA; www.irrc.org). In terms of community or activist pressures, few issues have arisen. Internally, upper management strongly supports GSCM at AMD. Part of AMD's EH&S management process is where EH&S personnel, in co-operation with executive management, determine strategic direction.

Initially, AMD's EPR section and the US-based Corporate Supply Management Group were responsible for designing and implementing GSCM. Expertise in both EH&S and supply-chain management practices required close co-ordination. As the programme has matured and additional potential risk areas have been identified, other groups such as AMD's facilities, quality and manufacturing organisations have been integrated into the process.

The major internal business driver for GSCM has been organisational risk management. A twofold strategy emerged when establishing AMD's programme: (1) to develop a methodology that effectively identified potential risk factors; and (2) to develop a programme that generated useful data that could be easily maintained. The intangible nature of risk and the associated benefits from GSCM make the overall value of these practices difficult to quantify (Carter and Dresner

2001). GSCM practices that avoid even a single supply interruption incident arising from an EH&S issue can offer a substantial return on investment by decreasing lost production time. As the practices mature and additional information is collected, AMD will work towards developing quantitative metrics to evaluate GSCM practices at AMD.

17.6.2 Evolution of Green Supply-chain Management at Advanced Micro Devices

GSCM at AMD has grown and evolved over the past few years in a manner similar to the environmentally sound supply-chain management methodology (Lamming et al. 1999). AMD targets those aspects of the supply chain with the potential to interrupt service or that represent a potential liability issue. The primary objectives of the programme are to identify potential risk to AMD and to work with suppliers to mitigate the risks. An equally important objective is to determine whether suppliers have sufficient resources and programmes to sustain safe and environmentally friendly operations in the communities in which they operate.

GSCM evolved from AMD's audit programme for waste-management service providers into an active EH&S component of supplier management. EH&S looked at various supply chain environmental management initiatives during this time (see the next section for a more detailed discussion). AMD's programme has expanded to include chemical suppliers and large construction contractors in the EH&S evaluation. Large construction contractors were added because of the high-risk nature of their work, especially with regard to safety concerns. AMD wants to be apprised of the environmental and safety records of these construction firms. For example, certain best practices should be undertaken when storing hazardous materials on a construction site. A review of requested information from the supplier will help to determine whether the company has sufficient knowledge of industry best practices and associated laws and regulations to act responsibly.

Foundry subcontractors and subcontract manufacturing suppliers are also included in the high-risk profile of the GSCM. Although AMD maintains facilities that perform all phases of the manufacturing process, some manufacturing processes have been outsourced as markets have grown and as manufacturing constraints have been reached. Historically, at AMD the selection of foundry and subcontract manufacturers was based largely on technical and manufacturing capabilities. EH&S concerns were not initially viewed as a significant risk issue for this supplier group. However, as GSCM at AMD grew, these suppliers were viewed as high-risk from a business-interruption perspective, and supplier EH&S programmes are becoming a part of the routine evaluation programme as well as a consideration in the initial selection criteria for these suppliers.

17.6.3 Quality, Supply-chain Management and Environmental Issues

Evolution of GSCM at AMD has some of its underpinnings in AMD's World Class Supplier (WCS) programme. The semiconductor manufacturing business is quality-driven, and, in the late 1980s, AMD recognised the need to integrate suppliers into the manufacturing process with the development of the Quality Vendor Programme

(QVP). The QVP initiated supplier partnering as a means to drive continuous improvement in AMD's supplier organisations while reducing risk and variability in AMD's manufacturing process. In the early 1990s, this programme was renamed the World Class Supplier programme, maintaining the original concepts behind the QVP. The WCS programme has grown into a standard business practice for AMD that includes processes for driving improvement, monitoring supplier performance, assessing the effectiveness of supplier business and quality systems, matching technology 'road maps' and recognition for supplier excellence. Because of the company's commitment to providing a safe work environment, EH&S has been involved in the WCS and other continuous improvement teams at AMD for the past decade.

The WCS teams consist of EH&S, supply management, quality, finance, facilities operations and technical staff. The teams review the materials and suppliers that are essential to the manufacturing process. These teams consider material quality and supplier relations and also work with suppliers to resolve any issues. As part of the WCS review process, the teams have quarterly meetings with major suppliers to address supply and quality issues and to provide them with feedback on how AMD perceives their performance. Performance is evaluated on service, technical ability, quality, costs and flexibility. Relevant environmental issues are also reviewed during this process (Sarkis 1999).

The WCS teams focus on the suppliers that are critical to the manufacturing needs of AMD or that are considered high-risk because of issues such as sole sourcing and intellectual property. For example, in the case of chemical suppliers, where there are a large number of suppliers, the WCS teams target companies that are the biggest suppliers in terms of material and dollar volumes and those that supply AMD with critical chemicals. These critical chemicals are not necessarily highly toxic but may be critical with respect to the manufacturing and production systems.

AMD has eight WCS teams that evaluate suppliers of a range of products, from speciality gases to consumables. Table 2 provides a listing of the teams and the types of supplier involved. Overall, five of the eight teams evaluate chemical suppliers, the other teams covering suppliers of specialised products critical to the manufacturing process. The WCS teams do not review suppliers that are not critical to the manufacturing process.

Teams are an essential part of the total quality management and continuous improvement philosophy at AMD. EH&S representation is common on many of these decision teams. In addition to the WCS teams, EH&S personnel contribute to manufacturing equipment improvement teams, facilities teams evaluating the planning and design of new construction projects and new-technology R&D teams evaluating potential EH&S impacts of future manufacturing technologies.

Table 17.2. World-class supplier team names and the types of suppliers targeted by the teams

Team name	Supplier type(s)
Wet Chemicals	Bulk commodity chemicals
Photolithography Chemicals	Specialty photo-resist and developer chemicals
Chemical-Mechanical Planarisation	Specialty planarisation chemicals
Consumables	Supplies consumed in the manufacturing process
Reticles	Reticles for imaging
Quartz	Quartz tube suppliers
Gases	Commodity and speciality gas suppliers
Silicon	Silicon wafer suppliers

17.6.4 Just-in-time Systems and Green Supply-chain Management

Another general business practice associated with GSCM practices is AMD's just-in-time (JIT) chemical delivery programmes that are in place at AMD's wafer manufacturing sites. AMD utilises a modified JIT programme that combines traditional JIT practices with the bulk purchase of certain commodity-type chemicals. The programme offers both environmental and business benefits. Increased handling and larger inventories of speciality chemicals result in potentially higher risks of spills and spoilage. JIT offers the advantage of a reduced inventory of expensive speciality chemicals and their associated inventory costs (Sarkis 1999).

While the chances or likelihood of an accident increase with more frequent deliveries, more deliveries also mean less efficient use of energy for transportation and handling. Where mechanically and economically feasible, AMD has moved away from JIT deliveries toward a preference for bulk delivery for certain commodity-type chemicals, making much more efficient use of transportation resources at a more favourable cost, particularly when compared with a delivery of chemicals in drums. Bulk deliveries decrease the amount of handling throughout the supply chain and consequently decrease the potential for accidents or spills. However, the trade-off with bulk deliveries is that the handling of larger quantities could result in larger spills in the event of an accident. In several cases, AMD and large chemical suppliers have used their combined expertise to design bulk chemical storage and handling facilities with improved EH&S features. These facilities automate storage and distribution of commodity-type chemicals, minimizing the physical handling of chemical containers and eliminating the need for disposal or recycling of chemical containers.

17.6.5 ISO 14000 and Green Supply-chain Management

Some companies, such as Ford and IBM are requiring or recommending that their suppliers be ISO 14001-certified.1 AMD has not taken this step and has no immediate plans to require certification as a condition of doing business. However, GSCM requirements are co-ordinated with ISO 14001 requirements and the global standards set by AMD. Although some suppliers to AMD are ISO 14001-certified, certification does not exempt a supplier from AMD's supplier auditing and assessment practices. Although ISO certification indicates the presence of an environmental management system, the certification is not indicative of the effectiveness of the system or of exemplary environmental performance.

17.7 The Practice of Green Supply-chain Management at Advanced Micro Devices

Thus far, the discussion has focused on a number of more general issues relevant to GSCM at AMD and the relationship to other organisational practices within the company. The following sections discuss GSCM practices at AMD. AMD's practices include both mature and relatively new elements.

17.7.1 Self-assessment and Auditing Questionnaires

The self-assessment questionnaire is the initial information-gathering instrument used by AMD. Questionnaires are sent to suppliers in the identified 'high-risk' groups, and responses are reviewed and scored relative to AMD's expectations for the industry and other suppliers. Survey questions are tailored for industry-specific or group-specific suppliers. For example, the questions asked of chemical suppliers differ from those asked of construction contractors for large projects. New survey instruments have been developed for each group to reflect the differences between supplier groups and different EH&S programme expectations.

AMD's survey development process takes into consideration the relative size of the suppliers and the amount of materials or services purchased by AMD. In general, the surveys are structured in two parts. Part 1 of the survey includes general questions about the company's EH&S management and commitment. Part 2 of the survey asks topic specific questions. For example, Part 2 of the chemical supplier assessment survey addresses emergency response and preparedness, distribution safety, pollution prevention, health and safety, product stewardship and process safety. AMD solicits feedback on the survey format and content from a few selected suppliers during pilot-testing prior to broad distribution. Pilot-testing helps to identify country-specific or region-specific language and idioms and other potentially confusing language that may lead to less useful responses. The feedback is used to improve the understandability of the questionnaire and, consequently, the usefulness of the data.

AMD uses available resources and past experience to develop survey questions. For example, the American Chemistry Council's Responsible Care™ programme was used to help design the questionnaire for chemical suppliers. For waste-management service providers there is an abundance of publicly available information related to environmental assessing and auditing of these service providers. Other questions are developed to identify potential sources of risk based on AMD's past experiences or expectations. AMD monitors changing assessment and audit trends through involvement in industry groups, review of published information and a network of industry contacts.

Questions on the self-assessment questionnaires are typically structured with multiple-choice responses for the supplier. The responses are structured to address a variety of situations for the supplier to select the most appropriate response. Each question is followed by an opportunity for respondents to comment on or clarify their responses. An example set of questions from AMD's chemical supplier EH&S assessment questionnaire is shown in Box 1.

Supplier responses are incorporated into a WCS review for that supplier. Additionally, as the WCS teams conduct technical assessments or audits of the suppliers' quality and technical capabilities, EH&S personnel may participate, depending on the score from the self-assessment. Responses from suppliers not included in WCS team responsibilities are compared with other suppliers in the same industry group. Suppliers not meeting AMD's expectations are asked either to clarify their responses or to provide additional information. AMD works co-operatively with suppliers to promote continuous improvement and to address areas that do not meet expectations. Under extreme circumstances, the relationship with a new potential supplier would not be pursued or the relationship with an existing supplier would be terminated.

For most supplier groups, periodic updates to supplier responses are incorporated into AMD's programme. Updated information helps AMD to identify areas of improvement for suppliers and to ascertain how supplier EH&S organisations have matured.   Supplier data is updated at varying frequencies, depending on the perceived risk to business interruption. For example, information on hazardous waste service provider is updated annually, whereas data on chemical suppliers is updated every two years.   However, an update to large construction contractor information occurs only when the contractor is proposing to bid on a new project for AMD.

17.7.2 Engaging Suppliers in Design for Environment, Health and Safety

One of the central ways that suppliers are involved with DfEH&S at AMD is through the material and manufacturing equipment review process. These reviews target chemicals and tools used in the manufacturing process and are conducted as part of the procurement process. For example, prior to purchasing a new chemical, whether for production or R&D, EH&S personnel review available information contained in the material safety data sheet and supplemental information from the chemical user and supplier. The review includes the proposed use, quantity, purpose, location of use and the type of health and safety protection systems proposed to manage the chemical and associated waste. Supplier information is critical at this juncture to ensure that proper systems are provided to protect employees and the environment. Several hundred new chemicals are reviewed annually.

Box 17.1. Example questions from the supplier assessment questionnaire of Advanced Micro Devices

▶ Is your company or site certified according to ISO 14001 or another environmental management system standard?
- ☐ No certification and not seeking certification
- ☐ No sites certified, but short-term company plans include certification
- ☐ Some sites are certified
- ☐ All sites certified

▶ Do all manufacturing sites have dedicated, full-time environmental, health and safety (EH&S) staff?
- ☐ There is no full-time EH&S staff at any of the manufacturing sites.
- ☐ Staff having other responsibilities, a contractor EH&S service or some combination of the two handles EH&S responsibilities.
- ☐ There is dedicated, full-time EH&S staff at some manufacturing sites.
- ☐ There is dedicated, full-time EH&S personnel at all manufacturing sites, with little or no corporate EH&S support.
- ☐ There is dedicated, full-time EH&S staff at all manufacturing sites supported by corporate EH&S staff.

▶ Are routine environmental and safety inspections, audits or assessments conducted at all manufacturing sites?
- ☐ No routine environmental and/or safety inspections, audits or assessments are performed at manufacturing sites.
- ☐ Environmental and safety inspections, audits or assessments are performed at manufacturing sites but with no regular frequency.
- ☐ Environmental and safety inspections, audits or assessments are performed on a regular frequency.
- ☐ EH&S audits or assessments are routinely performed with assigned responsibilities for corrective actions and follow-ups to resolution.

▶ Have there been any environmental or safety violations resulting in fines, penalties, compliance orders or similar actions over the past three years?
- ☐ There have been violations resulting in a total amount of $1,000 or more or there have been one or more compliance orders within the past three years.
- ☐ There have been violations resulting in a total amount of less than $1,000, or one compliance order within the past three years.
- ☐ There have been no violations within the past three years at any site owned or operated by the company.

▶ Does your company participate in voluntary industrial, government or community-based initiatives?
- ☐ There is no participation in voluntary industrial, government or community-based initiatives.
- ☐ There is some participation in voluntary industrial, government or community-based initiatives.
- ☐ There is participation in numerous voluntary industrial, government or community-based initiatives.

▶ Does your company monitor the EH&S performance of its suppliers and/or contractors?
- ☐ No performance criteria are placed on suppliers and/or contractors.
- ☐ EH&S performance criteria are being developed and documented.
- ☐ EH&S performance criteria have been developed and selectively implemented.
- ☐ Criteria have been effectively implemented and enforced for all suppliers and/or contractors.

▶ Are EH&S considerations a component of product or service design?
- ☐ No EH&S considerations are incorporated into product or service design.
- ☐ EH&S considerations are selectively incorporated into product and service design.
- ☐ EH&S considerations are consistently incorporated into product and service design.
- ☐ Design for EH&S concepts are fully integrated into product and service design.

▶ Has your company adopted and implemented global standards for EH&S performance that are applicable to all manufacturing operations worldwide?
- ☐ No global EH&S performance standards have been developed.
- ☐ Informal EH&S performance standards have been developed and implemented worldwide.
- ☐ Global EH&S performance standards have been adopted by the corporation and are in the process of being implemented.
- ☐ Global EH&S performance standards have been adopted and implemented.

AMD and its suppliers participate in the joint R&D of environmentally friendly technology and processes through International Sematech (www.sematech.org) and the National Science Foundation/Semiconductor Research Corporation Center for Environmentally Benign Semiconductor Manufacturing (www.cebsm.org). For example, AMD hosted a project in 2000 for International Sematech to research new wafer-cleaning technology. Working with equipment and material suppliers, AMD demonstrated the effectiveness of ozonated water for the removal of photo-resist residues.

17.7.3 Reduction of Packaging Waste at the Customer–supplier Interface

AMD and its suppliers actively pursue opportunities to reduce packaging waste and handling at the supplier–customer interface. For example, AMD switched from 55-gallon drums to 300-gallon totes to bulk tankers for several commodity-type chemicals to help reduce packaging waste. In addition to reducing waste, cost reductions have been realised as the quantity of chemicals purchased in bulk tankers has increased. When purchasing chemicals not suited for bulk quantities, AMD's preference is to purchase chemicals in returnable or re-usable containers.

17.7.4 Re-use and Recycling of Material

Owing to the ultra-high-purity requirement of materials used in the semiconductor manufacturing process, point-of-use recycling has shown limited success. Despite the manufacturing limitation on re-use of materials in the process, AMD has found other opportunities for re-using or recycling materials. For example, used sulphuric acid from the manufacturing process is re-used as a neutralising agent for corrosive waste-water as well as a neutralising agent in scrubbers for corrosive air emissions. The remaining sulphuric acid is collected and shipped off-site for re-use by other companies as an industrial-grade feedstock. In addition to chemical conservation and re-use, AMD also has active water conservation and re-use programmes as well as energy conservation programmes at each manufacturing site.

17.7.5   Life-cycle Assessment Cooperation from Suppliers

Historically, conducting a life-cycle assessment (LCA) on semiconductor products has not been an industry priority, because the product life expectancy is relatively short and semiconductor products are components of larger products. However, the need for LCA information on semiconductor products will become more crucial if the EU Impact on the Environment of Electrical and Electronic Equipment Directive passes the European Parliament. AMD has already experienced an increase in requests for LCA-type information from customers and anticipates requesting this type of information from its suppliers in the future.

17.7.6 Influencing Legislation in Co-operation with Suppliers

AMD relies on industry groups to influence legislation. Many of the groups include suppliers. One recent example of co-operation between manufacturers and suppliers is the case of perfluoroctyl sulphonates (PFOS). PFOS is a material used in small quantities in photo-resist and photo-developer solutions in the semiconductor manufacturing process, but it is better known for its vastly larger use in products such as Scotchguard™.

The US EPA proposed a broad restriction on the manufacturing of this chemical. Although replacement chemicals have been identified for some applications in the manufacturing process, alternative chemicals have not been identified for other applications. AMD and other semiconductor manufacturers and chemical suppliers affected by the restriction are working together to secure exemptions for critical uses where suitable replacements have not been identified.

17.8 Organisational Issues

Even though the GSCM policies and practices are becoming more pervasive and integrated into AMD's corporate operations and practices, the road has not been without some pitfalls. From an economic perspective, few barriers have developed. The monetary investment required to implement GSCM at AMD has been relatively minor. Most of the difficulties encountered in implementation and execution have been associated with integrating GSCM into existing operational practices.

Operationally, the movement towards GSCM has required a slight cultural change for the company. Integrating suppliers and EH&S into the decision-making and managerial processes was initially challenging. However, with the involvement of EH&S personnel in the WCS teams, integrating EH&S concerns into the supplier selection process has grown easier. Many of AMD's internal groups realise the value of EH&S considerations and understand the need for EH&S involvement and have been accepting of the change.

Chemical suppliers were also accepting of EH&S involvement because of similar expectations in their own industry. For example, many chemical manufacturers participate in the Responsible Care™ programme, which requires a continuous EH&S improvement commitment. Acceptance of an EH&S component for AMD supplier evaluations has not been a significant barrier.

The construction industry may not be as accustomed to supply-chain EH&S issues as the chemical industry, but AMD has had early success. The EH&S evaluation was successfully used to assist in the selection of the company to build AMD's facility in Singapore. The evaluation was used as a pre-bid tool to select companies that could bid on the project and it was also a factor in the final selection process.

Tailoring self-assessment questionnaires to different supplier groups required some initial effort. However, previously developed questions and publicly available information have been used to adapt questions for newly identified risk categories. AMD has avoided developing questionnaires for different-sized companies within each identified risk category. For example, the chemical supplier questionnaire has been used for companies with seven employees and for companies with 7,000 employees. Although some smaller companies have difficulty answering all the questions, they were able to complete the survey to the best of their ability.

AMD's programme is electronically available to all employees, and regular updates to management communicate the latest evaluation results. AMD is currently implementing an enterprise resource planning system that will help to centralise procurement information worldwide. This system will help to ensure that all suppliers in high-risk business areas are captured.

17.9 Future Issues

In concluding this case study, some of the plans and issues that AMD faces in the future include the following:

- AMD will extend GSCM practices to other supplier and commodity groups, after identifying and evaluating their risk in relation to AMD's manufacturing operations. Equipment maintenance and cleaning service suppliers may be the next groups of companies subject to GSCM practices at AMD. Typically, these organisations are smaller and do not have large numbers of staff dedicated to maintaining EH&S programmes. The questionnaire will need to be customised for these smaller suppliers.
- As additional data is collected, AMD will continue to investigate ways to measure and quantify the value of GSCM activities.
- AMD would like to expand its GSCM information management systems to facilitate internal access to data across functional units.
- AMD would like to expand its GSCM electronic information management systems to allow for more frequent and convenient communication between suppliers and AMD.

References

AMD (Advanced Micro Devices) (2000) 2000 Sustainability Progress Report (Sunnyvale, CA: AMD, www.AMD.com/).

Carter, C.R., and M. Dresner (2001) 'Purchasing's Role in Environmental Management: Cross-functional Development of Grounded Theory', Journal of Supply Chain Management 37.3: 12-26.

Lamming, R.C., P. Cousins, F. Bowen and A. Faruk (1999) 'A Comprehensive Conceptual Model for Managing Environmental Impacts, Costs and Risks in Supply Chains', in Proceedings of the Eighth Annual IPSERA Conference, Belfast and Dublin, 1999.

Sarkis, J. (1999) How Green is the Supply Chain? Practice and Research (working paper; Worcester, MA: Graduate School of Management, Clark University).

US AEP (US–Asia Environmental Partnership) (2002) 'Greening the Supply Chain', www.usaep.org/ctem/greening.htm, accessed 22 March 2002.

Part IV

Tools and Technology

18

Environmental Quality in the Supply Chain of an Original Equipment Manufacturer: What Does It Mean?

Menno Nagel

Delft University of Technology, Faculty of Mechanical, Maritime and Materials Engineering, Mekelweg 2, 2628 CD DELFT, The Netherlands, Email: m.h.nagel@3me.tudelft.nl or mnagel@dualcore.nl.

This chapter approaches the environmental quality question in the supply chain from an original equipment manufacturer's (OEM) perspective within the electronics and telecommunications industry. The OEM perspective of both a customer and the supplier typically consider such aspects as price, delivery, service, technology and quality playing on an ongoing role, while environmental quality is an additional and new factor. A novel environmental supply chain evaluation approach is introduced here on the basis of the eco-supplier development cycle, which embodies six steps. The activation and continuation of the cycle is executed with environmental performance tools, which collect the supplier data and generate an environmental performance per supplier. The environmental performance expresses the supplier's total production behavior. Based on their environmental performance, suppliers can be evaluated and a proposed price reduction can be derived. The link between environmental performance and proposed price reduction transposes environmental quality into a business perspective. In this scope, the results of a global assessment of 25 printed board production facilities are discussed and a conclusion is drawn. In the long term, this new approach offers a complementary tool for environmental management systems like ISO 14001.

18.1 Introduction

Customers, suppliers, competitors, shareholders, and governments all play roles in the corporate competitive environment. Multiple relationships exist between the company and these entities. Within this setting a company operates in terms of product sales, production, procurement, and regulatory activities. The customer-supplier relationship's core activities are sales and procurement. Relationships are cornerstones of the global operating economic process, because this process can be described in terms of the sum of customer-supplier relationships. From the perspective of the customer as well as the supplier, such aspects as price, delivery, service, technology and quality play an ongoing role, while environmental quality is a new factor. This chapter approaches the supplier or set of suppliers, the 'supply chain', from a customers' perspective, as the requesting party in the scope of environmental quality. Each company producing products or delivering services has its own set of suppliers. The set of suppliers of an original equipment manufacturer (OEM) clearly differ depending on industry, where different materials, supplies, and even practices are exhibited.

Management of the supply chain of an OEM in the telecommunications industry is a complex activity because an average telecommunication product contains roughly 10,000 different components. When considering factors of price, delivery, service, technology and quality, ranking and classification of suppliers typically occurs with ranges from excellent to poor. Indeed, now suppliers should similarly be ranked and classified with relation to environmental aspects (Nagel, 2001).

In current supply chain practices of many telecommunication OEMs, semiconductors, cables, printed boards, housings, capacitors and various types of subassemblies are required to assemble a telecommunication product. These components are a sum of base materials. The raw materials are procured through a supplier of the supplier. In some cases, like copper, the next chain can be outlined: copper extraction, pure copper production, lead-frame production for semiconductor devices and lead-frame preparation before use. It is clear that several customer-supplier relationships exist in this chain, see Figure 18.1 for a generic supply chain.

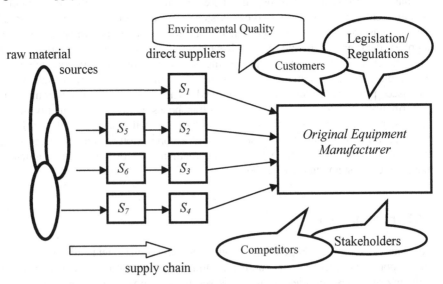

Figure 18.1. The supply chain as a sum of customer-supplier relationships

Environmental quality plays a role in each customer-supplier relationship. For example, supplier S_2 is the customer for supplier S_5. The entire supply chain of an OEM contains suppliers S_1 through S_4, i.e. from raw-material extraction to produced components. For the proposed supply chain approaches of OEMs, the contacts with the supply chain are limited mostly to the first tier of suppliers, i.e. S_1 to S_4. Introducing the concept of environmental quality in the OEM's direct supply chain, S_1 through S_4, may provide a substantial win-win opportunity from an environmental-business perspective. That is, the environmental load of a supplier's production facility can be linked to a proposed price reduction on the purchase turnover. Suppliers S_1 through S_4 generate environmental load in their different production processes, as well as the suppliers further down the chain. Each process step in the chain produces wastes, emissions and usable material and components. As well, each process step needs energy, auxiliary compounds, water, raw materials and/or subcomponents. This flows characteristics show that the supply chain can be approached from both an environmental and an economic perspective. Less use of materials and energy in a process step can constitute an environmental and an economic benefit. Each produced component can contain environmentally relevant substances or can use too much energy or can be non-recyclable. The introduction of the concept of environmental quality to each customer-supplier relationship in the chain offers an environmental-business opportunity when the suppliers' environmental performances are measured and integrated into the suppliers' negotiations. This chapter focuses on the environmental quality of processing methods for components *in* the supply chain. Furthermore, a global application of this new relative approach using performance analysis and metrics in the printed board industry will be described

18.2 An Environmental Supplier Development Approach

18.2.1 Introduction

The management of environmental quality in the supply chain can be driven from an organization's own corporate goals or from customers, competitors and/or legislation. Customers, competitors, stakeholders, legislation are external drivers for a company, while the corporate goals are internal drivers, like the realization of cost savings from an environmental perspective. When a customer of an OEM has specific questions relating to the material content of the delivered product, the questions should be answered directly or when for instance, the use of chromium in products is forbidden in Europe, the OEM should take action immediately. When the OEM carries out activities in compliance with its customer's request, and complies with the legislation, but does not study the backgrounds of these requests and laws, and anticipates and acts on future possibilities, the OEM adopts a reactive position. Independent of customer questions, regulations and laws, but linked to corporate goals, the major question above relating to the material content of products may be the trigger for a company to develop an environmental business strategy. To have in place its own environmental business strategy is a step leading to a more proactive position. An environmental supply chain strategy, a product strategy and a marketing strategy can be derived from a company's environmental business strategy. The links between costs and environmental impact should be a leading element in these strategies. Because mainly the supply chain determines the material content of the OEM's products, it emphasizes that a supply chain approach is necessary.

18.2.2 Environmental Loads, Metrics and the Eco-supplier Development Cycle

When suppliers' production processes are comparable, the environmental load per kilogram of produced component is typically comparable. For instance, suppliers A and B produce comparable printed boards. Supplier A uses 5 kilogram base materials and supplier B uses 7 kilogram base materials per 1 kilogram of printed board. Comparisons of A and B will probably show that supplier A has a better environmental performance than supplier B. This may also mean that supplier A has lower costs for the base materials and less solid waste. Less solid waste results in less waste handling costs, use of materials, auxiliary compounds, water and energy. They may also have congruent generation of air emissions and other solid and liquid wastes. These environmental load elements can be used to determine the environmental performance of a supplier's production facility. These environmental load elements form the basis for a supply chain management environmental performance tool. The generated environmental load (E_L) of a one kilogram component by the use of materials, auxiliary compounds, water, energy and packaging materials, can be argued to be inversely proportional to the environmental performance (E_P), which is general expressed by (18.1).

$$E_{P,SUPPLIER} \propto \frac{1}{E_L} \qquad (18.1)$$

Suppliers can be managed on environmental performance primarily because there are measurable, tangible metrics. Environmental supply chain management requires appropriate environmental performance management tools and metrics. In our view, an environmental performance tool for supplier assessments needs to contain at least two elements:

- A set of specified questions related to the use of materials, auxiliary compounds, water, energy, packaging materials, air emissions and waste, the so-called data collection process related to the seven environmental load elements.
- A model, which generates a quantitative environmental performance value that is useful to managers.

Environmental performance per supplier is a basic tool to help evaluate suppliers. Environmental performance can be linked to the supplier's purchase turnover, which can result in price reductions. Environmental quality needs to be integrated into the supply chain based on the supplier's environmental performance *and* the link to purchase turnover. Proposed price reductions linked to bad environmental performances trigger suppliers to improve themselves competitively. Assigning an economic cost (e.g. reduced price) to poor supplier environmental performance will encourage cost savings and better environmental performance.

Supplier development from an environmental perspective is defined as eco-supplier development, which is based on continual improvement. Eco-supplier development suggests two or more different measurable environmental situations of a supplier and the method of changing from environmental situation A to environmental situation B, see Figure 18.2.

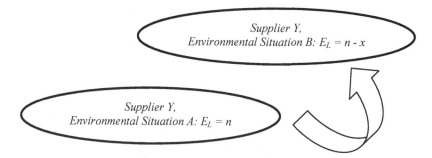

Figure 18.2. Eco-supplier development

The challenge for the OEM is *how* to activate suppliers in such a way that they initiate innovations in their processes and components from an environmental business perspective, which results in a reduction of the environmental load for the existing chain. Figure 18.2 shows the environmental load (E_L) situation A for supplier Y as $E_L = n$, while the environmental load in environmental situation B has been decreased by x to $E_L = n - x$. Eco-supplier development is a core competence in a supply chain policy. Eco-supplier development should be integrated into the supplier development cycle. A proposed eco-supplier development cycle embodies six steps as shown in Figure 18.3.

The first step in this proposed cycle is the determination and acquisition of supplier measurements. Environmental performance per supplier can be calculated and compared from these measurements, represented in the second and third steps of the cycle. Based on the environmental performance, proposed price reductions relating to the supplier's purchase turnover can be determined and negotiated with the supplier, the fourth step. This link puts environmental quality within the scope of a business perspective and results in an agreed price reduction, after negotiation, the fifth step. If the supplier has been classified as very bad and the proposed price reduction is 10%, the primary intention is not to discontinue business with the supplier, but to realize an agreed price reduction and on the basis of this to support the supplier with an eco-supplier development plan. These price reduction steps are clearly the "activation steps" of the cycle. Such a plan should contain actions for improvement such as reducing energy consumption by 5% at the same production level, the sixth step. The execution of an eco-supplier development plan is the supplier's responsibility, but encouragement and support by the end customer may help this development plan along. After, 3 or 4 years, for example, the supplier will be measured again and compared with its competitors. *The essence of the eco-supplier development cycle is to realize environmental improvements by price incentives in the scope of continual improvement.* The activation and continuation

of the eco-supplier development cycle cannot take place without environmental
performance tools.

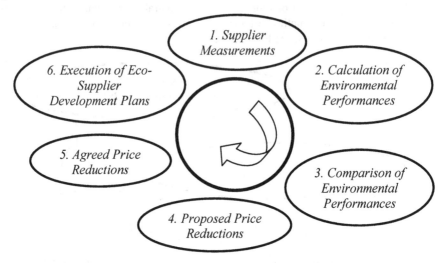

Figure 18.3. Eco-supplier development cycle

18.2.3 Environmental Process Modeling based on a Relative Approach

The production process environmental load contribution can be addressed from
both an absolute and a relative approach. The absolute approach makes a direct
link to environmental effects, like ozone depletion, greenhouse effect and so on,
while the relative approach assumes that a minimum use of materials, water,
energy and other resources always delivers an environmental benefit. From the
relative approach, a random supplier's production process may have five different
input flows and three different output flows, see Figure 18.4. The three different
output flows can be distinguished in two undesired and one desired output flow.
The desired output flow is the mass of manufactured products or components, m_{poc}.
These input and undesired output flows are defined as follows:

1. Input flow of base materials m_i: The desired component is produced from these
   materials for sale to the customer.
2. Input flow of auxiliary compounds m_{ac}: These chemical compounds are
   necessary to produce the desired component, but are not included in the
   component.
3. Input flow of water U_w: Water in combination with chemical compounds is
   necessary to produce the desired component, but is not included in the
   component.
4. Input flow of energy E: Energy is necessary to produce the desired component.
5. Output flow of air emissions E_m: The production of the desired component
   generates an undesired flow of air emissions.

6. Output flow of waste W_t: The production of the desired component generates a liquid and solid waste flow of water, chemical compounds, metals, plastics and paper, etc.
7. Input flow of packaging materials P_m: These packaging materials are used to transport the produced component from the production facility to the customer

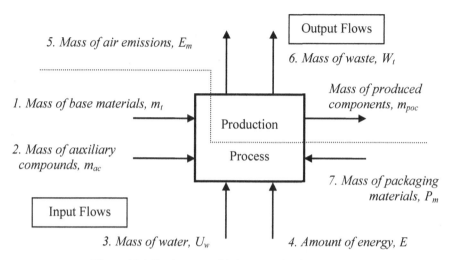

Figure 18.4. Environmental balance production process

All these input and output flows are a function of time. The rate of production depends on the rates of the use of materials, auxiliary compounds, water and energy, while the generated waste and air emissions per unit time are also linked to production rate. The production rate (R_p) and its relations to the other flows per unit time can be expressed by equations (18.2) through (18.8).

In these equations, I_1 through I_7 are defined as the environmental indicators, while k_1 through k_7 are constants, during a fixed period. The definition, dimension and determination of a constant depend on environmental research from a life-cycle perspective. The percentage of bromides in the resin of a laminate, for example, can determine k_1 and thus influence environmental indicator I_1. For a period, Δt, equations (18.2) through (18.8) can be rewritten, through integration. The result is equations (18.9) through (18.15), which show a linear system.

$$production \;\; rate: \;\; R_p = \frac{dm_{poc}}{dt} = I_1 \cdot k_1 \cdot \frac{dm_t}{dt} \tag{18.2}$$

$$R_p = I_2 \cdot k_2 \cdot \frac{dm_{ac}}{dt} \tag{18.3}$$

$$R_p = I_3 \cdot k_3 \cdot \frac{dU_w}{dt} \tag{18.4}$$

$$R_p = I_4 \cdot k_4 \cdot \frac{dE}{dt} \tag{18.5}$$

$$R_p = I_5 \cdot k_5 \cdot \frac{dE_m}{dt} \tag{18.6}$$

$$R_p = I_6 \cdot k_6 \cdot \frac{dW_t}{dt} \tag{18.7}$$

$$R_p = I_7 \cdot k_7 \cdot \frac{dP_m}{dt} \tag{18.8}$$

$$m_{poc} = I_1 \cdot k_1 \cdot m_t \qquad or \qquad I_1 = k_1 \cdot \frac{m_{poc}}{m_t} \tag{18.9}$$

$$m_{poc} = I_2 \cdot k_2 \cdot m_{ac} \qquad or \qquad I_2 = k_2 \cdot \frac{m_{poc}}{m_{ac}} \tag{18.10}$$

$$m_{poc} = I_3 \cdot k_3 \cdot U_w \qquad or \qquad I_3 = k_3 \cdot \frac{m_{poc}}{U_w} \tag{18.11}$$

$$m_{poc} = I_4 \cdot k_4 \cdot E \qquad or \qquad I_4 = k_4 \cdot \frac{m_{poc}}{E} \tag{18.12}$$

$$m_{poc} = I_5 \cdot k_5 \cdot E_m \qquad or \qquad I_5 = k_5 \cdot \frac{m_{poc}}{E_m} \qquad (18.13)$$

$$m_{poc} = I_6 \cdot k_6 \cdot W_t \qquad or \qquad I_6 = k_6 \cdot \frac{m_{poc}}{W_t} \qquad (18.14)$$

$$m_{poc} = I_7 \cdot k_7 \cdot P_m \qquad or \qquad I_7 = k_7 \cdot \frac{m_{poc}}{P_m} \qquad (18.15)$$

The mass of produced products or components, m_{poc}, during a period Δt, is described by seven linear equations. Each equation shows a multiplication of environmental indicator I and an environment load element m_t, m_{ac}, U_w, E, E_m, W_t, and P_m. Environmental indicators, I_1, I_2, I_3, I_4 and I_7 are defined as output-input indicators, while I_5 and I_6 are defined as output-output indicators. The environmental output-input indicators describe the relation between the produced output m_{poc} and the input flows. The environmental output-output indicators describe the relation between the produced output m_{poc} and the other output flows. The linear system of (18.9) through (18.15) can be transferred to a matrix and two vectors, expressed by (18.16). Component vector **P** of a produced mass of components, during period Δt, can be described by multiplying an environmental load matrix **E$_{LM}$** of the produced mass and an environmental performance vector **E$_P$** of the produced mass.

$$E_{LM} \cdot E_P = P \qquad (18.16)$$

Vector **E$_P$** represents the environmental performance of a mass of produced components m_{poc} during a period of time Δt and contains the seven environmental indicators I_1 through I_7. The operating ranges of environmental indicators I_1 through I_7, the environmental performance vector as a function of time, gives an environmental impression of the produced mass of products or components. The higher the ratios between the produced mass of products or components m_{poc} and the number of environmental load elements, the more efficient the production will be. This means in theory that environmental indicators I_1 through I_7 will operate between $0 \leq I_1, I_2, I_3, I_4, I_5, I_6, I_7 \leq \infty$, when $k_1, k_2, k_3, k_4, k_5, k_6, k_7 \geq 0$.

Regarding vector **E$_P$** in (18.16), the measured environmental indicators I_1 through I_7 should be compared with environmental reference indicators I_{1R} through I_{7R}, the 'best practices'. When in practice reference indicators I_{1R} through I_{7R} are chosen so that $I_{1R}, I_{2R} .....I_{7R} > I_1, I_2, ...I_7$, the ratios x_1 through x_7 between the measured and the reference indicators vary between $0 \leq x_1, x_2, x_3, x_4, x_5, x_6, x_7 \leq 1$. This results in expression (18.17).

$$x_n = \frac{I_n}{I_{nR}} \quad for \quad n = 1, 2, 3, 4, 5, 6, 7 \quad and \quad 0 \le x_n \le 1 \tag{18.17}$$

Vector E_P can also be based on ratios x_1 through x_7, see equation (18.18). When x_1 through x_7 are equal to 1, the norm of environmental performance vector, $\|E_P\|$, becomes 7, which represents the best performance value in terms of best practices. When x_1 through x_7 are equal to 0, the norm of the environmental performance vector, $\|E_P\|$, becomes 0, which represents the worst performance value in terms of best practices. A normalized environmental performance vector E_{PN} is shown in equation (18.19). The property of this vector is that its norm, $\|E_{PN}\|$, is equal to 1, when x_1 through x_7 are equal to 1. The normalized environmental performance, $\|E_{PN}\|$, can be determined from equation (18.19) and equation (18.20). Equation (18.21) is obtained when expression (18.17) is substituted in equation (18.20). Here, each measured indicator I_1 though I_7 is compared with its reference indicator I_{1R} through I_{7R}.

In expression (18.20) each ratio x_n has the same environmental weighting. This means in theory that each ratio x_n is multiplied by 1/7. In practice, however, it means that material consumption is equal to the consumption of auxiliary compounds, water, energy, etc. from an environmental perspective. Application of the same environmental weighting indirectly implies application of a quality approach, which means that the consumption of materials, auxiliary compounds, water, energy and packaging materials and the generation of air emissions and waste should be equal to the established *perfect* reference indicators I_{1R} through I_{7R}. If $I_1 = I_{1R}, I_2 = I_{2R}, \ldots\ldots I_7 = I_{7R}$, then $\|E_{PN}\| = 1$, which represents the best performance value. The operating range of the normalized environmental performance is given by (18.22). The operating range of $\|E_{PN}\|$ offers a simple solution with respect to supplier classification.

$$E_P = \begin{bmatrix} x_1 \\ x_2 \\ x_3 \\ x_4 \\ x_5 \\ x_6 \\ x_7 \end{bmatrix} \tag{18.18}$$

$$E_{PN} = \frac{1}{\sqrt{7}} \cdot \begin{bmatrix} x_1 \\ x_2 \\ x_3 \\ x_4 \\ x_5 \\ x_6 \\ x_7 \end{bmatrix}$$

(18.19)

$$\|E_{PN}\| = \sqrt{\frac{1}{7} \cdot \left\{ x_1^2 + x_2^2 + x_3^2 + x_4^2 + x_5^2 + x_6^2 + x_7^2 \right\}}$$

(18.20)

$$\|E_{PN}\| = \sqrt{\frac{1}{7} \cdot \left\{ \left[\frac{I_1}{I_{1R}} \right]^2 + \left[\frac{I_2}{I_{2R}} \right]^2 + \dots + \left[\frac{I_7}{I_{7R}} \right]^2 \right\}}$$

(18.21)

$$0 \leq \|E_{PN}\| \leq 1$$

(18.22)

$\|\mathbf{E_{PN}}\|$ can be applied to the supply chain of an OEM. $\|\mathbf{E_{PN}}\|$ expresses the environmental performance of a mass of produced components in a production facility during a period of time. If there are, for example 25 printed board suppliers that are to be assessed by means of a data collection process for each environmental load element, this metric provides the information that allows a normalized performance evaluation to be completed. That is, the suppliers can be easily benchmarked and classified. Table 18.1 contains an example of a supplier classification based on the environmental performance metric. If, for instance, the deviation of the assessed supplier, i.e. environmental indicators I_1 through I_7, is less than 10% of the reference indicators, the supplier is classified as E1. This means that $\|\mathbf{E_{PN}}\|$ operates between 0.9 and 1. In this way, each $\|\mathbf{E_{PN}}\|$ of a supplier can be redirected to an E-level and classified as good, sufficient, insufficient, bad and very bad, or other levels as desired by management and management policy.

18.3. A Global Application of an Environmental Performance Tool based on the Relative Approach in the Printed Board Industry

Through a case example, this environmental performance tool is applied to an OEM's printed board supply chain. The objective of this step is to establish normalized environmental performance for several printed board suppliers. In this scope, 25 suppliers' production facilities, A1 through A25, were selected for the execution of environmental assessments. These facilities are located in different regions around the globe and produce different kinds of printed boards. These 25 suppliers' production facilities were assessed with the aid of a well-organized procedure.

Table 18.1. Classification of supply chain

	Classification Supply Chain		
#	Environmental Indicators I_1 through I_7	$\|E_{PN}\|$	E-levels
1	0 through 10% deviation of I_{1R} through I_{7R}	0.9 - 1	$0.9 < E1 \leq 1$, good
2	10 through 20% deviation of I_{1R} through I_{7R}	0.8 – 0.9	$0.8 < E2 \leq 0.9$, sufficient
3	20 through 30% deviation of I_{1R} through I_{7R}	0.7 – 0.8	$0.7 < E3 \leq 0.8$, insufficient
4	30 through 40% deviation of I_{1R} through I_{7R}	0.6 – 0.7	$0.6 < E4 \leq 0.7$, bad
5	Larger than 40% deviation of I_{1R} through I_{7R}	0 – 0.6	$E5 \leq 0.6$, very bad

Before the start of the supplier assessments, the OEM compiled an overview of appropriate environmental contact persons for each facility. The facilities then received an introductory letter about the environmental activities, research and the OEM's supply chain strategy. This letter contained an explanation of the environmental-quality concept from a business perspective. This means in practice that the environmental assessment results are integrated in the business and the facilities will be classified as good, sufficient, insufficient, bad and very bad. The introductory letter also announced that by a certain date, the facility would be receiving a second letter plus a floppy disk containing an environmental survey. A purchaser, quality engineer, and environmental expert signed the introductory letter. Five weeks after the introductory letter, the second letter plus the environmental survey was sent to the suppliers' production facilities. This letter contained the same message as the first one, and was also signed by the same purchaser, quality engineer, and environmental expert. During the seven-week assessment period the environmental expert was available to answer questions and provide support. Most facilities contacted the environmental expert with remarks and questions. Both letters indicated that suppliers' facilities that did not respond to the environmental survey would be classified as very bad after the due date. In both letters, the OEM requested that the facilities send confirmation to the environmental expert about when they will be able to open the floppy disk containing the survey. During the assessment period, the environmental expert contacted the facilities to inquire about the status of the survey.

This procedure yielded a 100% result as *all* suppliers responded. Based on the procedure, supplier A_{13} was classified as very bad, which means the environmental indicators I_1 through I_7 will be established as 0, the normalized environmental performance becomes 0 and the proposed price reduction in the negotiations will be 10%. Supplier A_{22} exhibited comparable behavior. Supplier A_{22} was also classified as very bad and the proposed price reduction will also be 10%. Neither supplier exhibited supportive behavior. The other suppliers did respond to the questions of the data collection process. A study of the answers identifies inconsistencies in delivered supplier data. This means that some answers are not given or are unreliable. Different answers contradict each other in some cases. Another issue is that some suppliers did not read the explanation of the data collection process carefully. The mass balance provides insight into the suppliers' self-management behavior. The mass balance per supplier provides an initial impression of the inconsistency. In this case, independent of the inconsistency, the answers delivered were used to calculate indicators I_1 through I_7 for each facility.

Based on a set of *selected* reference indicators, which were provided by the suppliers A_{16} (I_{1R}), A_{25} (I_{2R}, I_{7R}), A_{23} (I_{3R}), A_{20} (I_{4R}) and A_{18} (I_{5R}, I_{6R}) and equation (18.21) the calculated normalized environmental performances vary between 0 and 0.66, see Table 18.2. The set of selected reference indicators form a reference supplier for the other suppliers. Supplier A_{12} has the highest performance, followed by suppliers A_{18} and A_{17}. The other suppliers have performances that vary between 0 and 0.53. Within this range, suppliers A_{11}, A_{13} and A_{22} have the lowest performances, while supplier A_{25} has the highest. But all these suppliers exhibit more than 40% deviation from the *selected* reference indicators. When the suppliers are ranked, using the classification from Table 18.1, all suppliers with the exception of A_{12}, A_{17} and A_{18} are classified as very bad, i.e. level E5. Suppliers A_{12}, A_{17} and A_{18} exhibit 34%, 38% and 36% deviation from the reference indicators respectively, which means a classification of bad, i.e. level E4. None of the suppliers can be classified as sufficient or good. These performances determine "environmental situation A" of the supply base. Furthermore, the normalized environmental performance can be integrated into the business by a link to a proposed price reduction. The result is that the supplier with the lowest performance receives the highest proposed price reduction of purchase turnover per supplier's facility (PT_S), for example, see suppliers A_{10}, A_{11}, A_{13} and A_{22}. *From a business perspective the five suppliers' facilities, which can deliver the highest cost savings, should have the first attention in the scope of the eco-supplier development cycle.* After an agreed price reduction with the supplier, a required "environmental situation B" can be established and eco-supplier development plans can be developed. The last column of Table 18.2 shows the "quality" of the mass balance of the suppliers. When a measure of inaccuracy is accepted within the range -15% through +15%, only suppliers A_5, A_6, A_8, A_{11}, A_{12}, A_{14}, A_{15}, A_{16} and A_{19} have a correct mass balance.

18.4 Discussion of Managerial Implications

The global application of the proposed environmental performance tool shows that the 25 suppliers' production facilities can be benchmarked by a numerical value, the normalized environmental performance. Even though all the assessed suppliers in Table 18.2 have a certified ISO14001 environmental management system, the OEM's supply chain manager cannot distinguish them in terms of good or bad. A comparison of 25 ISO14001 certified environmental management systems show that terms like environmental performance, environmental impact, continual improvement, etc. have been measured, interpreted and implemented in different ways. This emphasizes the fact that when all the suppliers' facilities have an environmental management system in place, they do not necessarily have the same metrics, nor do these systems guarantee good environmental performance.

Table 18.2. Calculated normalized environmental performances of assessed suppliers

Supplier	$\|E_{PN}\|$	Proposed Price Reduction	Classification	Region	Difference Δ (%)
A_1	0.3	7% of PT_s	E5: very bad	USA	+35.1
A_2	0.41	5.9% of PT_s	E5: very bad	USA	+18.3
A_3	0.39	6.1% of PT_s	E5: very bad	USA	+27.2
A_4	0.29	7.1% of PT_s	E5: very bad	Asia	+23
A_5	0.49	5.1% of PT_s	E5: very bad	USA	-1.1
A_6	0.24	7.6% of PT_s	E5: very bad	USA	+11.1
A_7	0.35	6.5% of PT_s	E5: very bad	Canada	+31.4
A_8	0.32	6.8% of PT_s	E5: very bad	Europe	-5.1
A_9	0.23	7.7% of PT_s	E5: very bad	Europe	+15.5
A_{10}	0.2	8% of PT_s	E5: very bad	Europe	-22.9
A_{11}	0.07	9.3% of PT_s	E5: very bad	USA	+2.9
A_{12}	0.66	3.4% of PT_s	E4: bad	USA	+0.1
A_{13}	0	10% of PT_s	E5: very bad	USA	-
A_{14}	0.41	5.9% of PT_s	E5: very bad	Europe	+13.3
A_{15}	0.46	5.4% of PT_s	E5: very bad	USA	+0.7
A_{16}	0.43	5.7% of PT_s	E5: very bad	USA	-2.8
A_{17}	0.62	3.8% of PT_s	E4: bad	USA	-89.8
A_{18}	0.64	3.6% of PT_s	E4: bad	USA	+36.3
A_{19}	0.4	6% of PT_s	E5: very bad	USA	-2.1
A_{20}	0.46	5.4% of PT_s	E5: very bad	Europe	-33.3
A_{21}	0.34	6.6% of PT_s	E5: very bad	Europe	+24.8
A_{22}	0	10% of PT_s	E5: very bad	Asia	-
A_{23}	0.38	6.2% of PT_s	E5: very bad	Asia	+48.3
A_{24}	0.37	6.3% of PT_s	E5: very bad	Asia	+20.5
A_{25}	0.53	4.7% of PT_s	E5: very bad	Europe	-98

Normalized environmental performance is a usable tangible metric within the scope of supply chain management.   Application of this environmental performance tool also shows, however, that only nine of the 25 assessed printed board facilities know what their mass balance is. This leads to possible difficulties in implementing a tool such as this where the data may need to be audited for accuracy purposes. The added strength of this environmental performance tool is that the accuracy of the supplier data can be checked. The accuracy may lead to requests and recommendations for improved measurement of environmental performance for these suppliers, as part of the cycle.

The data collection process related to the seven environmental load elements required a total of 39 questions to be answered. Answering these questions gives a detailed profile of the supplier's production behavior. Minimization of the questions means in general that the accuracy of the supplier data cannot be checked. From a pure environmental perspective, based on LCA (Life Cycle Assessment) method Eco-Indicator '95 (Nagel, 2001), the environmental load elements material and energy use are the most important to minimize. This means from a management perspective that the data collection process contains only eleven questions, which is easier to manage than 39 questions. Based on only one energy-oriented question the normalized environmental performance can be calculated. In this case the supplier's answer cannot be checked if the production location is not visited. Here, a link to a supplier's audit program of the OEM is an option for investigation. Based on one energy-oriented question, normalized environmental performances can be calculated and benchmarked for all kinds of suppliers' production facilities in different kinds of industry sectors, but the drawback is that the six other environmental load elements are not managed, which means no full application of the concept of environmental quality can occur.

From a supply chain management perspective, the right balance between the number of questions and the number of environmental load elements determines the quality of the normalized environmental performance and the quantity of managerial effort. Currently, the proposed price reduction has not been applied in practice. Integration of this aspect will mean that a change in thinking, handling and attitude of organizations and individuals should be realized.

18.5 Conclusion

This chapter has shown that environmental quality evaluation can be integrated into an OEM's existing supply chain by using an environmental performance tool. Application of this environmental performance tool has shown that suppliers can be ranked, classified and compared on the basis of environmental performance and proposed price reductions can be derived and used in the supplier negotiations. It also shows however, that only nine of the 25 assessed printed board facilities know what their mass balance is. So the added strength of this environmental performance tool is that the accuracy of the supplier data can be checked.

In this case, the environmental indicators and the normalized environmental performance were calculated independent of inaccuracies in supplier data. Inaccuracies in data are not a reason for not calculating environmental indicators and normalized environmental performance. In practice inaccuracies in data will be eliminated when the eco-supplier development cycle is activated and continued. Currently, all kinds of the data are available in a suppliers' production facilities for calculating environmental performance, but the suppliers have never aggregated the data in one profile. This means that there also seems to be a lack of management of the seven environmental load elements expressed in indicators from the own facility. The facility manages environmental load elements incidentally or partially, but not from an approach in which the overall facility is managed.

There is no management and research of the *seven environmental load elements* from a facility's perspective or from an OEM's supply chain perspective in the electronics industry. In the future, the business impact in terms of proposed price reductions can be expanded widely when suppliers deliver inaccurate data. In this case, it means that sixteen printed board facilities have no insight into their mass balance, which should result in a normalized environmental performance of 0, a proposed price reduction of 10% and a classification of very bad. The normalized environmental performances can be calculated and compared for the other nine printed board facilities, and proposed price reductions can be derived. The link to proposed price reductions is a major actor for realizing environmental improvements in the supply chain.

References

Nagel M.H., (2001), "Environmental Quality in the Supply Chain of an Original Equipment Manufacturer in the Electronics Industry" Ph.D. Dissertation, September 2001, ISBN 90-9015022-6.

19

Creating A Green Supply Chain: A Simulation and Modeling Approach

Khoo Hsien Hui[1], Trevor A. Spedding[2], Ian Bainbridge[3] and David M.R. Taplin[4]

[1]Department of Industrial & Systems Engineering, National University of Singapore,
1 Engineering Drive 2, Singapore 117576, g0203686@nus.edu.sg
[2]School of Management and Marketing, Faculty of Commerce, University of Wollongong,
Australia, spedding@uow.edu.au
[3]Cooperative Centre for Cast Metals Manufacturing, University of Queensland, Queensland
4072, Australia
[4]Visiting Professor of Systems Engineering, School of Engineering, University of
Greenwich, Chatnam Maritime, ME 4 4TB, U. K.

This paper presents a case study of a supply chain which is concerned with the distribution of aluminium metal, starting from raw material from a Metal Supplier to a Casting Plant, billets from the Casting Plant to the Component Producer, and finally, die-cast components from the Component Producer to the Market. The paper creates a green supply chain by integrating the concerns of transport pollution, marketing costs, time to market, recycling of scrap metal and energy conservation. Simulation and modelling tools are introduced to aid in the decision making process of distance selections and choices of transportation in the case study. Based on a series of user input selections, the simulation results are used to determine a range of optimal plant locations that will balance economical benefits (highest scrap values, least total costs, etc.) as well as environmental stewardship (least pollution).

19.1 Introduction

Business organizations are facing the increasing pressure of balancing marketing and environmental (green) performance. This is an issue which is becoming more important to the public (Shultz II and Holbrook 1999). In order to demonstrate good environmental management and sustainability, companies must learn to embrace a wide range of issues. Included are sustainable development, pollution and the community at large. The idea of green businesses forces the re-examination of the very purpose of a company's existence (Hick 2000). Adopting greener management practices as part of an enterprise's policy have increasingly turned into a major strategic thrust in business organizations which will carry well into the 21st century (Stead and Stead 2000). This calls for a new approach to performing business, from merely achieving economic profit to developing ecologically sensitive strategic management policies.   There are various approaches adopted by many enterprises in creating green enterprises, such as adopting eco-efficiency methods in the design of products (Hibbert 1998, Ottman 1999) or establishing industrial ecologies (Matthias 1999).

This paper looks at the creation of green enterprises from yet another perspective. The authors offer a unique approach by developing a simulation model for a supply chain that helps to achieve optimal performance (low marketing cost, fast delivery time, etc.) and least transport pollution. The conservation of energy and promotion of recycling scrap metal are also considered in the supply chain case study.

19.1.1 Supply Chain Management

Supply chain management usually takes into consideration issues of minimizing end cost (market cost), efficient logistical aspects, and timely delivery of goods (Cox 1999). However, at the beginning of the 21st century, a shift in focus can be observed. For example, business chain partners were formed to participate in implementing environmentally friendly practices that reduces waste and pollution (Melnyk and Handfield 1996).

This paper takes a look at the green or environmental concerns of a supply chain. One of the approaches adopted is by taking into consideration the levels of pollution from the various modes of transportation between plants.

19.1.2 Transport Pollution

The transfer of raw material and goods from one plant to the next in a supply chain occurs by various modes of transportation, namely – land, air or sea. The levels and types of transport pollution depend on the combination of two factors: the type of transportation and the distance travelled.

The pollution from diesel engine vehicles, such as heavy trucks, include gaseous pollutants of carbon monoxide (CO), oxides of nitrogen (NOx), particulate matter (PM) and volatile organic compounds (VOC). Some hydrocarbons (including VOC) from diesel emissions are carcinogenic. VOC are known to or are suspected of effecting human health. NOx is an invisible, toxic gas that can form fine aerosol particles or salts which can contribute to acid rain or fog. Fuel from engines that are not highly efficient may be emitted as particulate matter (PM). Toxic and cancer-causing chemicals can be carried by PM into the lungs. Moreover, ozone and PM are associated with adverse health and welfare effects, including respiratory illness, environmental damage and visibility problems, such as haze (United States Environmental Protection Agency 2000).

The impact of transport pollution can also be assessed in monetary terms; the cost of healthcare, the cost of days of work lost and the economic cost of premature deaths. In the UK for example, environmental economists have estimated the cost of air pollution from road transport at £19.7 billion a year (The Ashden Trust 1994). Many other researchers have discussed the problems of pollution in greater details (*e.g.*, Flachsbart 1999, Colvile *et al.*, 2001).

19.1.3 Sustainable Issues in Aluminium Production

In the aluminium production industry, cost effective methods are needed to address the issues of scrap metal generated from the production stages. This issue involves the Casting Plant which produces the aluminium billets, and the Component Producer, which produces die-cast or forged components. The locations of the plants have to be selected in a manner that would promote the convenience – in terms of distance travelled and monetary returns – of recycling scrap metal.

Another issue concerning sustainable development is the conservation of energy. An increasingly deregulated power industry is scrambling to keep pace with strong customer demand. This has forced many manufacturers to have a vested interest in reducing the costs of energy, which often represent a substantial portion of their total operating expenses (Quinn 2001). This issue is especially addressed in the Metal Supplier and Casting Plant, which consume a huge amount of energy for the melting of metal, necessitating the Casting Plant to locate close to the Metal Supplier so that molten metal may be supplied by the latter. (The theoretical amount of electricity required to melt 1 tonne of Al is 294 kWh [Street 1986]).

19.1.4 Simulation and Speed

Time and speed is crucial in today's fast paced competitive markets. Due to hyper-competition, enterprises that are not keeping pace with fluid marketing demands and changes may loose out to competitors with higher advantage of faster and speedier deliveries due to well-planned plant locations and good marketplace selections. Therefore computer simulation is a useful tool which offers a wide range of decision scenarios that saves time, energy and cost.

19.1.5 Simulation Software

ProcessModel (ProcessModel 2000) is a simulation software package which is commercially available for designing and improving systems. This software combines flowcharting technology and simulation to allow operations to be studied from a holistic view. It allows the testing of different options, or 'what if' scenarios to aid managers, planners and decision makers assess and analyse their company's processes or activities. By using activity charts to describe a sequence of actions (or decision making processes), the outcome of each action or decision becomes more transparent. The software allows users to create simulation models by using simple programming and mathematical formulas (also known as model logic).

The advantage of the simulation software and their applications for improving industrial processes to achieve more sustainable business results have been presented by various researchers. In a first example, Spedding et al. (1999) demonstrated how the *Tragedy of the Commons* was reinvented by companies which did not consider resource preservation and recycling activities as part of their business plans. In the second example, Khoo et al. (2001) used simulation tools to track production costs, pollution and waste levels, and measured a smelter company's performance based on two decisions – the first to allow the system to "run as usual", and next to implement more sustainable operations. In the third case, Taplin *et al.* (2001) suggested how simulation could be developed to integrate sustainable efforts or environmental management into a complete business model. These examples demonstrated how simulation tools facilitated the dissemination of information and the visual verification of making the right decisions, while saving time and costs.

19.1.6 Objective and Layout

This paper offers a unique simulation approach to aid the creation of a green supply chain, which consists of four plants, to achieve:

- A balance of low total market cost and low transport pollution
- Fast deliveries between plants
- Promotion of recycling of scrap metal
- Conservation of energy
- The use and application of simulation in decision making and for creating greener business practices.

The paper is laid out as follows.   The following section introduces the case study and introduces the four plants of the supply chain. The cost and pollution variables of the plants are also described.   Next, the details of the boundary distances of the plants, location settings and transport types within the supply chain are presented.   Four types of location cases are designed for the simulation study. The last few sections present the simulation, results and discussions.   Finally, the paper ends with a conclusion.

19.2   Case Study of a Supply Chain

The supply chain case study involves the distribution of metal, starting with metal ingots from the Supplier to the Casting Plant, billets from the Casting Plant (or Pilot Plant) to the Component Producer, and finally finished (die-cast or forged) components to the Market or end user. The Casting Plant produces billets made of aluminium mixed with various types of alloys. The billets are produced to fill a marketing niche where the demand for weight reduction in material is sought. This type of light metal provides a sustainable and environmentally friendly solution to improving energy efficiency in the aerospace, electronics and automotive industry. The Component Producer is a typical die-casting company that produces small precision components, with specialized design, tooling and net-shape production processes for the die-cast components. The supply chain, including the transport pollution and recycling activities, is shown in Figure 19.1.

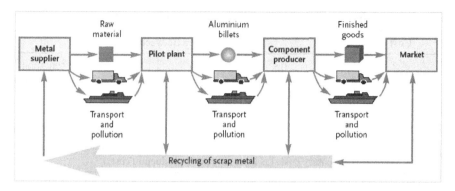

Figure 19.1. Supply chain case study

19.2.1 Marketing Time, Cost and Pollution Variables

The pollution and costing factors (inventory, scrap metal values and total costs) of the four plants within the chain are dependent on the following:

- Distances between plants
- Types or modes of transport between plants
- Amount of metal processed and transferred between the plants
- The choice of location for sending scrap metal

All calculated results of total transport cost, inventory cost and scrap metal values are performed according to the data displayed in Table 19.1. The transport pollution data used for the case study is displayed in Table 19.2.

Except for the Metal Supplier, the total accumulative cost of each plant is calculated according to the plant's respective distances from the previous one (*i.e.*, the distance of the Casting Plant reference to the Supplier, the distance of the Component Producer reference to the Casting Plant, etc.). The final total costs for the various plants are calculated in the following manner:

For the metal supplier, the total cost per tonne, $C^{MS}(total)$, is:

$$\text{CMS (total)} = \text{C(metal)} \tag{19.1}$$

Where C(metal) is the cost of metal per tonne.

For the casting plant, the total cost, $C^{CP}(total)$, is:

$$C^{CP}(\text{total}) = C^{MS}(\text{total}) + C^{CP}(\text{inventory}) - R^{CP}(\text{scrap}) + C^{MS-CP}(\text{transport}) + C^{CP}(\text{scrap}) + C^{CP}(\text{process}) \tag{19.2}$$

Where $C^{CP}(\text{inventory})$ is the cost of the casting-plant inventory, $R^{CP}(\text{scrap})$ is the return on casting-plant scrap value, $C^{MS-CP}(\text{transport})$ is the total cost of transport from the metal supplier to the casting plant, $C^{CP}(\text{scrap})$ is the total cost of transport to send scrap from the casting plant and $C^{CP}(\text{process})$ is the total metal process cost at the casting plant.

Table 19.1. Scrap values, inventory costs and transportation costs

Cost or Value	Distance Between Plants			
	0.5	500	500-1500	1500-5000
Scrap value (as a percentage of metal tonnage value)				
Casting Plant	10	2	2	2
Component Producer	10	2	2	2
Inventory costs (as a percentage of production costs)				
Casting Plant	1	5	8	10
Component Producer	1	5	8	10
Transportation costs (in Australian dollars per tonne per kilometre)*				
Forklift Truck	1.00	N/A	N/A	N/A
Truck	N/A	0.15	0.13	0.12
Rail	N/A	0.10	0.09	0.08
Ship	N/A	N/A	N/A	0.05

* Speed of transport: forklift truck, 10–25 km/h; truck, 70–80 km/h; train, approximately 50 km/h; ship, approximately 18 km/h.

Table 19.2. Vehicle pollution statistics

Source for figures on trucks and ships: QEPA 1999
Source for figures on rail transport and forklift trucks: US EPA 2000

Vehicle (g/l)	Pollutant (g/l)			
	NOx	VOCs	CO	PM
Trucks *	11.4	3.7	6.7	4.2
Rail **	59.75	4.315	9.1	4.315
Ship[†]	18,761.3	126.4	2.362	200
Forklift truck[‡]	37.6	0	20.3	1.6

* Capacity up to 20 tonnes and fuel consumption 50 l per 100 km
** Capacity up to 1,500 tonnes and fuel consumption 400–500 l per 100 km
[†] Generator power approximately 600 kW
[‡] Capacity up to 4 tonnes and fuel consumption 5–6 l per 18 km

Note: NOx = nitrogen oxides; VOCs = volatile organic compounds; CO = carbon monoxide; PM = particulate matter

For the component producer (component maker), the total cost, C^{CM}, is:

CCM(total) = CCP(total) + CCM(inventory) − RCM(scrap) + CCP–CM(transport)
+ CCM(scrap) + CCM(process)   (19.3)

Where C^{CM}(inventory) is the cost of the component-producer inventory, RCM(scrap) is the return on component-producer scrap value, C^{CP-CM}(transport) is the total cost of transport from the casting plant to the component producer, C^{CM}(scrap) is the total cost of transport to send scrap from the component producer and C^{CM}(process) is the total metal process cost at the component producer.

For the market (end-user), the total cost, C^{m}(total), is:

$$C^{m}(total) = C^{CM}(total) + C^{CM-m}(transport) \qquad (19.4)$$

Where C^{CM-m}(transport) is the total cost of transport from the component producer to market.

The possible modes of transportation are trucks (type 1), trucks and rail (type 2) or trucks and ship (type 3). It was decided that the mode of transport for a distance of 0.5 kilometres is by the use of forklifts (type 4). The reason for this decision is because in the actual situation, a great deal of time may be saved from the loading and unloading of material onto trucks.

The trucks used in the model are assumed to have a carrying capacity of 20 tonnes and the forklifts are assumed to have a carrying capacity of 4 tonnes. This means that the number of trucks and forklifts travelling in the model depends on the amount of metal travelling through the system. In selecting transport types of 2 and 3, the truck travelling distance to the rail or seaport is assumed to be 10 kilometres. The sample calculations for the four transportation selections are as follows:

The total transport cost for transport types 1 and 4, Cv(transport), is:

$$Cv(transport) = dv \times Cv(w,d) \times w \times nv \qquad (19.5)$$

Where dv is the distance travelled by vehicle v (here, v = truck or forklift truck), Cv(w, d) is the cost of vehicle v per tonne per distance, w is the amount of metal in tonnes and nv is the number of vehicles used.

The total transport cost for transport type 2, C_2(transport), is:

$$C_2(transport) = [d_{truck} \times C_{truck}(w, d) \times w \times n_{truck}] + [d_{rail} \times C_{rail}(w, d) \times w] \qquad (19.6)$$

Where d_{truck} and d_{rail} is the distance travelled by truck and rail, respectively, C_{truck} (w, d) and C_{rail} (w, d) is the cost of truck per tonne per distance and the cost of train per tonne per distance, respectively, and ntruck is the number of trucks used.

Similarly, for transport type 3, the total cost is:

$$C_2(transport) = [d_{truck} \times C_{truck}(w, d) \times w \times n_{truck}] + [d_{ship} \times C_{ship}(w, d) \times w] \qquad (19.7)$$

The 'time to deliver' is the amount of travel time spent in delivering the goods between the plants. The total time to deliver, *t(total)*, is:

$$t(total) = [s_v \times d^{MS-CP}] + [s_w \times d^{CP-CM}] + [s_x \times d^{CM-m}] \tag{19.8}$$

Where s_v, s_w and s_x are the travel speeds of vehicles v, w and x, respectively (v, w, x = truck, forklift truck, train or ship); d^{MS-CP} is the distance between the metal supplier and the casting plant; d^{CP-CM} is the distance between the casting plant and the component producer; and d^{CM-m} is the distance between the component producer and the market.

19.2.2 Boundary Distances and Location Cases

The boundary distance of the supply chain is defined as the approximate distance between the Metal Supplier and potential Market (end user). These distances are selected on the basis of a country such as Australia with a number of major cities. Within a city the distance between two plants is not likely to exceed 50 kilometres. If for example some plants are in one city and some in another, then the distance is between the cities are approximated to be 1,000 to 2,500 kilometres. If one or more of the plants are located off-shore, or in another country, then the condition may be stretched to 5,000 or to 10,000 kilometres. It is most practical that the locations of at least two consecutive plants are situated side by side, due to the following two reasons:

i) Promotion of scrap metal recycling activities
The scrap metal is generated from both Casting Plant and Component Producer, and yields the highest value when being sent to the next closest plant. Scrap metal may be sent only to the Supplier, Casting Plant or Market.

ii) Savings on transportation costs and least pollution
For the case where two plants are located at a "close" distance of approximately 0.5 kilometres, forklifts may be selected as the transport mode between the plants. Otherwise, the alternate transport types may be selected.

Therefore, the distances and transport selections that are tested for the simulation runs are designed according to the following four location settings. They are location cases A, B, C and D, shown in Figure 19.2.

19.3   Simulation Model and Results

Computer simulation is used to provide a fast and efficient method of capturing the outcomes of decisions. The function and purpose of the simulation model is to trace and compute the costing and pollution variables (from Tables 1 and 2) throughout the supply chain from beginning to end. The simulation model accepts the distances between plants as user input values and generates the cumulative total costs and pollution levels, based on the distances entered.

It is designed to first establish the location of the Casting Plant with reference to the Metal Supplier. The next user input selections determine the location of the Component Producer with reference to the Casting Plant, and finally the location of the Market with reference to the Component Producer. If "far" distances are selected, user input choices are also made for the different types (1, 2 or 3) transportation.

The simulation model is shown in Figure 19.3. An example of the user input pop-up menu provided by the model is displayed in Figure 19.4.

19.3.1 Simulation Runs

Based on user input selections, the simulation model generates the sets of results. A total of 40 simulation runs are performed, all taking into account the four location cases and transport types. The simulation entries are shown in Tables 19.3(a) and (b). In each simulation run, the amount of metal entered into the system is 1,000 tonnes.

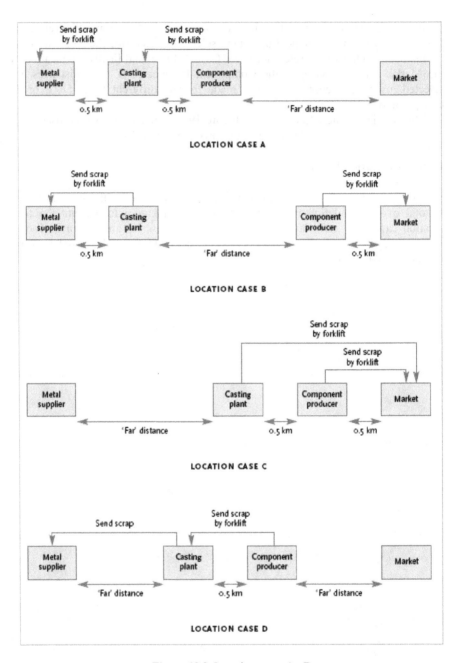

Figure 19.2. Location cases A - D

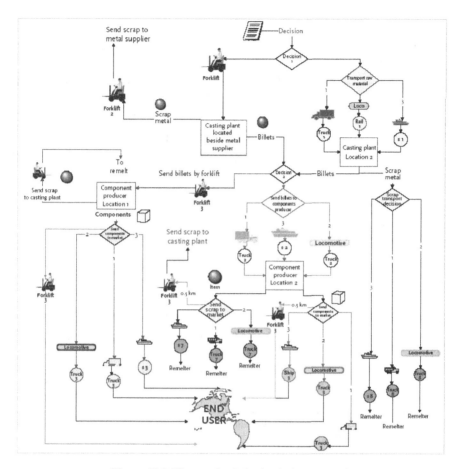

Figure 19.3. The supply chain simulation approach

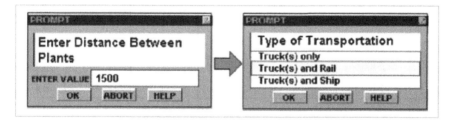

Figure 19.4. User pop-up simulation menus

Table 19.3. User selection of 40 simulation entries

Run	Distance			Transport		
	1	2	3	1	2	3
Location Case A						
1	0.5	0.5	625	4	4	1
2	0.5	0.5	1250	4	4	1
3	0.5	0.5	1250	4	4	2
4	0.5	0.5	2500	4	4	1
5	0.5	0.5	2500	4	4	2
6	0.5	0.5	5000	4	4	2
7	0.5	0.5	5000	4	4	3
Location Case B						
8	0.5	625	0.5	4	1	4
9	0.5	1250	0.5	4	1	4
10	0.5	1250	0.5	4	2	4
11	0.5	2500	0.5	4	1	4
12	0.5	2500	0.5	4	2	4
13	0.5	5000	0.5	4	2	4
Location Case C						
14	0.5	5000	0.5	4	3	4
15	625	0.5	0.5	1	4	4
16	1250	0.5	0.5	1	4	4
17	1250	0.5	0.5	2	4	4
18	2500	0.5	0.5	1	4	4
19	2500	0.5	0.5	2	4	4
20	5000	0.5	0.5	2	4	4
21	5000	0.5	0.5	3	4	4

Table 19.3. User selection of 40 simulation entries (continued)

Run	Distance			Transport		
	1	2	3	1	2	3
Location Case D						
22	625	0.5	625	1	4	1
23	625	0.5	1250	1	4	1
24	625	0.5	1250	1	4	2
25	1250	0.5	625	2	4	1
26	1250	0.5	625	1	4	1
27	625	0.5	2500	1	4	1
28	625	0.5	2500	1	4	2
29	2500	0.5	625	2	4	1
30	2500	0.5	625	1	4	1
31	1250	0.5	1250	1	4	1
32	2500	0.5	2500	2	4	2
33	1250	0.5	2500	1	4	1
34	1250	0.5	2500	2	4	2
35	2500	0.5	1250	1	4	1
36	2500	0.5	1250	1	4	1
37	2500	0.5	5000	1	4	2
38	2500	0.5	5000	2	4	2
39	5000	0.5	5000	2	4	3
40	5000	0.5	5000	3	4	3

19.3.2 Simulation Results

Figures 19.5 and 19.6 display the Total Transport Costs and Total Pollution respectively. Figures 19.7 displays the Total Marketing Costs of the supply chain. The 'Total Time To Deliver' is displayed in Figure 19.8.

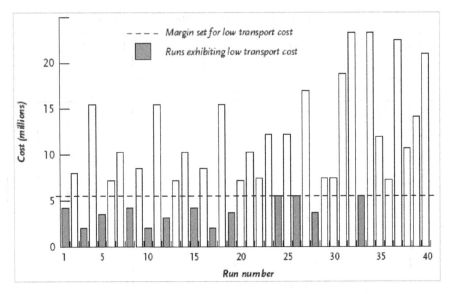

Figure 19.5. Total supply chain transport costs

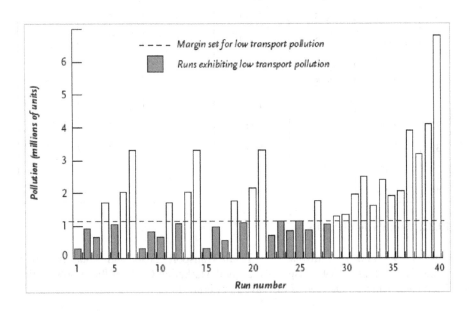

Figure 19.6. Total supply chain transport pollution

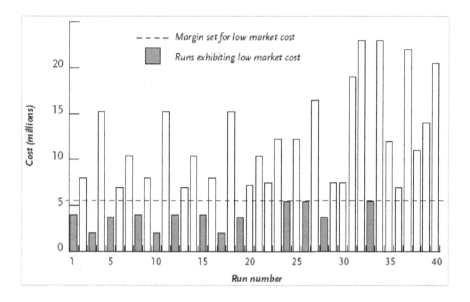

Figure 19.7. Total supply chain market costs

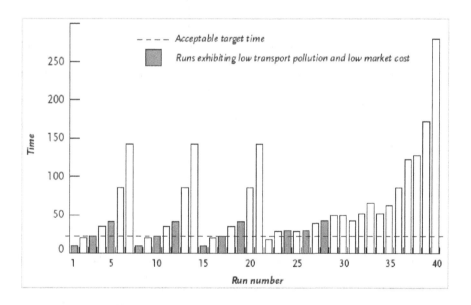

Figure 19.8. Total supply chain 'time to deliver'

From the simulation result, a margin is selected for "low" values of total cost and pollution. The final assessment of the best range of locations from the simulation runs is based on the following:

The "low" values of Total Transport Pollution results are:

▶ Run Nos. **1, 2, 3, 5, 8, 9, 10, 12, 15, 16, 17, 19, 22, 23, 24, 25, 26** and **28.**
The "low" values of Total Market Cost are:

▶ Run Nos. **1, 3, 5, 8, 10, 12, 15, 17, 19, 24, 26, 28** and **33.**
The matching numbers for balancing minimum costs and pollution are:

▶ Run Nos. **1, 3, 5, 8, 10, 12, 15, 17, 19, 24, 26** and **28.**

With reference to Table 3, run no. **1** is a Case A location setting with a distance of 625 kilometres between the Component Producer and Market. The transportation selected between the last two plants is type 1 (trucks only). Run nos. **3** and **5** refers to a Case A location setting with a type 2 (trucks and rail) transportation mode between the "far distance" plants. Run nos. **8, 10** and **12** suggests a Case B location setting where the distances between the Casting Plant and Component Producer are 625, 1,250 and 2,500 kilometres respectively.

Run nos. **15, 17** and **19** suggest the Case C location settings. The "far distances" selected for no. **15** is 625 kilometres, for no. **17** is 1,250 kilometres, and for no. **19** is 2,500 kilometres. For run no. **15**, the mode of transportation selected between the Metal Supplier and Casting Plant is type 1 (by trucks only). As for run nos. **17** and **19**, the mode of transportation selected between the first two plants is type 2 (by trucks and rail). Run nos. **24** and **26** proposes the Case D location setting, where the trade-off of least costs, least pollution and distance between the plants can be found in the combined distances of 625 and 1,250 kilometres, as well as the use of trucks and rail. Finally, run no. **28** also proposes the Case D setting. This time, the "far distances" are selected as 625 kilometres for the first two plants and 2,500 kilometres for the last two plants. The transport modes selected are type 1 from the supplier to the Casting Plant, and type 2 from the Component Producer to the Market.

The Total 'Time To Deliver' displayed in Figure 8 depicts the minimum total travelling time between plants, subjected to the distances and transport type selected. A "target time" is selected to meet the requirements of "just in time" demands. Based on the 'target time' margin, run nos. **5, 12, 19, 24, 26** and **28** are omitted from the selections, leaving run nos. **1, 3, 8, 10, 15** and **17** for further consideration.

19.3.3 Conservation of Energy

Within the selected run nos., the next issue addressed is the conservation of energy. At the Metal Supplier, melting of metal is performed to transform the aluminium into slabs or ingots of the right shapes and sizes. The metal melting activity consumes a high amount of energy. At the Casting Plant, these slabs or ingots of aluminium metal are melted again for the billet production process.

In order to conserve energy, it was suggested that *molten metal* is transferred by truck from the Metal Supplier to the Casting Plant. This suggestion was reasonable for travel distances within 50 kilometres. Therefore the case A and B settings (run nos. **1, 3, 8** and **10**) are the final selections.

This suggestion saves energy and cost for the Casting Plant. In the drive towards establishing sustainable development, this type of energy calculation is treated as an essential part of the model. An ideal condition requires 294 kWh (kilowatt-hours) for processing 1 tonne of aluminium. This is the energy equation used for the Pilot Plant plus an additional estimate of 1% to compensate for metal losses. The energy requirement of the Component Producer is estimated to be 25% of that consumed by the Casting Plant.

The energy savings for the Casting Plant are shown in Figure 19.9. The results show that the location of the Component Producer imposes no difference in the amount of energy spent.

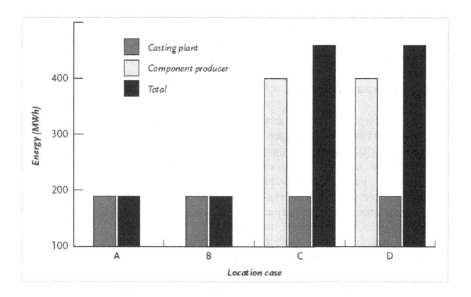

Figure 19.9. Energy differences for the casting plant and component producer

The energy savings are also evaluated for the Metal Supplier. On an average basis, the energy consumption is quite high for melting metal, usually an average value of about 15 kWh for every 1 kilogram of metal. However, some very efficient plants may get down to 13.5 kWh per kilogram. This type of energy difference is definitely a significant value for metal production on an annual basis. By adopting this type of energy savings for the Metal Supplier, the energy savings for the supply chain can be up to about 90%. These types of energy conservation methods are important factors for enterprises that want to demonstrate good corporate citizenship. It involves caring for future generations who may face shortage of energy supply (Denton, 1998). An additional environmental bonus due to better energy efficiency is less burning of coal and less greenhouse emissions.

19.4   Final Results and Discussions

Further analyses of the selected locations are performed based on the marketing needs and demands of aluminium components. The end user, or Market will demand a guaranteed "just in time" supply, or impose a price reduction due to the need to maintain stocks. It is therefore necessary for the model to reflect this demand in terms of the final transportation time, or an offsetting price reduction for the maintenance of stocks, at or adjacent to the market. Whilst the model is capable of taking these factors into account in terms of real, or input details, it cannot take into account "perceived", or attitudinal resistance to the model logic.

The final range of locations were confirmed to be run nos. **1, 3, 8** and **10**. These locations guarantee that, in the supply chain, recycling is promoted, marketing costs are low, delivery time is shortened and transport pollution is minimized. The rest of the discussion is concerned with the following three areas: i) green supply chain, ii) the use of simulation and iii) reliability and limitations of the simulation model.

19.4.1 Green Supply Chain

It was mentioned earlier that enterprises are facing the increasing pressure of balancing marketing performance with environmental issues. These issues are creating new challenges in businesses such as energy preservation and pollution abatement, not only as a precondition for long-term survival, but also as an ingredient for long-term profitability (Gifford, 1997; Miller, 1999). Therefore companies have created networks of suppliers to build common understandings and learning about waste reductions and operational efficiencies in the delivery of existing products and services (Cox, 1999).

In alignment with sustainable business requirements, energy conservation was adopted for the supply chain by selecting near distances to allow the transfer of molten metal between plants. The design of the plant location settings also promoted the returns of scrap metal recycling and optimal marketing performance (cost and time), as well as low transport pollution. In this manner, environmental, marketing and customer needs are all satisfied, given the strict logic of the model.

19.4.2 The Use of Simulation

The case study has demonstrated the advantages of simulation as a powerful tool for making decisions. The various decision scenarios enabled outcomes (pollution and costs levels) that do not incur any losses in time or expenses. These tools have been tested for its application in other parts of industry for creating greener plants and environmental management purposes (Spedding et al., 1999; Khoo et al., 2001; Taplin et al., 2001).

In testing the outcomes of a series of decisions, quantitative results are required due to the ability to make the comparisons of costs and benefits easier. This approach has demonstrated the potential of advanced technology in playing a significant role in sustainable development and creating more socially responsible business enterprises – not because of potential breakthroughs to replace natural resources or air, but because technological advances can facilitate dissemination of information, enhancement of communication and visual verification of making the right decisions (Saemann, 1992; Christensen, 1999).

19.4.3 Reliability and Limitations of the Simulation Model

The model was created to reflect true-to-life outcomes of a series of decisions. The robustness of the simulation model depends on several factors:

- *Availability of data* The outcome of the simulation results relied heavily on the accuracy of the data collected, which were treated as the model's *variables*. In the case study, the authors worked closely with the owner of the plant to ensure that the data (*e.g.*, costs, distances, etc) used for the simulation reasonably reflected the actual case.
- *Design of the model* The methodology used for designing the model has a direct effect on its ability to represent the true-to-life events. The simulation model incorporated simple mathematical formulas to form the model's logical actions (also known as the system's *behaviour*). Again, the authors worked closely with the plant owner to inspect the model's logical actions (this is known as "walk through approach") to ensure that the model was fit for its intended purpose.
- *Limitations* The simulation model and its results were based on *ideal* conditions. In order to make the system more realistic, unexpected events could be programmed within its logic, such as, cases of machine or vehicle breakdowns, human intervention (in handling material), changes in customer orders or the fluctuation of market demands. These cases would interrupt the schedule for the delivery of goods to the end user. In order to include these factors within the simulation study, a more intensive research on transportation, marketing demands and the production line of each plant has to be carried out. This type of work also requires a higher level of programming, and perhaps, a more sophisticated simulation package. Other areas, such as the option for selecting more fuel-efficient transport types and their associated costs, could also be incorporated.

19.5 Conclusion

The article has demonstrated the potential of simulation tools to created a green supply chain which balances the concerns of transport pollution, marketing costs, time to market, recycling of scrap metal and energy conservation. The simulation tools allowed the testing of a series of decisions and their outcomes, while saving costs and time. The final selected plant locations have demonstrated how a supply chain consisting of four plants could tackle transport pollution and energy issues, while meeting customer demands. Part of the rationale for this supply chain design is the recognition that the Earth is a legitimate stakeholder and the reduction of pollution (CO, VOCs, NOx, and PM) serves to help maintain its natural climate.

The results of the simulation study were derived from an ideal situation. It is suggested that further developments could be made, such as, the study of material ordering, production time and inventory control – from supplier to distributor to end users – in the pursuit of establishing a green supply chain. The same approach and methodology could be applied to other supply chain cases, provided that the relevant data is available, and that the operations of the real system can be modelled.

References

Ashden Trust (1994). How Vehicle Pollution Affects Our Health. London: Ashden Trust.

Christensen, J. (1999). Seeing a Forest to Save the Trees. New York Times.

Colvile, R.N., Hutchinson, E.J., Mindell, J.S. and Warren, R.F. (2001). The transport sector as a source of air pollution, Atmospheric Environment, 35(9), 1537-65.

Cox, A. (1999). Power, value and supply chain management. International Journal of Supply Chain Management, 4(4), 167-75.

Denton, T. (1998). Sustainable development at the next level. Chemical Market Reporter, 253(7), 3-4.

Flachsbart, P.G. (1999). Human exposure to carbon monoxide from mobile sources. Chemosphere Global Change Science, 1(1-3), 301-29.

Gifford, D. (1997). The value of going green. Harvard Business Review, 75, 11-2.

Hardin, G. (1968). Tragedy of the Commons. Science, 162(1243), 48.

Hibbert, L. (1998). Sustainable activity. Professional Engineering , 11, 32-3.

Hick, S. (2000). Morals maketh the money. Australian CPA, 70, 72-3.

Khoo, H.H., Spedding, T.A., Tobin, L., Taplin, D. (2001). Integrated Simulation and Modeling Approach To Decision Making and Environmental Protection. Environment, Development and Sustainability, 3(2), 93-108.

Matthias, R. (1999). Strategies for promoting a sustainable industrial ecology, Environmental Science & Technology, 33(13), 280-82.

Melnyk, S. and Handfield, R. (1996). Greenspeak. Purchasing Today, July, 32-6.

Miller, W. H. (1998). Citizenship: a competitive asset, Industry Week, 247(15), 104-8.

Ottman, J. A. (1999). How to develop really new, new products. Marketing News 33(3), 5-7.

ProcessModel (2000) Software & Services (ProcessModel, Inc.).

QEPA (Queensland Environmental Protection Agency) (1999). Sustainable Queensland, www.epa.qld.gov.au.

Quinn, B. (2001). Manufacturers squeeze out more energy efficiency. Pollution Engineering, 33(1), 23-4.

Saemann, R. (1992). The environment and the need for new technology: empowerment and ethical values. Columbia Journal of World Business, 27, 186-93.

Shultz II, C.J. and Holbrook, M.B. (1999). Marketing and tragedy of the commons: a synthesis, commentary, and analysis for action. Journal of Public Policy and Marketing, 18(2), 218-29.

Spedding, T.A., Khoo, H.H., Taplin, D. (1999). An integrated simulation approach for teaching industrial ecology. Proceedings of the HKK Conference. Waterloo: Canada.

Stead, J.G. and Stead, E. (2000), Eco-enterprise strategy: standing for sustainability. Journal of Business Ethics, 24(4), 313-29.

Street, A. C. (1986). The Diecasting Book. 2nd Edition, Portcullis Press.

Taplin, D.M.R., Spedding, T.A., Khoo, H.H. (2001). Environmental security: simulation and modelling of Industrial Process Ecology. Paper presented at the Strasbourg Forum, Council of Europe, Strasbourg: France.

US EPA (US Environmental Protection Agency) (2000). Air Pollutants. www.epa.gov.

20

Computer-aided Resource Efficiency: How Software and a Common Data Format can Enhance the Assessment of Environmental Impacts and Costs Along the Supply Chain

Severin Beucker[1] and Claus Lang-Koetz[2]

[1]Fraunhofer Institut fuer Arbeitswirtschaft und Organisation Competence Center Innovationsmanagement, Nobelstr. 12, 70569 Stuttgart, Germany, severin.beucker@iao.fraunhofer.de
[2]Institute for Human Factors and Technology Management (IAT), University of Stuttgart, Nobelstr. 12, 70569, Stuttgart, Germany, claus.lang-koetz@iao.fraunhofer.de

20.1 Developments in E-business and its Environmental Impacts

In recent years traditional economics and business have been radically changed and their turnover accelerated by the use of information and communications technology (ICT) (Wirtz 2001). New methods of conducting business such as 'e-commerce' and 'e-procurement' are leading to a change in structure, from traditional commerce towards new business strategies using ICT, especially in the field of business-to-business (B2B) and business-to-consumer (B2C) relations.

From an economics perspective the acceleration and globalisation of sales and procurement have manifold positive effects, such as the ability to compete in markets and prices. B2C sales applications make use of single-product shipping, including air and surface transport. Worldwide procurement in combination with low transportation costs as well as dense and fast transportation networks guarantee a permanent supply of goods. Together, cheap and fast transportation allow rapid changes in supply structures and cut-backs in storage capacities in favour of just-in-time (JIT) delivery (Westkämper 2003).

Seen from an environmental perspective, the effects of e-commerce and e-procurement are difficult to assess and are not yet clear in terms of their amplitude. First results from case studies show that the potential environmental effects of e-commerce lead back to traditional environmental issues such as packaging materials and transport. Matthews and Hendrickson (2001) compared different logistic networks for the delivery of books in the USA according to their energy consumption and use of packaging materials. Williams and Tagama (2001) did a similar study for the Japanese market, connecting the environmental effects of book retailing to population density.

20.1.1 E-commerce and E-procurement Influences on Transportation and Packaging Material

Supplier and production structures in a globalised economy are complex networks and have always been connected to manifold environmental impacts, such as deforestation, air-borne and water-borne emissions from the extraction of raw materials and so on. With the intense use of fossil-based energy for production and transportation, environmental impacts have accelerated. E-commerce is, among other things, a result of a widespread communication network combined with cheap and fast transport. Transportation in particular contributes to various environmental impacts such as emissions, energy use and consumption of land surface. The effects depend on factors such as the means of transportation (of the carrier), the distance covered, the weight and volume of the product and the workload of the carrier. The environmental impacts of transportation are, in many cases, diffuse and difficult to attribute, and only a few of the effects, such as higher traffic rates, cause immediate and direct local impacts.

One of the more visible and calculable ecological and financial outputs is caused by packaging material from the delivery of products. Direct costs can arise from packaging material for disposal. However, some costs are hidden, such as additional working time for handling of packaging material, losses in quality or even loss of employee productivity as a result of injury. Past studies have shown that environmental impacts emerging from packaging materials can be connected to manifold effects, such as energy and water consumption as well as toxic effects from combustion or disposal (for an overview on related studies, see UBA 1997).

Taking the expected growth in e-commerce and e-procurement into account, a need for a combined financial and environmental assessment for the purpose of balanced decision-making with regard to producing companies is becoming more and more obvious. Cost-saving potentials and the possibility of achieving reduced environmental impacts can be drawn from such an assessment. In addition, the interests of different stakeholders regarding environmental matters as well as the goal of attaining a more sustainable economic system, declared as necessary by the environmental policies of many countries, can be a strong motivation for achieving an enhanced assessment of production processes and products.

To combine environmental reliability with financial reliability requires solid environmental data and information on a company's production and its supply chains. Whereas the financial assessment of a production company and its suppliers can be achieved by various methods of price calculation, an assessment of environmental performance is more difficult because of the sensibility of production and product-related data. Often, data on the environmental impacts of a product or a production process contain information on product or process-specific specifications and are therefore considered as trade secrets. Regardless of such difficulties, the need for information calls for new assessment methods and efficient support for such methods to allow companies to weight and assess their production as well as that of their suppliers with regard to financial and environmental restrictions.

So far, only a few studies have focused on the combined assessment of environmental and financial effects along the supply chain (see e.g. Cohen 2001;

Williams and Tagama 2001). Although in the past few years a lot of methodological work has been conducted on environmental management systems (EMS) in companies, the efficient assessment of a company's production and its supply chains still requires technical and applicable concepts and methods for its realisation. The approach taken in Section 20.2 is to describe the methodology of such a combined assessment, bringing financial and environmental information together. In Section 20.3 we will focus on how software can support such an assessment and how use of a common data format for product data can help in gathering environmental data in a company and along its supply chain. In Section 20.4 we provide a summary and highlight conclusions to be drawn.

20.2 The CARE Project: Objectives and Approach

The research project known as CARE (Computer Aided Resource Efficiency Accounting for Medium-Sized Enterprises)[1] consists of a scientific core project and three case studies performed with the companies Toshiba Europe GmbH (computer assembly), Nolte Möbel (furniture) and Muckenhaupt & Nusselt GmbH & Co. KG (cable-based communications).

The objective of the CARE project is to develop a financial and environmental information system for management and decision-making in companies. The approach uses the concept of resource efficiency accounting (REA), which combines cost accounting with environmental impact data on production processes and the product life-cycle.

20.2.1 Methodological Background

The REA approach was developed by the Wuppertal Institute for Climate, Environment and Energy (see Orbach *et al.* 1998). Its objective is to give a company and its management new insights into and aggregated information on financial and environmental effects for the assessment of a company and its supply chain. REA can be applied at different levels within a company and its supply chain. The different levels require specific data and information on material and energy flows as well as environmental impact data. The basis of the financial assessment within REA is a process-oriented cost-accounting approach that allocates unspecific costs, such as residue costs or costs for operating supplies, to the originating process or product.

---

[1] The CARE project was publicly funded by the German Ministry for Education and Research. Scientific partners in the project were the Institut für Arbeitswissenschaft und Technologiemanagement (IAT), at the University of Stuttgart, and the Wuppertal Institute for Climate, Environment and Energy, and Synergitec a consulting company in Freiburg. For further information, see the project website, at http://www.bum.iao.fraunhofer.de.

For the environmental impact assessment, the REA uses material intensity (MI) values.[2] They are available for many resources, materials and standardised transportation and production processes. MI values focus on the material and energy consumption in five different categories related to the production of a material or product:

- Water
- Air
- Soil
- Biotic Material
- Abiotic Material

MI values do not assess the toxicity of a material. They can be seen as indicators that give consolidated and aggregated information on environmental impacts.

REA can be applied at different levels within a company and its supply chain. Different assessment tasks, such as the assessment of the inputs and outputs of a company, its processes and its products, can be conducted. The results of REA can be evaluated and displayed in various ways, depending on the company's assessment and management systems. An applicable and aggregated way to sum up the results of the assessment process is to calculate financial and environmental key figures that can relate to specific production processes or products (e.g. see the aggregation of key figures in Fig. 20.1: The combination of material, cost and environmental impact data in resource efficiency accounting).

A specific form of visualisation of the results from REA is a portfolio presentation that arranges different processes or products relative to each other according to their financial and environmental impacts. Figure 20.1 shows a portfolio presentation from a case study of Toshiba Europe that compared sets of packaging materials used by different suppliers for the delivery of key components for computer notebook production.

To evaluate key figures or portfolio presentations within REA, specific cost and environmental impact data on the company, its production processes and products are needed. The different levels of REA with their specific data requirements are listed as follows:

- Company level REA: the application of REA at the company level requires detailed data on the consumption of substances, materials, items and energy over a defined time-frame. All data must be sorted according to input and output flows.
- Process level REA: at the process level, the same input and output data as for the company level is required, but in greater detail and on a per-process basis. Cost data on material, personnel, machinery and so on is needed at the same level of detail.

---

[2]The MI values were developed by the Wuppertal Institute and count the material intensity of a product, material or service in metric tonnes. The MI is calculated as the mass-equivalents caused by the production of the product or component in the five categories listed in the text. For further information on the method and its use, see publications by F. Schmidt-Bleek, at www.factor10-institute.org.

- Product level REA: costs and material consumption determined at the process level are allocated to specific products. Data generated from the process level REA is supplemented by MI values, including information on the environmental impacts of preliminary and downstream processes of the supply chain and product life-cycle.

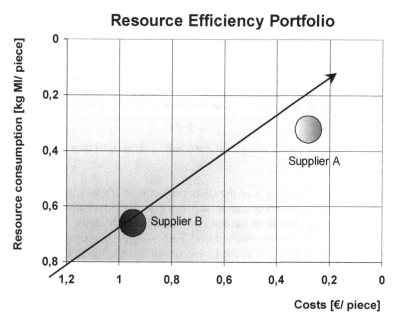

Figure 20.1. Assessment and weighting of different sets of packaging materials in a resource efficiency portfolio at Toshiba Europe

All data used for the different levels of REA application should be fact-based data. A possible source of such data may be a company's enterprise resource planning (ERP) system.[3]

Of the three levels, the product level is the most sophisticated and places the greatest demands on data quality and data structure. It thus also places the maximum demand for the testing of REA in companies.

Figure 20.2 shows the general use and combination of different data on costs, masses and environmental impacts as combined in REA. The combination of cost and MI data from different steps of the production process and product life-cycle allow various levels of assessment at the levels of REA listed above.

---

[3]An ERP system can be considered as a the backbone of a company's information system, containing data on materials, production planning, purchasing and sales, cost accounting and so on. An example of a widely used ERP system is SAP/R3 (for further information on the ERP System SAP/R3 see http://www.sap.com).

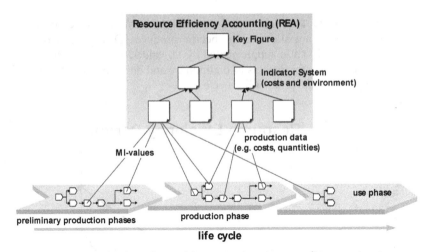

Figure 20.2. The combination of material, costs and environmental impact data in resource efficiency accounting (REA)

One of the main objectives of the CARE project is to advance and test REA at its different levels of application and to support that implementation with software that requires existing data from the company and its supply chain. The specific task of the Institut für Arbeitswissenschaft und Technologiemanagement (IAT; Institute for Technology Management) of the University of Stuttgart is to develop concepts and solutions to support the different levels of REA with data on costs, mass and energy flows as well as environmental impact data as described above. The different levels of REA with their specific needs for data, as well as possible solutions, will be discussed in the following sections.

20.2.2 Material Flow Analysis as Basis for Resource Efficiency Accounting

Material flows describe the conversion and dislocation of raw materials, operating supplies, semi-finished goods or products and thus provide the basis of corporate activity in most manufacturing companies. Material flows, with its corresponding costs, can be analysed and assessed by material flow analysis. Material flow analysis focuses on the modelling, analysis, assessment and control of material and energy flows in a company and aims at identifying potentials for optimisation. It requires activity-based and contemporary data on material input and output and can be applied to single processes, whole companies or production networks (see Bullinger and Beucker 1999).

In addition to material and energy flows, the costs for materials, activities and so on can be evaluated. Different cost-accounting approaches such as activity-based costing (see Schaltegger, Burrit 2000) or flow-cost accounting (see Strobel and Loew 2001) can be supported by this method. For the application of different levels of REA, material flow analysis can be used to create a detailed database. The data required for REA is described in the following section.

20.2.3 Data Requirements of Resource Efficiency Accounting

As mentioned in Section 20.2.1, the different levels of REA require specific data. The application of REA at the product level was identified as the most sophisticated case of REA application. Its data needs can therefore serve as the case of maximum demand for the application of REA in companies. For the application of REA at the product level, data and information are required on the:

- Structure of production processes: the structure of production processes, including the connections between the individual processes, can often be identified from process diagrams or from work plans describing product ingredients and how they have to be assembled. Work plans often also contain information on personnel expenses and machinery as well as allocations by workplaces.
- Materials, substances and energy (input) consumed per process: such data can be found in bills of material, work plans and routings or in production orders. Product-related material inputs to and outputs from processes are often available in the bills of materials. Other outputs, such as waste, often have to be recorded by measurement.
- Output per process: outputs of energy, waste and waste-water from a process are not usually measured but can be estimated if necessary. Some companies record such measurements with use of separate software systems.
- Mass and cost allocation per process: precise mass and cost allocations per process and the outputs of each process have to be made. In particular, mass allocations for outputs are often not available and therefore have to be assessed.
- MI: MI values for all materials and energy used in the production and life-cycle of the product have to available for the completion of the environmental impact assessment. In the CARE project this data is provided by the Wuppertal Institute.

Many of the types of data specified above can be found in a company's ERP system. Hence, ERP systems and the data contained within them are a valuable source of information in support of the application of REA. Possibilities for the use of ERP systems and their data for REA will be discussed in the next section.

20.3 Data and Software to Measure and Assess Environmental Performance

All assessment of a company's environmental and financial performance starts with reliable data and information on the production itself. Prices and material consumption at various levels of specification are necessary for a transparent classification of processes consuming materials and producing waste. Within the classification and specification of such data, ERP systems play a significant role. They are widely used for production planning and control and therefore can contain valuable information for the environmental and financial assessment of a company's production processes and products.

A survey conducted among 151 ERP software companies from Germany, Switzerland and Austria in 2001 evaluated the environmental functionalities of ERP systems (see Rey *et al.* 2002). The goal of the survey was to assess to what extent ERP systems can support different tasks of environmental management. Among other questions, the availability of data with environmental relevance was queried. This data was divided into the following types:

- Description of materials in order to classify them according to the substances they contain
- Official categories of hazardous materials
- Criteria on how different kinds of waste can be disposed of
- Behaviour of residuary materials when exposed to the environment
- Material and energy flows in manufacturing processes
- Substance data such as physical and chemical properties or chemical formulae
- Consumption of energy and emissions of operating supplies such as oils and lubricants
- Water, energy and waste balances for companies, plants, processes or products, and the amount of emissions such as waste and waste-water

Of the 151 questionnaires sent out, 10% were returned. They showed that many ERP systems can already be used for the classification of material data, for the monitoring and controlling of waste according to disposal criteria and for the administration of dangerous goods. However, not many of the systems can be used for the monitoring and control of material, energy and water consumption (see Fig. 20.3).

The lack of possibilities for controlling material and energy flows (see the point the graph crosses the lower vertical axis) indicates that ERP systems and their data are not yet sufficiently developed to be used for the identification of optimisation potentials. It also points out possible further developments of ERP systems and their functionalities. The ability to calculate material consumption and waste production and their corresponding costs and the ability to assign these costs to the places where they were caused within the company could open up new possibilities for financial and environmental performance measurement.

Nevertheless, ERP systems can serve as a valuable data source for applying REA within a company: Most ERP systems available today contain a major portion of the data required to conduct REA as specified in Section 20.2.3: details of work plans and routings, production orders, bills of materials, and material properties from the master data set can all be used to carry out a material flow analysis. Although this data does not allow detailed evaluations of material and energy flows, it can be used to evaluate material losses, resource consumption in general and the way in which waste was created and disposed of.

The usefulness of data from ERP systems will always depend on the specific data quality in a company as well as on technical restrictions arising from specific software. Data may be available in different formats and structures depending on the software used in the company. However, the gathering of additional data (manually or from other information technology [IT] systems) is still necessary in most cases. Hence, the IT structure and IT systems of the company play an important role in data acquisition.

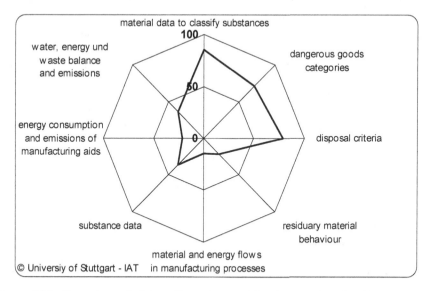

Figure 20.3. Percentage of data with environmental relevance in enterprise resource planning (ERP) systems; results are from a survey of 151 ERP software companies conducted in Germany, Austria and Switzerland. (Note: the outer octagon corresponds to 100 %, the inner octagon to 50 %, and the centre of the octagons (the origin of the six axes) to 0 %.)

To conduct REA, data from ERP systems and other sources can be used in environmental management information systems (EMIS) for assessment by means of calculation and evaluation. EMIS are a group of specific software tools for the interpretation of environmental data and information as seen from different assessment foci.[4] EMIS can be grouped into four different categories (see Bullinger and Jürgens 1999):

- Environmental and environmental law databases
- Systems for the organisational support of environmental management
- Systems to perform material flow management
- Systems to conduct life cycle assessments (LCA)

For the purpose of performing REA, EMIS with a focus on material flow management or LCA can be used, as they have functionalities for conducting a

---

[4]Although the term 'environmental management information system' generally refers to organisational–technical systems for systematically obtaining, processing and making environmentally relevant information available in companies (Page and Rautenstrauch 2001), the CARE project defines EMIS as software systems that support the tasks of an EMS.

material flow analysis (for the advantages and disadvantages of EMIS categories when applied to REA, see Beucker *et al.* 2002).

EMIS can be used to conduct a material flow analysis or to calculate resource efficiency according to the REA method. Results calculated can then be displayed in the ERP system for evaluation and visualisation. Data exchange between ERP systems and EMIS is still a complicated process that often requires manual data transfer and/or costly adaptation of the software being used. A standard data format for data exchange between ERP systems and EMIS would simplify automatic data exchange between ERP systems and EMIS. This would enhance the ability of ERP systems and EMIS to interact.

To date, a standard data format for such a data exchange has not yet been defined. The IAT and Fraunhofer-Institut für Arbeitswirtschaft und Organisation (IAO) worked together with software developers to define a data format that serves as a pre-standard for data exchange. This data format was published as a so-called 'publicly available specification' (PAS).[5] The goal is to facilitate the data transfer between different ERP systems and EMIS. Such a data transfer would have several advantages:

- Data access would not depend on communication or complicated workflows inside the company.
- It would decrease data redundancy: The data already collected and stored in the ERP system could be used by the EMIS without additional expenses to regather the data in the EMIS.
- Checks regarding the consistency of the same data stored in multiple locations in different systems would not be necessary.

The PAS has to be seen in the broader context of the development of standards for business and environmental data. Recent developments in e-business show the attractiveness of XML (extensible mark-up language) for the standardisation of data in exchange processes such as product catalogues or sales transaction (see Kelkar *et al.* 2001, 2002; Schmitz *et al.* 2001). Regarding the standardisation of environmental data, current developments can be found mainly in the research area. Researchers on the Ecoinvent project are developing a common data format, based on XML, for LCA data.[6] The Environmental Mark-up Language (EML) initiative has started a discussion of the standardisation process for environmental data, using XML, but so far it has focused on corporate environmental reports and spatial environmental data.[7]

---

[5]A PAS can be considered as a first step in the standardisation process. A PAS is a normative document representing a consensus reached within a working group but is not the result of an international agreement in the way that standards of the International Organisation for Standardisation (ISO) are. For further information, see the ISO website, at www.iso.ch.

[6]See the website of Ecoinvent, at www.ecoinvent.ch.

[7]For further information, see the website of the EML initiative, hosted by the Technical University of Berlin, at www.xml-eml.de.

20.3.1 Specification Development for Data Exchange Between Environmental Management Information Systems and Enterprise Resource Planning Systems

As mentioned above, a PAS can be considered as a first step in the standardisation process. The PAS on data exchange between EMIS and ERP systems has been developed in a three-step process (see PAS 1025 and Wohlgemuth et al. 2004):

- Business scenarios that require intercommunication between ERP systems and EMIS have been collected.
- These business scenarios have been generalised to describe different applications for the proposed data exchange between ERP systems and EMIS.
- The data to be transmitted and a generic data format have been described by means of XML.

Interest groups determine the type of data to be communicated. Thus, business scenarios that include the specific tasks of environmental management and the requirements of the people involved in such management formed the starting point for determining a basic structure for the PAS. Environmental data is needed by different interest groups inside and outside a company. Examples of such business scenarios are as follows:

- The environmental manager needs environmental data to carry out an internal environment and safety report or to conduct an input–output analysis
- The controlling department would like to know where costly material and energy losses occurred in order to improve the material efficiency
- The cost centre requires the exact costs of materials used and disposed of in order to control those costs accordingly
- The public requests yearly reports on how the company treats the environment, as set down in the external environmental report
- A yearly environmental cost report has to be sent to the national German statistics agency

Data transfer between different software systems can take place via interfaces, allowing direct access to external software systems. Data can be read from or written into a system, and certain functionalities can be accessed from outside. Interface files for data transfer are generated when data is written as output into a data file from a software system. Standardisation of such a data format can lead to the universal exchangeability of data between different systems.

Examples of data to be transmitted are:

- Product mass, energy and costs
- Production structure, material flow diagrams, bills of materials, routings, work plans and the work centres involved
- Results calculated in the EMIS such as performance indicators and graphical evaluations (e.g. charts and Sankey diagrams)

The PAS for data exchange between EMIS and ERP systems includes a data format for the data to be transmitted: this format will be described in XML technology.

The data format consists of the data structure (the order of elements, tree hierarchy, etc.), the allowed data types (such as integers, strings, floats, etc.), the definition of units (kilograms, metres, euros, etc.), the description of system boundaries or reference bases or reference time-periods.

20.3.2 Going Beyond the Factory Gate: The Environmental Product Specification

Assessing environmental impacts inside the company requires data mainly from the company itself. However, if one takes the preliminary phases of the supply chain and the use phase of the product into account the problem becomes more complex and may involve many different partners. The PAS consortium is focusing on the data exchange between EMIS and ERP systems, but research projects at IAT, and Fraunhofer-IAO in co-operation with industry have shown the need for an environmental product specification (EPS) as an extension to the PAS to secure environmental data transfer along the supply chain. [8]

The production of goods is most often split into production sequences. These complex production sequences can be described by a net structure (see Steinaecker 2001). As a material is produced and carried on from one company to another it raises the need for a detailed description of its characteristics to be passed along also. This description may include information directly needed in the production process, such as weight, ingredients, components or a bill of materials. It may also include data that may be described as environmental data, such as disassembly information for the recycling process, declarations of hazardous substances, guidelines for disposal or quantitative information on environmental impact. Usually, the data needed for the production process is transmitted to the next member of the production net, whereas environmental data is not. Such data might already have been generated in one company but not sent to the next member of the supply chain. Thus, valuable data does not reach all interested parties. For example, such data may be of value to a company's decision-making process if raw material is to be purchased or when new machinery is to be acquired. The simultaneous passing on along the supply chain of both data directly needed in the production process and environmental product data leads to increased transparency. This would be in the interest of many companies, consumers and also the public in general who require more specific data on product-related environmental impacts.

The passing on of data needed for the production process can help not only in co-ordinating business activities in the development of products but also in scheduling production orders. Such data sharing has become a regular business practice for many companies (e.g. in supply chain networks; see Steinaecker and

---

[8]   For instance, see the description of the European Union EPSILON (Environmental Planning and Control Systems in Logistic Networks) project in Jürgens et al. (1999).

Kühner 2001) and between firms collaborating in the development of products.[9] The recipient of the data may be another company (B2B data transfer) or the customer at the end of the supply chain (B2C data transfer).[10] The data has to be adapted according to the recipient's needs. The addition of environmental information would simplify the process. The product specification could be read by the ERP system of another company processing data needed for the production process or environmental data. It could also be used by software for consumers to evaluate the environmental impacts of a product.

The combination of data needed for the production process and environmental data is referred to as an 'environmental product specification' (EPS) and can be seen as an extension of the PAS for data exchange between EMIS and ERP systems as described above. Such an EPS could be used for intra-company and inter-company data exchange, as shown in Figure 20.4.

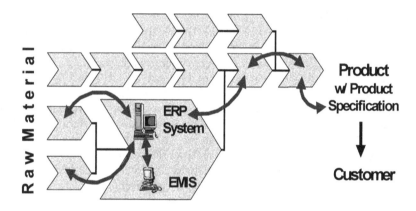

Figure 20.4. Data transfer between different members of the production chain (Grey symbols represent companies, black lines material flows and arrows symbolise possible data transfer between companies, inside a company or to the customer. Note: ERP means: enterprise resource planning, EMIS means Environmental Management Information System)

The process of passing on an EPS will find widespread acceptance only if such data is put into a common data format that is agreed on by many interest groups. This data format could form the basis for data files available in electronic form. Data could then be transferred automatically from one information system to another, as already achieved with product catalogues and sales transaction data in e-business, as described in the opening text of this section on software.

---

[9] Many partner companies use the Standard for the Exchange of Product Data (STEP) while collaborating on the development of new products. This standard is specified in ISO 10303 and is continuously being improved (go to www.iso.org).

[10] Eco-labels already inform interested customers about environmentally friendly products but they do not contain quantitative information on environmental impacts. For an overview of existing eco-labels, go to www.labelinfo.ch or www.eco-labels.org.

20.4 Summary and Conclusions

The changes and acceleration of economic activities and business, driven by technological developments in e-business, create financial benefits for many companies as well as having environmental effects. Owing to their complex production structures, with worldwide supply chains, companies face great difficulty in assessing the effects of their activities, particularly in the environmental field.

The CARE project is developing an information system for companies to integrate financial accounting and environmental impact assessment. The objective is to help organisations gain new insights into the environmental and financial effects of their production and supply chains.

The methodological background for the assessment is provided by the REA approach and by MI values, developed by the Wuppertal Institute for Climate, Environment and Energy.

The specific task of the IAT, at the University of Stuttgart, is to develop concepts and solutions to support the different levels of REA with data on costs and on mass and energy flows as well as with environmental impact data.

Calculations at the different levels of REA can be facilitated by the use of production data available in the ERP system of the company concerned. Such data might be available in bills of materials, work plans or in production orders. Although ERP systems can provide most of the data needed to perform REA, another group of software systems, the EMIS, can be used for the assessment of financial and environmental impacts by means of material flow analysis.

The exchange of data between ERP systems and EMIS and the interpretation of that data can be eased by a publicly available specification (PAS) for exchanging environmentally relevant data between ERP systems and EMIS. The PAS defines a data format for the exchange of data between the two systems and therefore enhances the ability to carry out a combined financial and environmental assessment of a company. We also postulate that to extend the assessment process to the supply chain an additional data structure, called an 'environmental product specification' (EPS), will be required for inter-company exchange of environmental data. An EPS can be seen as an extension of the PAS for data exchange between EMIS and ERP systems, applicable along the whole supply chain.

In conclusion, the integration and combination of financial and environmental data in companies and along their supply chain may be made possible by the use of consistent data formats at the different levels of production and along the supply chain.

References

Beucker, S., C. Lang and U. Rey (2002) *Betriebliche Umweltinformations-systeme und ihre Funktion für die Ressourceneffizienzrechnung* (project report of the CARE [Computer Aided Resource Efficiency Accounting for Medium-Sized Enterprises] research project; Stuttgart, Germany: CARE, http://care.oekoeffizienz.de).

Bullinger, H.-J., and S. Beucker (2000) *Stoffstrommanagement: Erfolgsfaktor für den betrieblichen Umweltschutz, Tagungsband zum 3: Management Produktion und Umwelt* (Stuttgart, Germany: Fraunhofer Informationszentrum Raum und Bau (IRB) Verlag).

Bullinger, H.-J., and G. Jürgens (1999) *Betriebliche Umweltinformationssysteme in der Praxis, Tagungsband zum 2: Management-Symposium Produktion und Umwelt* (Stuttgart, Germany: Fraunhofer Informationszentrum Raum und Bau (IRB) Verlag).

Cohen, N. (2001) 'The Environmental Impacts of E-Commerce', in L.M. Hilty and P. Gilgen (eds.), *Sustainability in the Information Society, Part 1: Impacts and Applications* (Umwelt-Informatik aktuell; Marburg, Germany: Metropolis): 41-52.

Jürgens, G., D. Schnapperelle and J.v. Steinaecker (1999) 'EPSILON: Ein System zur Unterstützung einer umweltorientierten Planung in Unternehmensnetzwerken', in C. Rautenstrauch and M. Schenk (eds.), *Umweltinformatik zwischen Theorie und Industrieanwendung, 13: Internationales Symposium 'Informatik für den Umweltschutz' der Gesellschaft für Informatik (GI)* (Marburg, Germany: Metropolis): 46-54.

Kelkar, O., B. Otto and V. Schmitz (2001) *Spezifikation openTRANS V1.0* (Stuttgart, Germany: openTRANS, www.opentrans.org).

Kelkar, O., J. Leukel and V. Schmitz (2002) 'Towards Extended Price Models in XML Standards for Electronic Product Catalogs', in *Proceedings of the 4th International Conference on Enterprise Information Systemes (ICEIS 2002), April 2002, Ciudad Real, Spain*: 937-945.

Mathews, S., Hendrickson, C. (2001) Economic and Environmental Implications of Online Retailing in the United States' in Hilty, L.M. and Gilgen, P. (eds.) Sustainability in the Information Society, Part 1: Impacts and Applications (Umwelt-Informatik aktuell; Marburg, Germany: Metropolis Verlag): 65-72.

Orbach, T., Liedtke, C., Duppel, H. (1998) 'Umweltkostenrechnung: Stand und Entwicklungsperspektiven', in Döttinger, K., Roth K., Lutz, U. (eds.), *Springer Loseblattsystem Betriebliches Umweltmanagement: Grundlagen, Methoden, Praxisbeispiele* (Berlin, Germany: Springer, 8th edn): 1.02, 1-27.

Page, P., and C. Rautenstrauch (2001) 'Environmental Informatics: Methods, Tools and Applications in Environmental Information Processing', in C. Rautenstrauch and S. Patig (eds.), *Environmental Information Systems in Industry and Public Administration* (Hershey, USA and London, UK: Idea Group Publishing): 2-11.

PAS 1025 (2003) Deutsches Institut für Normung: *Austausch umweltrelevanter Daten zwischen ERP-Systemen und betrieblichen Umweltinformationssystemen* (Berlin, Germany: Beuth-Verlag).

Rey, U., C. Lang and S. Beucker (2002) *ERP-Systeme und ihr Datenangebot für die Ressourceneffizienz-Rechnung* (project report of the CARE [Computer Aided Resource Efficiency Accounting for Medium-Sized Enterprises] research project; Stuttgart, Germany: CARE, http://care.oekoeffizienz.de).

Schaltegger, S., and R. Burritt (2000) *Contemporary Environmental Accounting: Issues, Concepts and Practice* (Sheffield, UK: Greenleaf Publishing).

Schmitz, V., O. Kelkar, T. Pastoors, C. Hümpel and T. Renner (2001) *Spezifikation BMEcat V1.2* (Essen, Germany: BMEcat, www.bmecat.org).

Steinaecker, J.v. (2001) *Ein Informationsmodell zur Modellierung und Planung von netzwerkartigen Produktionsstrukturen* (doctoral thesis, Fakultät Konstruktions- und Fertigungstechnik der Universität Stuttgart, Heimsheim, Germany: Jost-Jetter Verlag).

Steinaecker, J.v., J., and M. Kühner (2001) 'Supply Chain Management: Revolution oder Modewort?', in O. Lawrenz, K. Hildebrand and M. Nenninger (eds.), *Supply Chain Management: Konzepte, Erfahrungsberichte und Strategien auf dem Weg zu digitalen Wertschöpfungsnetzen* (Wiesbaden, Germany: Vieweg Verlag, 2nd edn).

Strobel, M., and T. Loew (2001) 'Stoff- und energieflussorientierte Kostenrechnung', in *Handbuch Umweltcontrolling* (Munich, Germany: German Ministry for the Environment, and Federal Environmental Protection Agency): 523-36.

UBA (Umweltbundesamt) (1995) Methodik der produktbezogenen Ökobilanzen: Wirkungsbilanz und Bewertung (report UBA/23; Berlin, Germany: UBA).

UBA (Umweltbundesamt) (1997) Materialien zu Ökobilanzen und Lebensweganalysen (report UBA 26/97; Berlin, Germany: UBA).

Westkämper, E. (2003) Einführung in neue Organisationsformen im Unternehmen. in Bullinger, H.-J., Warnecke, H. J. Westkämper, E. (Eds.): *Neue Organisationsformen im Unternehmen. Ein Handbuch für das moderne Management* (Berlin, Germany: Springer Verlag).

Williams, E., and T. Tagama (2001) 'Energy Analysis of E-commerce and Conventional Retail Distribution of Books in Japan', in L.M. Hilty and P. Gilgen (eds.), *Sustainability in the Information Society, Part 1: Impacts and Applications* (Umwelt-Informatik aktuell; Marburg, Germany: Metropolis): 73-80.

Wirtz, B. W. (2001) Electronic Business (2nd edn, Wiesbaden, Germany: Gabler Verlag).

Wohlgemuth, V., Niebuhr, C., Lang, C. (2004): Exchanging Environmental Relevant Data between ERP-Systems and Industrial Environmental Management Information Systems using PAS 1025, in *Proceedings of the 18th International Conference for Environmental Protection*, volume 1, pp. 183-193, Cern Geneva, Switzerland: October 21-23, 2004.

21

E-commerce Solutions to Environmental Purchasing

Steve V. Walton[1] and Chris Galea[2]

[1]Goizueta Business School, Emory University, 1300 Clifton Road, Atlanta, GA 30322-2710, USA, Steve_Walton@bus.emory.edu
[2]Department of Business Administration, St. Francis Xavier University, P.O. Box 5000, Antigonish, NS B2G 2W5 Canada, cgalea@stfx.ca

The dot-com bubble that burst in March 2000 marked the end of an amazing run of unbelievable hype. It turned out, for example, that profits actually *do* matter and that first to market is often the same as first to bankruptcy. But in the same way that the impact of electronic commerce (e-commerce) on business was never as large as the hype promised, the impact of e-commerce on business in general and purchasing and materials management in particular is larger than the post-bubble pessimism suggests.

By the same token, the dot-com frenzy promised improved environmental performance by consolidating shipments, by substituting one-truck trips for multiple car trips and by 'dematerialising' products. Some of these turned out to be empty promises, but e-commerce can significantly improve the environmental performance of purchasing and materials management. In this chapter we describe 'environmental purchasing' and demonstrate how e-commerce can be used to make purchasing even more environmentally sound.

21.1 What is Environmental Purchasing?

At the highest level of abstraction, environmental purchasing is buying 'products or services that have a lesser or reduced effect on human health and the environment when compared with competing products or services that serve the same purpose' (Executive Order 13101, US GPO, Office of the President of the United States, 1998). In the USA, the Pacific Northwest Pollution Prevention Resource Centre (PPRC) refers to this concept as 'affirmative purchasing'.[1] Environmental purchasing seeks to find both environmental and operational performance gains in the activities needed to acquire products and services for a buying organisation. Governments at all levels (city, county, state and national)

---

[1]See the web page on the website of the Pacific Northwest PRRC, at www.pprc.org/pprc/pubs/topics/envpurch.html, accessed 26 October 2002.

and in many locations (from Santa Monica, CA, to Belfast, Northern Ireland, UK) have adopted environmental purchasing; many governments, including the US government through Executive Order 13101, give preference to suppliers that further the goals of environmental purchasing.

Environmental purchasing represents an opportunity to a business whether the business intends to sell to a government or not. One study into the role of purchasing in environmental management found sufficient evidence to offer a proposition suggesting that there is a positive relationship between the success of the environmental projects of a business and the quality of its products and processes. As importantly, the study further suggests that this improved quality is associated with decreased operating cost (Carter and Dresner 2001). This finding is significant, because it demonstrates a real link between the environment and other components of operations strategy. Operations strategy usually involves setting relative priorities on the components of cost, quality, time and flexibility. Environmental purchasing specifically, and environmental operations generally, require that this list of strategic components be expanded to explicitly include the environment so that the definition of operational success is also expanded to include environmental performance.

To expand the definition offered in Executive Order 13101, we can regard environmental purchasing as having two distinct facets: the purchase of environmental products and services, and the acquisition of products and services that give an environmental advantage. The purchase of environmental products and services, for example, involves all those activities associated with evaluating, selecting and managing a provider of solid-waste management services. This facet would also include things such as the hiring of an architect to design an environmentally friendly building. In addition, it includes the hiring of consultants or other experts to help the company pursue accreditation in different environmental certification schemes, such as LEED™, of the US Green Building Council (US GBC),[2] or the ISO 14000 series, of the International Organisation for Standardisation (ISO 1996). Finally, the purchasing of environmental products and services includes the buying of tradable emissions credits. Reuters reports, for example, that the sulphur dioxide market is a US$4 billion market.[3] In fact, the now-open Chicago Climate Exchange (CCX)[4] has set a goal of reducing greenhouse gas emissions by 2% from the 1999 level and to further reduce emissions by 1% per year after that (*The Economist,* 19 October 2002: 60.).

---

[2] The US GBC has developed the Leadership in Energy and Environmental Design (LEED™) programme, which sets guidelines for designing and constructing environmentally friendly buildings. LEED™ emphasises five design areas: building site selection and erosion control; water efficiency; energy and atmosphere; materials and resources; and indoor environmental quality.

[3] See Reuters, at http://news.cnet.com/investor/news/newsitem/0-9900-1028-20341252-0.html, accessed 26 October 2002.

[4] See the website of the Chicago Climate Exchange, at www.chicagoclimatex.com/, accessed 17 November 2005.

Members of the CCX include DuPont, Ford, International Paper, Mexico City and the City of Chicago (see www.chicagoclimatex.com).

The acquisition of products and services that have an environmental advantage is more complex. The advantage may be embedded in the purchased product itself, as is the case, for example, with items that have a higher percentage of post-consumer recycled content than comparable items, or items with significant bio-content (i.e. bio-based products). Alternatively, the environmental advantage could be embedded in the process used to make the purchased product; for example, the purchased product could be made by an organisation with a commitment to using renewable energy. This, of course, means that an opportunity exists to improve the product's environmental advantage by improving the buying company's process; for example, by buying volatile organic compounds (VOCs) such as acetone in smaller containers the buying company can reduce the free release through evaporation of the VOC. Table 21.1 shows several more examples of each type of environmental advantage attainable in purchased items and the corresponding improvement opportunities for use of environmental processes.

Traditional paper-based methods or e-commerce methods of purchasing can be applied to environmental purchasing. There appear to be certain efficiencies in using e-commerce, though. In the following sections we explore how e-commerce can be used to accomplish the goals of environmental purchasing, and provide an example both of unsuccessful and of successful deployment of e-commerce for environmental purchasing.

Table 21.1. Measures of environmental advantage gained through purchased products

Environmental advantage embedded in the product	Environmental advantage embedded in the supplier's process	Improvement opportunities embedded in the supplier's process
Increased recyclable or re-usable content	Increased process energy efficiency	Decreased energy use
Recyclability	Decreased process scrap rates and increased yields	Decreased scrap rates and increased yields
Remanufacturability	Optimised delivery frequency and volume	Reduced amount of packaging waste
Reduced off-gas	Lower process emission levels	Lower process emissions

21.2 How does E-commerce Enter into Environmental Purchasing?

Companies can use e-commerce to dramatically improve the environmental performance of the services and products they offer while, at the same time, often reducing purchasing and operating costs and increasing customer satisfaction. There are three main ways in which companies can achieve this. The first is to take a traditionally paper-based service and shift it to an online process. A second approach is to use e-commerce to refine and improve an existing process. Last, e-commerce can be used to develop whole new approaches to purchasing that can release previously untapped areas of value while improving environmental performance (this idea is discussed in more detail in Metter *et al.* 2006).

21.2.1 Shifting Paper-based Systems to Online Processes

In some industries, there is a natural fit between a firm's traditional paper-based services and the ability of e-commerce to deliver dramatically increased customer service at substantially lower costs. An example of this can be found in the airline industry.

Over the past five years most airlines in North America have shifted from issuing paper tickets to electronic tickets (commonly referred to as e-tickets.) The shift has been dramatic. For example, United Airlines currently issues more than two thirds of its tickets in electronic form; Continental, over 50%. Air Canada wants all its customers to shift to e-tickets; in fact, it now charges a Can$30 ($US20) surcharge for any customer who has the audacity to request a paper ticket!

The main driver behind this shift by airlines has clearly been cost reduction. However, there are also direct environmental benefits in addition to increased customer satisfaction. Eliminating millions of paper tickets obviously reduces paper use—a not insubstantial impact especially when one considers that most paper tickets are hard to recycle because of the mixture of paper and carbon paper used. In addition, all the paper and energy used in mailing the tickets is vastly reduced or eliminated. In contrast, e-tickets have virtually no additional environmental impacts, save, perhaps, the marginal paper used by the passengers that choose to print their itinerary.

E-tickets also improve customer satisfaction. United Airlines has conducted surveys that suggest that 90% of customers that have used e-tickets prefer them to paper tickets. The reasons for this are somewhat self-explanatory. First, e-tickets eliminate one worry of harried travellers: tickets that are lost or forgotten at home. Moreover, e-tickets can be issued instantly, thus eliminating the worry that 'the tickets won't get to me in time'. Last, any changes to e-tickets are much easier to implement, as no new tickets have to be issued. Delta Air Lines, Inc. has recently taken the e-ticket concept one step further and offers electronic boarding passes. A traveller can access his or her itinerary from the web, check in for the flight and electronically secure a boarding pass that can then be printed locally.

21.2.2 Using E-commerce to Improve Existing Processes

The shift from paper-based purchase orders to electronically exchanged purchase orders results in cost savings, for a number of reasons. Inventories can be reduced, fewer orders are lost, delivery cycle times can be greatly improved and so on. Such improvements naturally also lead to increased customer satisfaction.[5]

There are also positive environmental impacts that can flow from using e-commerce to further the goals of environmental purchasing. Electronic purchasing tends to be much less energy-intensive and material-intensive than are paper-based models. Also, the benefits of, for example, reduced inventories have positive environmental effects such as the reduction in storage space and materials handling required. Again, however, a caveat needs to be added in that only a full life-cycle assessment (LCA) can actually reveal whether the net gain is positive. For example, just-in-time (JIT) deliveries are a natural extension of electronic purchasing, but Sarkis (1995) and Angell and Klassen (1999), for example, suggest that the move to JIT systems leads to an increased environmental impact as a consequence of the more frequent deliveries that result from their use. To complicate matters further, Angell and Klassen (1999) also suggest that centralised purchasing is likely to lead to a fuller life-cycle perspective on purchasing, but that a less centralised purchasing organisation is likely to better capture the environmental concerns of its customers.

21.2.3 Using New E-commerce-based Systems to Create New Environmental Value

Portions of current supply chain design are inherently wasteful, resulting in negative environmental impacts. Take, for example, products found defective by the consumer. Most supply chains are designed to move materials and information forward, toward the end-customer; very few are designed to take products back. As a result, defective products are usually thrown away rather than returned to the manufacturer for rework. Even worse, products returned by customers because they do not fit or are the wrong colour to match the carpet often cannot be sold as new and thus enter into this wasteful chain despite the fact that there is nothing wrong with the actual item.

However, redesigning the supply chain to take products back from the customer could result not only in cost savings but also in enhanced environmental performance. For example, defective products could be taken back from customers and, rather than discarded, sold to a liquidator. Likewise, products could be redesigned for easy disassembly, making it possible for the still-functioning parts to be re-used or sold through different channels. Another option is to have a different manufacturer do the rework, especially if the original manufacturer is far from the point of sale or return. Each of these other disposition methods extracts residual value from the product for the company. In addition, however, positive environmental benefit is realised. This comes in the form of the decreased amount

---

[5] We will offer few examples in this section because later in this chapter, in Section 21.4, we will discuss in detail how Delta Air Lines, Inc. and Interface-LLC developed an e-commerce-enabled system for chemical purchasing and management that has dramatically reduced cost and decreased environmental impact.

of material going to landfill and the subsequent reduction in the amount of virgin materials needed. In addition, much of the embedded energy of any newly manufactured products is 'saved'.

A number of examples of such reverse logistics processes are emerging. Both Volkswagen and BMW have set up automobile *dis*assembly plants in Germany. Here, cars that have reached the end of their useful lives are taken apart and their various components re-used or recycled. Needless to say, the cars from these companies are now being designed not only for ease of manufacturing but also for ease of disassembly. Such reverse manufacturing requires the exchange of much information, which is often supplied over the Internet or on proprietary data networks.

Another example emerges from the electronics and consumer products industry. A start-up company called 180Commerce has developed Returns Management Platform™. The company provides the information infrastructure that allows manufacturing companies to offload the reverse logistics process to 180Commerce as their preferred service provider. Sarkis *et al.* (2002) provide a detailed discussion of reverse e-logistics and available service providers.

21.3 E-commerce and Poor Environmental Purchasing Outcomes

In the frenzied heyday of the dot.com decade, the benefits of e-commerce reigned supreme. Besides creating previously unimagined, if not temporary, wealth, e-commerce also seemed to hold the key to achieving sustainability. For example, instead of driving to shopping centres we would now shop online, with thousands of individual car trips being replaced by efficient door-to-door shipping networks. Much free time would be made available, time that we could use to slow down our frenzied lives. All would benefit.

Few questioned the underlying assumptions behind these claims; such was the euphoria surrounding e-commerce and the rise of online interaction. However, Galea and Walton (2002) looked at the online grocer Webvan and the effects of people purchasing their groceries online. Galea and Walton came to the conclusion that online grocery purchasing was not only environmentally much more harmful but also socially regressive. In Webvan's case it also proved to be economically unsustainable as the company eventually went bankrupt, in spite of an initial investment of over US$1.2 billion, the second largest venture funded investment into a company, behind only Amazon. In this section we look further at the Webvan example and analyse why e-commerce purchasing in that case was not environmentally positive. (for a full description of Webvan's service model, see Galea and Walton 2002).

Three main arguments were put forward concerning the environmental gains to be had from grocery delivery: reduced emissions, reduced use of fuel, and reduced packaging. At first glance, such a system would seem to have nothing but positive environmental effects. Who could argue against replacing thousands of personal car trips with one efficient delivery by truck? Surely the vast increases in efficiency that Webvan's e-commerce purchasing system created would also result in greater environmental performance? The deeper analysis shown in Galea and Walton's study proved otherwise.

Environmentally, Webvan's last-mile delivery model (i.e. the process by which the final consumer received the groceries) consumed approximately 50% more fuel and emitted more than 25 times as much air pollution (e.g. carbon monoxide, oxides of nitrogen and particulate matter) compared with personal grocery purchasing. This was because Webvan's trucks were much more polluting than are cars and although they did displace some car trips they also had to travel further because the distribution centres had to be located on the outskirts of town. Moreover, many people combine grocery shopping with other necessary car trips (such as going to work); hence the number of actual trips displaced by Webvan's trucks was much less than originally anticipated. Furthermore, Webvan's infrastructure model added to the pressures of urban sprawl, with the concomitant environmental problems associated with such sprawl.

The hopes of reducing packaging also proved to be misplaced. Webvan hoped to reduce packaging by delivering the groceries in re-usable plastic totes, which reduced the need for bags. These, however, proved impractical; Webvan eventually moved to the use of plastic bags inside the totes, eliminating packaging reduction as a possible environmental advantage.

Clearly, online purchasing models are not, *ipso facto*, environmentally superior. Much depends on the design of the system and whether the whole-system effects have been thoroughly evaluated. Moreover, models that depend on 'last-mile delivery' face added difficulty. This is especially true when the e-commerce purchasing system simply replaces a current process rather than creating a whole new approach to managing a customer need.

The move to electronically enabled purchasing has other unintended consequences. Sarkis *et al.* (2002) point out that moving to electronic procurement as part of a forward e-logistics strategy requires deploying more computer technology. Computers are required at the buying site, the selling site and to create the information infrastructure to connect them. A News.com report quotes data from the US Energy Information Administration (EIA) that estimates that the Internet information technology (IT) sector accounts for between 8% and 13% of US electricity use (Konrad 2001). This, of course, means that powering the infrastructure for electronic procurement accounts for a significant portion of the emissions related to electricity generation.

Further, Sarkis *et al.* (2002) point out that a move to e-commerce solutions generally increases the rate of disposal of computer equipment. They quote studies that estimate that 15 million personal computers are disposed of each year. This represents a significant issue, because these devices are 'the source of many toxic substances used in manufacturing and heat treatment, such as lead, mercury, cadmium, chromium and bromine' (Sarkis *et al.* 2002: 39).

21.4 E-commerce and Improved Environmental Purchasing Performance

Companies can also use e-commerce as an enabler to overhaul or streamline existing processes. In this section, an example of how e-commerce significantly improved environmental purchasing is highlighted by the case of Delta Air Lines, Inc. and Interface-LLC. Delta replaced an inefficient chemical-purchasing procedure with an integrated management system that improved operational performance, reduced chemical usage and met stringent environmental reporting requirements. E-commerce enabled and enhanced the outcome of this innovative design.[6]

Before constructing a new model of chemical purchasing, Delta lacked a centralised method to match amounts bought with those used, emitted and disposed of, which created significant regulatory reporting problems. After Delta paired with Interface-LLC it was able to create an electronically enabled chemical management system that saved nearly US$1 million via reduced inventories, lowered insurance premiums and fewer instances of expired shelf-life materials in fiscal years 1996 and 1997 and that continues to pay off.

From 1994 to 1997, Delta undertook a re-engineering effort it dubbed '7.5'. It was designed to lower Delta's cost measure (cost per available seat mile) from 10.8 cents to 7.5 cents. Delta reduced staff through early retirement and refocused on core businesses by outsourcing ancillary services. In 1994, Delta realised that its chemical management system was inadequately tracking, managing and reporting chemical use and disposal at the Technical Operations Centre at Hartsfield International Airport in Atlanta, GA. The two independent events of '7.5' and Delta's realisation about the weaknesses in its chemical management system spurred it to respond.

Prior to 1995, Delta's chemical-purchasing system was not integrated with point-of-use monitoring. This limited company-wide visibility so that managers could not match chemicals bought and used with amounts discarded. Delta researched how other companies handled purchasing to craft a template for redesign. The redesign effort targeted three main goals: to manage chemicals better, to capture all necessary data concerning chemical use and to do it at a lower cost. Delta recognised it needed an integrated chemical management system. This entailed switching from basing its purchasing decisions on unit cost to deciding according to total cost of ownership, the idea being that a company pays more than the per-unit purchase price when it buys materials and supplies. Total cost includes expenses associated with procurement, handling, warehousing, environmental fees, late deliveries, poor quality, incomplete deliveries and inventory costs. Table 21.2 shows some of the components of total cost that Delta considered.

---

[6]This example builds on material gathered to develop the academic case presented in the book *Successful Service Operations Management* (Metters *et al.* 2006).

21.4.1 Chemicals at Delta Air Lines, Inc.

Chemical management at Delta was highly complex before integration because of the large scale of operations and number of suppliers, the decentralised use of chemicals bought via a centralised purchasing process, and exacting environmental regulations. Table 21.3 illustrates the scope, listing some ground-based activities, with examples of waste-streams and chemicals used in each. The complexity was partly attributable to the difficulty of monitoring and managing the content of waste-streams. Adding a wrong substance into the mix could make hazardous what would otherwise be harmless, resulting in higher disposal costs.

Delta spends about US$16 million on chemicals annually, including US$1.8 million for cabin-cleaning chemicals such as window cleaner and chewing-gum remover. It buys about 1,500 different stock-keeping units (SKUs): 1,000 routine items and 500 rare-use or special-order items. Per-unit costs range from 67 cents for a cleaner, to about US$10,000 for three pounds of brazing powder used for the metal build-up on landing-gear components. Some 30% of these SKUs have a limited shelf-life, ranging from 6 months to 24 months. On expiration, most of these products become hazardous waste. The material cost for expired chemicals was about US$250,000 in 1996. Chemicals were stored in general use at the Technical Operations Centre, requiring significant space.

Delta's chemical-vendor database included 350 suppliers. Many of the products could have been obtained through distributors, thus reducing the size of the supplier base, but this would have added another administrative layer. Delta therefore decided to buy most chemicals from manufacturers, which bypassed the middlemen but kept the vendor base large.

Decentralised use of chemicals caused the centralised buying process to work poorly. 'Maverick buys' were particularly problematic. For example, mechanics working on ground-service equipment often found it easier to go to a local hardware shop to buy supplies such as upholstery cleaner than to submit a purchase order. Managers struggled to track uses, quantities and locations of chemicals. Delta also lost some of its leverage as a large buyer.

The rules for holding large quantities of chemicals were strict. For example, the 1986 Community Right to Know Act sets a threshold of 10,000 pounds of hazardous chemicals; any facility that purchases or possesses more than this amount is mandated to report to the US Environmental Protection Agency (EPA). Other acts mandate tracking use and emissions. Delta's large inventory commanded compliance with these and other regulatory bodies, each with its own data collection and reporting requirements.

21.4.2 A New Way to Manage Chemicals

In 1995, a long-time chemical supplier collaborated with Delta to form a new company, Interface-LLC. The creative thinking exemplified by Delta and Interface-LLC would not have been possible without the capabilities embodied in electronic data exchange. Interface-LLC agreed to purchase Delta's chemical inventory and relocate it off-site. Then, acting as the 'gatekeeper', Interface-LLC would sell the chemicals back to Delta by using the existing information infrastructure along with new environmental chemical-tracking software developed for Delta and implemented by Interface-LLC. Although this may seem inefficient, it worked brilliantly. It immediately solved Delta's problem of simplifying the supplier base while minimising the use of middlemen: Delta went from 350 suppliers to one. Furthermore, relocation of the chemicals released for Delta 30,000 square feet of space for maintenance, in a building valued at US$30 per square foot.

Table 21.2. Delta Air Lines, Inc. examples of the total cost of ownership of various cost components

Component	Example
Quality	•Labour cost of: – Inspecting incoming materials – Returning expired chemicals – Correcting defects not detected until later, during maintenance
Delivery	• Cost of process disruptions from not having the right chemical to do the job
Inventory	•Cost of storage space required to stock chemicals •Insurance costs for holding significant quantities of hazardous chemicals •Labour cost to stock chemicals
Compliance	• Cost of record-keeping • Cost of reporting requirements •Potential US EPA fines from failure to comply with regulations
Material	• Unit price for chemical material purchase
Inventory carrying	• Cost to handle, label and shelve material[a]
Labour	• Cost of mechanic to transfer contents of larger container into smaller container • Cost of relabelling and so on to meet application need and US OSHA employee-right-to-know or HAZCOM requirements[b]
Safety and risk	• Assigned costs (relative to potential for occupational exposure later, immediate risk to Delta property and future liability for cleanups from mismanagement) associated with handling and exposure to chemicals based on health and flammability rating and studied number of transfers<6pt added after>
Disposal	• Extra costs associated with disposal of expired material

[a]Generally, this is 11% of unit price for Interface-LLC and 18.5% of unit price for Delta
[b]OSHA = Occupational Safety and Health Administration

Interface-LLC absorbed all contracts Delta had negotiated with its suppliers. It agreed to deliver routine requests within three hours and expedited orders within two hours. It guaranteed a 95% fill rate (the proportion of orders a company receives that can be met with on-site inventory). Interface-LLC opened a small distribution centre 2 miles from the Technical Operations Centre. Delta transmitted orders to Interface-LLC via its existing requisition system (a database management programme), facsimile, electronic data interchange (EDI) or a web-based electronic catalogue.

Table 21.3. Waste-streams at Delta Air Lines, Inc.

Activity	Description	Waste-streams	Chemicals
Airport Customer Service	Provide baggage handling, cabin cleaning, security	Waste from cabin services, 'blue water' from lavatories	De-icer and cabin cleaners (e.g. chewing-gum remover)
Line maintenance	Conduct repairs to aircraft on the ramp	Replaced parts, tyres, oil, hydraulic fluid	Solvents such as acetone, de-greasers and paint
Hangar maintenance	Conduct major repairs, overhauls, maintenance	Solvents, oils, paints, plating-shop waste, paint strippers	Solvents and cleaners such as acetone, sealant and hydraulic fluid
Ground service equipment	Provide and maintain ACS vehicles and equipment	Solvents, oils, paints, paint strippers	Solvents such as acetone, de-greasers and engine oil

From SUCCESSFUL SERVICE OPERATIONS MANAGEMENT 2nd edition by METTERS/KING-METTERS/PULLMAN. © 2006. Reprinted with permission of South-Western, a division of Thomson Learning: www.thomsonrights.com. Fax 800-730-2215.

Implementing the programme had its challenges. Unexpectedly, some people at Delta resisted relinquishing responsibility for chemical purchasing and management. Continued operational success eventually persuaded sceptics by demonstrating the value that could be gained by the new design, and most problems were overcome within six months.

But these operational gains were only one part of the total benefit Interface-LLC generated for Delta. Interface-LLC offered other value-added services, many relying on e-commerce, including streamlining the management of the materials safety data sheet (MSDS), bar coding and tracking chemicals delivered to shops throughout the operation, developing an 'approved chemicals' list and negotiating with manufacturers. Interface-LLC also extended the scope of control to include the sourcing of safety products. Because the structure of the contract allowed for joint cost savings, Interface-LLC was motivated continually to find innovative ways to reduce expenses. For example, prior to integration, grease was bought in five-gallon buckets, 35% of which became waste. Interface-LLC shifted the purchase to 14-ounce tubes, thus finding ways to increase material efficiency, with a corresponding increase in environmental efficiency, by using more appropriate packaging.

The chemical management programme also gave Interface-LLC a new tool to attract business. It was able to approach other airlines and demonstrate clearly the success of its innovative environmental purchasing system. Interface-LLC now services Northwest and Comair as well as Delta and is negotiating several other chemical management contracts.

21.5 Conclusions

E-commerce can be a powerful way to improve the performance of environmental purchasing. But the gains possible are not automatic; they can only be achieved by looking beyond the technology and considering the business goals of the processes being automated. As the Webvan case clearly demonstrates, application of good technology to a poorly conceived business objective leads to poor business performance. In the case of Webvan it also led to poor environmental performance. Past research has shown that the achievement of supply chain gains from technology requires integration of the technology with the supply chain process (Walton and Marucheck 1997). One would expect the same to hold true for the relationship between e-commerce and environmental purchasing improvements.

Furthermore, environmental purchasing will be successful only if the business measurement and evaluation systems used are changed. If reduction in VOC use, for example, is not a stated and measured goal, there is little business incentive for an organisation to pursue VOC efficiencies through e-commerce-enabled environmental purchasing. At a minimum, the environmental improvement goals should be linked to appropriate cost drivers; as the Delta case shows, reduction of VOC purchases can have a direct link on business performance.

One significant challenge for electronically enabled environmental purchasing is the availability and diffusion of the technology required to implement it. For large companies the technological barriers to entry into e-commerce-based environmental purchasing are low, but for smaller companies, which tend to be less environmentally aware to begin with, this barrier to entry may be more significant. For example, adoption of EDI has been very slow for smaller businesses. Often, small to medium-sized enterprises (SMEs) will adopt EDI only at the insistence of a larger trading partner. This may be true for environmental purchasing through e-commerce as well.

More research and study needs to be put into environmental purchasing and how e-commerce can facilitate it, but we believe that there are significant business and environmental gains to be made through the use of e-commerce for environmental purchasing.

References

Angell, L.C., and R.D. Klassen (199) 'Integrating Environmental Issues into the Mainstream: An Agenda for Research in Operations Management', *Journal of Operations Management* 17.5: 575-98.

Carter, C.R., and M. Dresner (2001) 'Purchasing's Role in Environmental Management: Cross-functional Development of Grounded Theory', *Journal of Supply Chain Management* (Summer 2001): pp 12-28.

Community Right to Know Act (1986), 42 U.S.C. 11001 et seq. (1986), (Washington, DC: US Government Printing Office).

Galea, C., and S. Walton (2002) 'Is E-commerce Sustainable? Lessons from Webvan', in J. Park and N. Roome (eds.), *The Ecology of the New Economy* (Sheffield, UK: Greenleaf Publications): pp 100-109.

ISO (International Organisation for Standardisation) (1996) *ISO 14000 System of Standards: Environmental Management Systems: Specification and Guidance for Use* (Geneva, Switzerland: ISO).

Konrad, R. (2001) 'Server Farms on Hot Seat amid Power Woes', news.com, http://news.com.com/2100-1017-257567.html?legacy=cnet, accessed 13 November 2002.

Metters, R., K. King-Metters, M. Pullman and S. Walton (2006) *Successful Service Operations Management* (Mason, OH, USA: Thomson South Western Publishing).

Min, H., and W. Galle (1997) 'Green Purchasing Strategies: Trends and Implications', *International Journal of Purchasing and Materials Management* (August 1997): pp 10-18.

Sarkis, J. (1995) 'Supply Chain Management and Environmentally Conscious Design and Manufacturing', *International Journal of Environmentally Conscious Design and Manufacturing* 4.2: 43-52.

Sarkis, J., L. Meade and S. Talluri (2002) 'E-logistics and the Natural Environment', in J. Park and N. Roome (eds.), *The Ecology of the New Economy* (Sheffield, UK: Greenleaf Publications): pp 35-51

The Economist (2002) 'Trading Hot Air: A New Approach to Global Warming', *The Economist*, 19 October 2002: 60.

US GPO (US Government Printing Office) Office of the President of the United States (1998) *Executive Order 13101: Greening the Government through Waste Prevention, Recycling and Federal Acquisition* (19 September 1998; Washington, DC: US GPO).

Walton, S.V., and A.S. Marucheck (1997) 'The Relationship Between EDI and Supplier Reliability', *International Journal of Purchasing and Materials Management* (Summer 1997): 30-25.

Walton, S.V., R. B. Handfield and S.A. Melnyk (1998) 'The Green Supply Chain: Integrating Suppliers into Environmental Management Processes', *International Journal of Purchasing and Materials Management* (Spring 1998): pp 2-12.

Contributor Biographies

Maher Ajam is a Ph.D. student at Salford University – Manchester UK. He received his B.E. Degree in Civil Engineering and Masters Degree in Engineering Management from American University of Beirut in 1998 & 2001 respectively. Currently he is working as an assistant Project Manager in D.G. Jones & Partners. His research interests are in Information Technology utilisation in Project Management, ND modeling, E-commerce in Engineering Management and other related Engineering Management topics. Has some published articles in these areas.

Ian Bainbridge has degrees in metallurgy and commerce from Melbourne and is currently undertaking a PhD degree at the University of Queensland. He has been involved in the aluminium industry for over 40 years and presently operates his own metallurgical consulting business specializing in aluminium and magnesium casting and specialty products production.

Severin Beucker holds a degree in Environmental Engineering and has been researcher at the Institut für Arbeitswissenschaft und Technologiemanagement (IAT), University of Stuttgart and Fraunhofer-Institut für Arbeitswirtschaft und Organisation (IAO) since 1999. His areas of expertise are corporate sustainability management and environmental management information systems. He is project co-ordinator of IAT's participation in the CARE project.

Frances Bowen is an Associate Professor in the Strategy and Global Management Area of the Haskayne School of Business, University of Calgary. Her research interests include corporate environmental management, particularly as explained by the resource-based and behavioral views of the firm. Dr. Bowen's research has been presented at more than 20 international conferences and published in academic journals such as *Journal of Management Studies, Production and Operations Management, International Journal of Operations and Production Management* and *British Management Journal.* When this research was conducted, Dr. Bowen was a Research Officer at the School of Management, University of

Bath partially funded by the Engineering and Physical Sciences Research Council and London Underground Ltd.

Geoff Christopherson has been involved in transport management, logistics and supply chain management up to senior operational level, including Director General of Movements and Transport in the Department of Defense, since 1958, and was a member of the Commonwealth Transport Industries Advisory Council from 1986 to 1989. He was made a Member of the Order of Australia for services to Defense transport operations in 19889 and from 1990-91 he was Defense Research Fellow at RMIT. After his retirement from the Australian Army in December 1991 he took up an appointment as a member of the Higher Education academic staff in the Faculty of Business at RMIT. During his time at RMIT Geoff Christopherson at various times was Program Leader for the logistics management group, Acting Director, Post Graduate programs and Acting Head of the School of Marketing.   In 1996 he was an exchange senior lecturer at the University of Huddersfield in the United Kingdom for one semester. In 2001-2002 he was a member of the Commonwealth Department of Transport and Regional Services Freight Transport Logistics Industry Action Agenda Steering Committee, and chaired the Education and Training Working Group of that Committee.

Paul Cousins is Professor of Operations Management and CIPS Professor of Supply Chain Management at Manchester Business School, The University of Manchester. His career to date has spanned a range of business sectors from industry (Westland Helicopters & Sikorsky Aircraft), consulting (A.T.Kearney) to academia (University of Bath and The University of Melbourne, Australia). It has led him to work in a variety of countries, including the UK, Europe, USA, and Australia. He has obtained over £1.5m in research grants and has conducted research and consultancy work for numerous firms across a range of industrial sectors at national and international level.

Frank Ebinger studied Business Administration at the University for Applied Science of Fulda and the University of Kassel. From 1996 to 2003, he worked as research associate and scientific coordinator at the Institute for Applied Ecology (Öko-Institut). He holds a PhD in business administration from the Carl von Ossietzky Universität Oldenburg, Germany. Since 2003 he works as Assistant Professor at the Institute of Forestry Economics, University of Freiburg. His main research interests are in the fields of Strategic Management, Relationship Management, Corporate Social Responsibility, and recently in NGO-strategies.

Adam Faruk is the Assistant Director of the Ashridge Centre for Business and Society at the Ashridge Business School. His interests include corporate governance, socially responsible investment, sustainability reporting, and changing relations between business, government and civil society. When this research was conducted, Dr. Faruk was a Research Officer at the School of Management, University of Bath partially funded by the Engineering and Physical Sciences Research Council and London Underground Ltd.

Chris Galea is a father, educator, outdoor enthusiast, builder, sailor and entrepreneur. He currently teaches at the Gerald Schwartz School of Business at St Francis Xavier University in Antigonish, Nova Scotia. He was also part of the founding faculty of the Sustainable Enterprise Academy at the Schulich School of Business at York University in Toronto. Much of his doctoral and current research is in the area of management learning as it relates to sustainability. Chris lives by the ocean surrounded by land he cherishes and people he cares about.

Yong Geng is an associate professor, associate director in Institute for Eco-planning and Development at Dalian University of Technology, P. R. of China. He got his Ph.D in chemical engineering at Dalhousie University in Canada. His main research interests are industry ecology and integrated water resource management.

Maria Goldbach studied business administration and economics in Germany and France. From 2000 to 2003 she was a research assistant at the Chair of Production and the Environment, Institute of Business Administration, University of Oldenburg, Germany. She holds a PhD in business administration from this university. Her main research areas are supply relationships and networks as well as organisational implications in cost management in supply chains. She is now studying Medicine at the University of Witten/Herdecke, Germany.

Jeremy Hall is an Associate Professor and Fellow of the International Institute for Resource Industries & Sustainability Studies (IRIS) at the Haskayne School of Business, University of Calgary, Canada. He teaches Business Strategy and Sustainable Development, Innovation Management, Corporate Strategy, and Entrepreneurship. His research interests include sustainable development innovation, stakeholder ambiguity, radical technology development and inter-firm innovation dynamics. He has conducted international research in the agricultural, energy, aerospace, retailing, chemicals, forestry and consultancy sectors, with support from Genome Canada and the Social Sciences and Humanities Research Council (SSHRC) of Canada. Output from this research has been published in MIT Sloan Management Review, R&D Management, Harvard Business Review (Latin American editions), Journal of Cleaner Production, Journal of Business Venturing, Greener Management International and Industrial & Corporate Change, among others. Before joining the University of Calgary, Dr Hall was a Research Fellow and Lecturer at SPRU—Science and Technology Policy Research, University of Sussex, UK, where he also received his doctorate in 1999.

Lic. Sc. **Kirsi Hämäläinen** works as a coordinator of sustainable development for the city of Tampere, Finland. Se is doing her Ph.D. studies in corporate environmental management and focusing on intentional development of industrial ecology practices on the basis of the network theory in her research. She has completed a licentiate's degree in corporate environmental management in University of Jyväskylä, School of Business and Economics.

Burton Hamner is an international development consultant specializing in sustainable industry, at the facility and policy levels. He has Master's degrees in

Business Administration and in Marine Affairs from the University of Washington and has worked in 15 countries for clients such as the World Bank, United Nations, US Agency for International Development and leading corporations. He has been a visiting professor at the Asian Institute of Management (AIM) in Manila and the Universidad del Pacifico in Lima. Burton produces the website, www.cleanerproduction.com and has published academic articles about sustainable industry in a range of journals.

John Harris is environmental manager at Eli Lilly Incorporated. He has been a Health, Safety and Environmental professional for a number of years. He has been active in a number of professional societies and has recently chaired a number of committees for the Global Environmental Management Initiative (GEMI). Currently at Eli Lilly he is involved in management of a number of green supply chain management projects and manager of their environmental management systems.

Marilyn M. Helms is Sesquicentennial Endowed Chair and Professor of Management at Dalton State College (DSC). In addition, she is the director of DSC's Centre for Applied Business Studies, working particularly on research projects, seminars and training programmes for area manufacturing industries. Prior to coming to Dalton State College, Dr. Helms taught at the University of Tennessee at Chattanooga and directed their Institute for Women as Entrepreneurs. She teaches production and operations management classes as well as classes in quality. She holds a doctoral degree in business administration, a masters degree in business administration and a bachelors degree in of business administration, all from the University of Memphis. She is a certified fellow in production and inventory management (CFPIM) and a certified integrated resources manager (CIRM) at the American Production and Inventory Control Society (APICS). She has published numerous journal articles and cases. She has been awarded grants from the US Department of Education, the Coleman Foundation and the Southern Regional Education Board. She was awarded the Fulbright Teaching and Research Award and taught at the University of Coimbra, Portugal, in 2000. Her current research interests include just-in-time implementation, manufacturing strategy and supply chain management. She has local and regional consulting experience and has spoken to international and national groups.

Aref A. Hervani holds a PhD in natural resource economics from West Virginia University, Morgantown, WV. His dissertation, *Oligopsony Elements in the Recycled Wastepaper Market: Existence and Implications*, led to his interest in recycling and reverses logistics for recycling. He also holds a Master of Arts degree in Economics from Ohio State University, and a Bachelor of Arts degree in Economics from University of South Carolina, Columbia, SC, as well as a minor in International Studies. Prior to coming to Chicago State University, Dr. Hervani taught at Dalton State College, Miami University, West Virginia University and Otterbein College. His experience as a research consultant included projects on the economics of food security and population growth as well as recycling for the newsprint industry. His research interests include environmental degradation, recourse depletion, recycling and oligopsony power and factor market performance. He is a member of the American Agricultural Economic Association (AAEA) and the American Economic Association (AEA).

Booi Hon Kam is Associate Professor of Transport Planning and Director of Operations, School of Management, RMIT University. An alumnus of the University of California at Los Angeles, Booi Kam holds a MS degree in Engineering Systems and a Ph D in Urban Planning. Booi's research interest straddles a diversified range of disciplines, having published in areas on postgraduate research supervision, systems engineering, helicopter design, housing and real estates studies, transport and land-use modelling, traffic accidents analysis, travel survey research, green logistics, privatization of agricultural plantations, and e-marketing.

Hsien H. Khoo received her degrees from the Nanyang Technological University of Singapore. She is currently working as a Research Engineer with the Department of Industrial Engineering. Her area of research involves green business practices and sustainable manufacturing.

Richard Lamming is Director of the School of Management at Southampton University. Formerly, between 1991 and 2003, Professor Lamming held the CIPS Chair of Purchasing and Supply Management at the School of Management, University of Bath, where he founded the Centre for Research in Strategic Purchasing and Supply (CRiSPS). Prof Lamming has been deeply involved with the UK Chartered Institute of Purchasing and Supply. In 1995 he was awarded the CIPS' highest honour, the Swinbank Medal, for outstanding services to the Purchasing profession. He was the founding Editor of the European Journal of Purchasing and Supply Management, (1994-2000: Elsevier Scientific, UK) and first Chairman of the International Purchasing and Supply Education and Research Association, (IPSERA: 1993-1995). In October 2002 he was awarded the Hans Ovelgonne Award by the International Federation of Purchasing and Materials Management for his outstanding contribution to Research and Development in the field of Purchasing and Materials Management.

Claus Lang-Koetz holds a Diplom-Ingenieur degree in environmental engineering and an MSc in water resources engineering and management. He has been a researcher at the Institut für Arbeitswissenschaft und Technologiemanagement (IAT), University of Stuttgart, and Fraunhofer-Institut für Arbeitswirtschaft und Organisation (IAO) since 2000. He works in the field of environmental management information systems.

Toufic Mezher is a Professor of Engineering Management at the American University of Beirut, Faculty of Engineering and Architecture. He received the B.S. degree in Civil Engineering for the University of Florida in 1982, Gainesville, the Mater of Engineering Management degree and the Doctor of Science in Artificial Intelligence and Human Factors from George Washington University in 1987 and 1992 respectively. His current research interests are in Technology

Management, Knowledge Management, Total Quality Management, Environmental Management, Human Resource Management, Building Intelligent Decision Support Systems, and other related Engineering Management topics. He has many published articles in these areas.

Menno Nagel was born in Soest, The Netherlands on October 23, 1961. He was awarded his BSc degree in telecommunication technology in 1985. His study of electrical engineering started at Delft University of Technology in 1985 and he received his MSc degree in 1990. He joined AT&T on January 1, 1991 as a Member of Technical Staff of Bell Laboratories, and was engaged in component reliability for integrated circuits for a three-year period. His second assignment in March 1994, involved supply chain management for interconnection devices, component reliability for optical connectors and controlled impedance issues for printed boards. He undertook a special assignment on modeling and simulating telecommunication networks during 1996. He became a Senior Engineer Design for Environment in October 1996. In this research position, he researched, developed, evaluated, validated and implemented the concept of environmental quality relating to the supply chain of an Original Equipment Manufacturer. During this research period, he was working towards a Ph.D. degree in environmental supply chain management at Delft University of Technology. He was promoted to a Distinguished Member of Technical Staff of Lucent Technologies, Bell Laboratories in September 1998. Lucent Technologies acclaimed him as an "Environmental Hero and Champion" for the planning of the future in 1998, 1999 and 2001. He received his PhD degree in September 2001. From August 1, 2002 he operates as an Assistant Professor in Design & Life Cycle Engineering at Delft University of Technology. Since January 1, 2005 he is also manager of Delft Center for Engineering Design.

Lutz Preuss teaches sustainability and business ethics at the School of Management of Royal Holloway College, University of London, and holds a PhD from King's College London. His research interests lie in the fields of managing for sustainability, business ethics and corporate social responsibility. Dr Preuss is author of 'The Green Multiplier', published by Palgrave in 2005. His research has also been published in academic journals, such as the Journal of Business Ethics, Business Strategy and the Environment or Business Ethics: A European Review. He has been involved with EBEN-UK, the UK association of the European Business Ethics Network, ever since it inception in 1994. From 2002 to 2005 he served as Treasurer and Membership Secretary on the Executive of EBEN-UK.

Purba Rao is holder of the Don Benigno Toda, Jr. Chair in Business Management at the Asian Institute of Management. She teaches quantitative management and environmental management. She is currently its Management Research Report Coordinator. Prof. Rao has had extensive consulting experience with San Miguel Corporation, the Plaza Fair/Fairmart Business Houses, the Philippine Transmarine Carriers, Inc., and the Indo Phil Textile Company. Prof. Rao is the author of Environmental Management Systems in Southeast Asia – Towards a Green Millennium (AIM, 2000) Her research has been published in a variety of international journals including Asia Pacific Journal of Economics and Business, International Journal of Operations and Productions Management (IJOPM), Philippine Business and Environment, International Journal of Systems Sciences, Journal of Mathematical Science. Prof. Rao is a Fellow in Management (equivalent to Doctorate) of the Indian Institute of Management - Calcutta. She holds a Master of Science in Applied Mathematics from the College of Science, University of Calcutta (1971) and Bachelor of Science at the Presidency College - Calcutta (1969).

Ed Rhodes is a senior lecturer in the Department of Design and Innovation within the Open University. He also serves as Director of the Centre for the Analysis of Supply Chain Innovation and Dynamics (CASCAID). He has published extensively in the field of supply chain management and innovation. His particular interests are in implementing new technologies, using systemic processes in order to elucidate all aspects of supply systems including the importance of networks, social capital, governance, the role of the aftermarket, lead companies power, and strategic thinking to understand the evolution of supply chains. His work in this book draws directly from his extended theory of treating supply chains as total product systems. This can be applied to any supply system to support the criticality of greening each component within the system, thus making the entire system more sustainable.

Joseph Sarkis is Professor of Operations and Environmental Management at Clark University in Worcester, Massachusetts. He earned his Ph.D. in Management Science from the University of Buffalo. He has a wide variety of research and teaching interests ranging from corporate environmental management to operations management topics. He has published extensively in a wide variety of outlets with over 170 publications. He serves on a number of editorial boards of internationally recognized journals. He is currently editor of Management Research News.

Uwe Schneidewind studied business administration in Cologne/Germany and Paris/France. From 1991 to 1992 he was Business Consultant at Roland Berger & Partner in the field strategic environmental management. From 1993 to 1997 he worked as researcher at the Institute for Ecology and the Economy at the University of St. Gall. Since 1998 he is full professor of Environmental management at the University of Oldenburg. Since 2004 he has served as President of the university.

Kosmas Smyrnios holds the position of Professor and Director of Research in the School of Management, at RMIT University. Kosmas is Associate Editor of the *Family Business Review*, and he is also on the editorial Board of a number of academic journals. Since the awarding of his PhD, Kosmas has developed an extensive applied research record with over 70 international and national refereed publications in different disciplines, including marketing, psychology, physics, management, and accounting. Kosmas has established international credentials in the family business and SME areas, having been involved in a number of prominent national and international research projects. Kosmas is a Foundation

Board Member of the International Family Enterprise Research Academy (IFERA). In 1998 and 2001, he was awarded prizes for the Best International Research Papers at the 9th and 12th World Family Business Network Conference in Paris and Rome, respectively. Professor Smyrnios is also frequently called upon to provide expert media commentary on pertinent matters relating to SMEs.

Trevor Spedding is the Head of School of Management and Marketing in the Faculty of Commerce at the University of Wollongong. Before joining the university, he was Medway Chair of Manufacturing Engineering in the Medway School of Engineering at the University of Greenwich. He was Associate Professor in the School of Mechanical and Production Engineering at Nanyang Technological University (NTU) in Singapore. Trevor has a Honours Degree in Mathematics and Statistics and obtained a PhD for research concerned with the statistical characterisation of engineering processes. He is chartered engineer and statistician and has worked as a consultant and conducted short courses for several prominent companies in the UK, Europe, USA and Singapore.

Robert Sroufe is an Assistant Professor in the Department of Operations and Strategic Management at the Carroll School of Management, Boston College. His research is primarily focused on environmental management systems and his interests include supply chain management, design for environment, environmentally responsible manufacturing, and performance measurement. His work has been published in the *Production and Operations Management Society, Journal of Operations Management*, the *European Journal of Operations Research, International Journal of Operations and Production Management*, and *the International Journal of Production Research*.

David M.R. Taplin is presently a visiting Professor of Industrial Process Ecology at the Medway School of Engineering, Intelligent Systems Centre, University of Greenwich, England. In 1997 he was appointed as Professor of Sustainable Manufacturing at Nanyang Technological University, Singapore. Later he became a Professorial Associate in the Faculty of Technology at Brunel from (1998-2002). He was also a Dean of Central Queensland University (CQU) in Australia. At CQU he established the School of Industrial Ecology and Built Environment (SIEBE), the new initiatives in Process Engineering and Light Metals (PELM) and in Sustainable Product Integrated Engineering (SPINE), respectively. At that time he was appointed CQU Honorary Professor of Industrial Ecology and Adjunct Professor of Asset Sustainability at Queensland University of Technology.

Gregory Theyel is an Assistant Professor of Strategy at California State University. His research interests are in firm strategy and technology management, particularly collaboration and regional resources for technological innovation in industries including biotechnology, renewable energy, chemicals, and electronics. His publications have appeared in journals such as *Strategic Management Journal* and *International Journal of Operations and Productiont Management*. He has also been a management consultant for 15 years assisting corporations and governments develop strategy and environmental policy.

Philip Trowbridge, P.E. is Advanced Micro Devices (AMD) Manager of Corporate Responsibility and is based in Austin, Texas. Prior to assuming his current role, Mr. Trowbridge was a member of AMD's Extended Producer Responsibility staff and the Environmental Health and Safety (EHS) department. He has more than 20 years of experience in the environmental field and has worked on supply chain EHS issues for the last six years.

Anette von Ahsen is a research assistant at the Chair of Environmental Management and Controlling, Duisburg-Essen University, Campus Essen. Her recent research and publications include the analysis of integrated quality and environmental management as well as environmental reporting.

Rhett Walker is a Professor of Business in the School of Business at La Trobe University, and is also Associate Dean (Regions) of the Faculty of Law and Management. His research and teaching interests include customer service and the nature of service mindedness; the marketing and management of services, particularly in business-to-business contexts; and competitive market positioning. His research papers have been presented at, and published in the proceedings of, many academic conferences nationally and internationally, and his work has been published in a variety of leading scholarly journals including the *European Journal of Marketing*, the *Journal of Marketing Management*, the *Journal of Services Marketing,* the *International Journal of Service Industry Management,* the *International Journal of Entrepreneurship and Innovation Management, Marketing Intelligence and Planning,* and *Advances in International Marketing.* He is the co-author of 3 textbooks one of which, *Services Marketing – An Asia-Pacific Perspective*, is the leading text in this field throughout Australasia.

Steve Walton is Associate Professor in the Practice Decision and Information Analysis at the Goizueta Business School at Emory University. His current research interests include uses of electronic commerce technologies for environmental and supply chain management, environmental impacts of supply chain management and applications of qualitative research methods. He has published in the *Journal of Operations Management, European Journal of Operational Research, International Journal of Operations and Production Management, International Journal of Purchasing and Materials Management.* He has also served as consultant for a number of large and medium sized corporations.

James Warren is a lecturer in the Department of Design and Innovation within the Open University. His main interests are focused on environmental emissions control and sustainable practices in transport planning and policy. His previous work has included analysis of supply systems within transport, use of taxation and road pricing to alter mobility trends, as well as the study of transport policy in developing countries in urban areas. With Ruth Carter and Ed Rhodes, he is the editor of a major new reader in the field of supply chains and total product systems.

Qinghua Zhu is an associate professor in Institute for Eco-planning and Development at Dalian University of Technology, P. R. of China. She received her bachelor and master degrees in electrical engineering, and her Ph.D in systems engineering. Her main research interests are integrated green supply chain management and solid waste management.

Index